Plant Genetics and Biotechnology in Biodiversity

Plant Genetics and Biotechnology in Biodiversity

Special Issue Editors

Rosa Rao
Giandomenico Corrado

MDPI • Basel • Beijing • Wuhan • Barcelona • Belgrade

MDPI

Special Issue Editors

Rosa Rao
Università degli Studi di Napoli
"Federico II"
Italy

Giandomenico Corrado
Università degli Studi di Napoli
"Federico II"
Italy

Editorial Office
MDPI
St. Alban-Anlage 66
Basel, Switzerland

This is a reprint of articles from the Special Issue published online in the open access journal *Diversity* (ISSN 1424-2818) from 2017 to 2018 (available at: http://www.mdpi.com/journal/diversity/special_issues/plant_genetics_biotechnology)

For citation purposes, cite each article independently as indicated on the article page online and as indicated below:

LastName, A.A.; LastName, B.B.; LastName, C.C. Article Title. *Journal Name* **Year**, *Article Number*, Page Range.

ISBN 978-3-03842-003-3 (Pbk)
ISBN 978-3-03842-004-0 (PDF)

Contents

About the Special Issue Editors

Rosa Rao is Full Professor of Plant Genetics at the University of Naples "Federico II" (Italy). Her research interests include the molecular characterization of crop biodiversity, along with food authentication and multidimensional approaches for studying defences against plant enemies.

Giandomenico Corrado (MSc, PhD) is currently Researcher at the University of Naples "Federico II" (Italy). He got his PhD at the University of Leeds (UK). His research activities include the molecular characterization of genetic diversity in landraces, with emphasis on the elucidation of defence mechanisms against biotic stress.

Preface to "Plant Genetics and Biotechnology in Biodiversity"

To increase both sustainability and productivity in agriculture, crop science, from pre-breeding to production management, needs to make better use of genetic diversity. For instance, adapted and exotic genetic resources can greatly contribute to generate varieties with enhanced adaptation to different climatic conditions and with improved plasticity in response to stressful factors. These features are central to reducing vulnerability to stress and supporting more diversified and sustainable agro-ecological systems.

Scientific and technical advances allow the generation of large quantities of genotypic and phenotypic data at a fraction of the cost required decades ago. Similarly, novel biotechnological approaches are revolutionizing the way we can edit or delete specific sequences of the plant genome. The impact of these innovations is wide-ranging and includes the possibility to improve our understanding and exploitation of plant genetic resources for food and agriculture (PGRFA). These new technologies are expected to transform the study of PGRFA into a more technologically intensive, information-rich science, in which the integration and interpretation of different kinds of data will be central to fully unlock the potential of genetic diversity for agriculture.

Considering that the number of research articles on these topics is rapidly increasing, this Special Issue (SI) was designed to collect perspectives and experimental studies that cover contemporary advances in plant genetics and biotechnology for plant diversity in agriculture. While the usefulness of molecular tools to describe genetic polymorphisms and the importance of generic resources in agriculture has been largely covered in the literature, we felt the need to highlight how the management, description, and use of agricultural diversity is changing in the backdrop of molecular genetics and biotechnology. To this aim, we included both reviews, which summarize the current state of understanding, and experimental articles, which provide empirical cases on the description and exploitation of different kinds of plant genetic diversity. The SI was then organized into three parts. The first includes three thematic reviews on the scientific, technological, and legal advances in plant diversity and agriculture. The second part consists of three contributions on specific examples of the exploitation of different sources of genetic diversity (i.e., landraces, mutant populations, and wild-gene pools) in crops (i.e., tomato, barley, and wheat, respectively). The third part is made of six research articles on the study of molecular and/or phenotypic diversity to address basic or applied questions in trees (olive and grape), as well as herbaceous crops (maize, rice, and wheat) and wild species. In collecting the contributions, we were pleased to note that the variety of topics reflects well the wide-ranging changes made possible by recent progress in plant molecular genetics and biotechnology. This SI was also launched to honour the memory of Prof. Scarascia-Mugnozza, and an editorial highlights his contribution to the study of plant genetic resources and his incessant activity to foster national and international cooperation for the protection of PGRFA.

Finally, we were delighted by the positive response of many colleagues, who enthusiastically contributed to this SI. We owe to them, to the anonymous reviewers, and to the Editorial Office special gratitude for having ensured the overall quality of the SI and that the project did come to completion on schedule.

Rosa Rao, Giandomenico Corrado

Special Issue Editors

diversity

MDPI

Editorial

Special Issue: Plant Genetics and Biotechnology in Biodiversity

Giandomenico Corrado and Rosa Rao *

Dipartimento di Agraria, Università degli Studi di Napoli, via Università 100, 80055 Portici (NA), Italy; giandomenico.corrado@unina.it
* Correspondence: rao@unina.it

Received: 21 March 2018; Accepted: 23 March 2018; Published: 27 March 2018

Abstract: The rapid progress and increasing affordability of novel investigation tools in plant genetics and biotechnology offer previously inaccessible opportunities for the exploitation of plant genetic diversity in agriculture. The Special Issue was lunched to highlight how new technologies are improving both genotyping and phenotyping methods, thus allowing us to uncover crop diversity and use genetic variability for plant breeding with remarkable precision and speed. Three thematic reviews report on scientific, technological, and legal advances in plant diversity and agriculture. Three contributions provide specific examples of the exploitation of different kinds of genetic resources, ranging from landraces to mutant populations. Six research articles are illustrative examples of the study of molecular and/or phenotypic diversity to address basic or applied questions in different plant species. Finally, this SI was also launched to honor the memory of Prof. Gian Tommaso Scarascia Mugnozza and a dedicated Editorial acknowledges his work in plant breeding and biodiversity protection.

Keywords: genetic diversity; agro-biodiversity; landraces; crop; breeding; agriculture

---

The worldwide loss of biodiversity is a concern of increasing importance, and the field of agriculture is no exception. If we only take into account the number of cultivated plant varieties, it is evident that we are witnessing an unprecedented loss of agricultural genetic resources: more than 90 percent of crop varieties have disappeared from farmers' fields in less than a century [1]. The rise of intensive farming is arguably the major driver of agro-diversity loss. This system of cultivation is virtually a "technological standard" in many industrialized economies and also essential to support a rapidly increasing world population. Input-intensive agricultural systems benefit from uniformity and, in every setting, they associated with a rapid displacement of "traditional" crops and livestock. Industrial agriculture proved to be effective in supplying a large variety of affordable and abundant food, yet its multiple negative consequences (e.g. biotic and abiotic stress vulnerability, dependence on chemicals, yield stagnation, habitat degradation, decreasing profits for small farm-based communities, etc.) have largely remained unaddressed. In addition, the loss of agricultural diversity irredeemably reduces the options and resources for long-term sustainability and prosperity.

Agriculture is an indispensable element in the food production of any economy, and therefore should contribute to both conservation and the sustainable use of biodiversity. If necessary, this goal should be pursued beyond the constraints imposed by the profit motive. There is a growing consensus that whenever a reliable and safe food supply is guaranteed, intensive agricultural systems should move towards more diversified agro-ecological production systems. Reducing the use of off-farm chemical inputs, safeguarding biodiversity, and stimulating beneficial interactions between different species are all part of an integrated strategy that includes a more profitable exploitation of plant genetic diversity, and, in the narrower sense, the use of already existing or improved landraces. For instance, landraces can compete with modern cultivars in terms of output, especially under durable environmental stress and with reduced chemical input. Moreover, landraces may contribute to improved health thanks to their higher content of beneficial bioactive compounds (e.g. antioxidants, glycosides, flavonoids, etc.).

This Special Issue (SI) was lunched to highlight how the rapid advances and increasing affordability of novel investigation tools in plant genetics and biotechnology offer previously inaccessible opportunities. New technologies are improving both genotyping and phenotyping methods, allowing us to uncover crop diversity and exploit genetic variability for plant breeding with remarkable precision and speed.

The SI includes three thematic reviews that guide readers through the main recent scientific [2], technological [3], and legal [4] developments for the characterization, conservation, and use of plant genetic resources. The source of diversity as well as the approaches to its use can be different. Three more specific contributions present the possibilities, limitations, and achievements related to improving crops by exploiting different kinds of genetic resources, landraces in the case of tomatoes [5], wild gene pools for wheat [6], and mutants for barley [7]. Research articles also offer varied examples of the study of molecular and/or phenotypic diversity in trees (olive and grape) [8,9] as well as herbaceous crops (maize, rice, and wheat) [10–12] and wild species [13]. These papers illustrate how recent knowledge about DNA-based analytical tools is converted into useful applications for the description, understanding, and utilization of the genetic diversity.

This SI was also launched to honor the memory of Prof. Gian Tommaso Scarascia Mugnozza. We were pleased to include an Editorial that acknowledges not only his work in plant breeding and diversity, but also his incessant activities "to protect, conserve and utilize biodiversity properly, and in particular genetic resources" [14,15]. Prof. Scarascia Mugnozza was an active promoter of international collaborations, and we would like to pay tribute to him, remembering especially his memorable quote: "as the importance of plant genetic resources will be more acknowledged, partnership will not be any easier, it will be more necessary." Our closing wish is that this SI and its content would also serve to strengthen partnerships and promote a future alliance of ideas and goals in order to study, protect, and exploit biodiversity for the agricultural challenges that lie ahead [16].

Finally, we thank all the authors for their important contribution to this SI. We would also like to acknowledge the various submissions that could not be published. Regrettably, part of the editors' work is to make a decision based on the reviewers' comments—for which we are deeply grateful—and their own opinion. We also wish to thank the staff members at the MDPI editorial office (in particular Ms. Wei Zhang) for their support.

References

1. Food and Agriculture Organization (FAO). *Building on Gender, Agrobiodiversity and Local Knowledge: A Training Manual*; FAO: Rome, Italy, 2005; p. 18.
2. Corrado, G.; Rao, R. Towards the genomic basis of local adaptation in landraces. *Diversity* **2017**, *9*, 51. [CrossRef]
3. D'Agostino, N.; Tripodi, P. NGS-based genotyping, high-throughput phenotyping and genome-wide association studies laid the foundations for next-generation breeding in horticultural crops. *Diversity* **2017**, *9*, 38. [CrossRef]
4. Sonnino, A. International instruments for conservation and sustainable use of plant genetic resources for food and agriculture: An historical appraisal. *Diversity* **2017**, *9*, 50. [CrossRef]
5. Carbonell, P.; Alonso, A.; Grau, A.; Salinas, J.F.; García-Martínez, S.; Ruiz, J.J. Twenty years of tomato breeding at EPSO-UMH: Transfer resistance from wild types to local landraces—from the first molecular markers to genotyping by sequencing (GBS). *Diversity* **2018**, *10*, 12. [CrossRef]
6. Ceoloni, C.; Kuzmanović, L.; Ruggeri, R.; Rossini, F.; Forte, P.; Cuccurullo, A.; Bitti, A. Harnessing genetic diversity of wild gene pools to enhance wheat crop production and sustainability: Challenges and opportunities. *Diversity* **2017**, *9*, 55. [CrossRef]

7. Terzi, V.; Tumino, G.; Pagani, D.; Rizza, F.; Ghizzoni, R.; Morcia, C.; Stanca, A.M. Barley developmental mutants: The high road to understand the cereal spike morphology. *Diversity* **2017**, *9*, 21. [CrossRef]
8. Veloso, M.M.; Simões-Costa, M.C.; Carneiro, L.C.; Guimarães, J.B.; Mateus, C.; Fevereiro, P.; Pinto-Ricardo, C. Olive tree (*Oleaeuropaea* L.) diversity in traditional small farms of Ficalho, Portugal. *Diversity* **2018**, *10*, 5. [CrossRef]
9. Porceddu, A.; Camiolo, S. Patterns of spontaneous nucleotide substitutions in grape processed pseudogenes. *Diversity* **2017**, *9*, 45. [CrossRef]
10. Palumbo, F.; Galla, G.; Martínez-Bello, L.; Barcaccia, G. Venetian local corn (*Zea mays* L.) germplasm: Disclosing the genetic anatomy of old landraces suited for typical cornmeal mush production. *Diversity* **2017**, *9*, 32. [CrossRef]
11. El-Namaky, R.; Bare Coulibaly, M.M.; Alhassan, M.; Traore, K.; Nwilene, F.; Dieng, I.; Ortiz, R.; Manneh, B. Putting plant genetic diversity and variability at work for breeding: Hybrid rice suitability in West Africa. *Diversity* **2017**, *9*, 27. [CrossRef]
12. Nigro, D.; Fortunato, S.; Giove, S.L.; Mangini, G.; Yacoubi, I.; Simeone, R.; Blanco, A.; Gadaleta, A. Allelic variants of glutamine synthetase and glutamate synthase genes in a collection of durum wheat and association with grain protein content. *Diversity* **2017**, *9*, 52. [CrossRef]
13. Gross, C.L.; Fatemi, M.; Julien, M.; McPherson, H.; Van Klinken, R. The phylogeny and biogeography of *Phyla nodiflora* (Verbenaceae) reveals native and invasive lineages throughout the world. *Diversity* **2017**, *9*, 20. [CrossRef]
14. Pagnotta, M.A.; Noorani, A. The contribution of professor GianTommassoScarasciaMugnozza to the conservation and sustainable use of biodiversity. *Diversity* **2018**, *10*, 10. [CrossRef]
15. Scarascia Mugnozza, G. *The protection of biodiversity and the conservation and use of genetic resources for food and agriculture: Potential and perspectives*; F.A.O.: Rome, Italy, 1995; Volume 19.
16. Foley, J.A.; Ramankutty, N.; Brauman, K.A.; Cassidy, E.S.; Gerber, J.S.; Johnston, M.; Mueller, N.D.; O'Connell, C.; Ray, D.K.; West, P.C. Solutions for a cultivated planet. *Nature* **2011**, *478*, 337. [CrossRef] [PubMed]

diversity

MDPI

Review

Towards the Genomic Basis of Local Adaptation in Landraces

Giandomenico Corrado * and Rosa Rao

Dipartimento di Agraria, Università degli Studi di Napoli, via Università 100, Portici (NA) 80055, Italy;
rao@unina.it
* Correspondence: giandomenico.corrado@unina.it

Received: 27 September 2017; Accepted: 2 November 2017; Published: 4 November 2017

Abstract: Landraces are key elements of agricultural biodiversity that have long been considered a source of useful traits. Their importance goes beyond subsistence agriculture and the essential need to preserve genetic diversity, because landraces are farmer-developed populations that are often adapted to environmental conditions of significance to tackle environmental concerns. It is therefore increasingly important to identify adaptive traits in crop landraces and understand their molecular basis. This knowledge is potentially useful for promoting more sustainable agricultural techniques, reducing the environmental impact of high-input cropping systems, and diminishing the vulnerability of agriculture to global climate change. In this review, we present an overview of the opportunities and limitations offered by landraces' genomics. We discuss how rapid advances in DNA sequencing techniques, plant phenotyping, and recombinant DNA-based biotechnology encourage both the identification and the validation of the genomic signature of local adaptation in crop landraces. The integration of 'omics' sciences, molecular population genetics, and field studies can provide information inaccessible with earlier technological tools. Although empirical knowledge on the genetic and genomic basis of local adaptation is still fragmented, it is predicted that genomic scans for adaptation will unlock an intraspecific molecular diversity that may be different from that of modern varieties.

Keywords: genomics; differentiation; genome-environment association

1. Crop Landraces

Public awareness on the importance of biodiversity conservation is strengthening over time [1]. Climate change, pollution, environmental disasters, loss of natural habitats, environmental degradation, and overexploitation of resources regularly make front-page news. Without taking into consideration the impact of the measures implemented to avoid biodiversity loss, large attention is generally given to wild species, especially those at risk of extinction [1,2]. Agricultural biodiversity (i.e., the variety and variability of animals, plants, and microorganisms that are used directly or indirectly for food and agriculture [3]) is largely regarded as a subset of biodiversity. However, agriculture and biodiversity are closely tied. Their mutual dependence is crucial not only to ensure yield today, but also to contribute to a more resilient, sustainable agriculture. This includes the development of solutions for water-saving technologies and for minimizing the detrimental effects of global climate change on crops [4,5].

Plant genetic resources for food and agriculture (PGRFA) are the central components of agricultural biodiversity because they constitute the primary elements of the production process. Crop improvement relies on genetic diversity. Taking into account the trends and efforts of modern breeding, the main part of genetic diversity of cultivated species is expected to be found in traditional varieties, also known as landraces. It is not easy to provide an all-purpose definition of landraces because of their complex nature. Different classifications have been proposed in the literature [6–8]. For instance, the apparently simple distinction between autochthonous and allochthonous landraces is not easy to put in practice, considering that it is difficult to clearly identify geographical boundaries and define in quantitative terms a "recent introduction". In spite of different definitions, human management is integral to the development and maintenance of landraces. Moreover, there is a consensus that plant landraces are dynamic populations that may be particularly adapted to certain environments. Other features that are usually attributed to landraces are yield stability, adaptability to sustainable farming, and resilience to stress, although these characteristics are expected to be evident especially in low-input agricultural systems [6–8]. For instance, submergence survival in rice landraces was retained as a beneficial trait in local ecosystems [9]. Similarly, a rice variety originating from regions with poor soil (e.g., phosphorus-deficient lowlands) was used to isolate the phosphorus-starvation tolerance 1 (PSTOL1) gene, which is absent in modern varieties [10,11].

While landraces are widely employed in low-income countries, in high-income countries they are associated with traditional or amateur farming, and niche products [12]. In advanced economies, landraces usually receive attention because of the consumers' perception of food production of higher quality. Traditional agricultural products have a prominent role in supporting social, historical, and cultural identity, and are becoming increasingly appealing [13]. Trust, transparency, uniqueness, and authenticity are central drives of today's consumers, especially those who have a wealth of resources at their fingertips. The perceived authenticity of a food product is usually connected with its origin and culture, including traditional cooking and specific industrial transformations [14].

The myriad of landraces, which constituted the cultivated genetic material for millennia, has been progressively displaced by modern cultivars in almost all the agricultural settings of the high-income countries [15]. The genetic erosion of cultivated material in southern Italy was estimated to be over 70% in terms of collected samples [16]. Modern agriculture, including plant breeding, has been frequently evoked to explain the incessant erosion of PGRFA [17]. In essence, plant breeding represents a fast evolutionary process to develop improved varieties, and in the last decades, it has been largely based on elite breeding pools [18]. Although new genetic diversity has introduced exploiting crop wild relatives, gene flow, or mutation, plant breeding is usually accompanied by loss of allelic diversity. Trait uniformity and stability are an essential target for breeders, not only for technical reasons related to current cropping systems, but also for protection purposes. Moreover, elite breeding populations move towards a reduced heterozygosity because of inbreeding and the random fluctuation in gene frequency (genetic drift) associated with small, effective population sizes [18].

Plant domestication and breeding have greatly increased food quality by removing unpleasant characteristics (e.g., excessive bitterness, sourness, pungency, toxic compounds, spikiness, hairiness, etc.) and favored others (e.g., sweetness, attractive color, relative amount of flesh or pulp, regular fruit shape, etc.) [19]. Although for different vegetables (e.g., tomato, summer squash, pepper, etc.), fruit size, color, or shape present a morphological variety absent in wild species [20], plant breeding may have unintentionally diminished fruit quality in exchange for production traits [21–23]. Selection for high yield may have reduced the relative amount of main components of fruit taste, such as sugar, aldehydes, and volatile organic compounds (VOC) [24]. Some Italian tomato landraces displayed higher level of metabolites related to fruit quality [25] and a potentially useful phenotypic variability that deserves a further genetic characterization [26]. Landraces can also have superior technological aptitude, which can be exploited for typical, highly-valued products, such as the San Marzano tomato tins and the Portuguese high quality maize bread [27,28].

2. Landraces as a Source of Local Adaptation

Excluding the cultural value, resilience, nutrition content, sensorial value, and compliance to low-input farming are traits that are controlled often by multiple genes. In landraces, these features constitute the bulk of their local adaptation (LA) because they are the target of main farmer-mediated evolutionary forces.

LA is a process by which a population becomes better suited to its local environment than other members of the same species [29]. Although LA is frequently linked to climate, there is ample evidence that plants adapt to different environmental elements, including biotic and non-climatic abiotic factors [30,31]. In ecology, LA is usually measured as the difference in fitness between a population in its environment or growing elsewhere, or by comparing the fitness of a local and an introduced population in one environment [29]. In agriculture, fitness may not necessary be a trait of interest, and LA usually relates to crop yield, or more generally, to a phenotype in response to certain environments and agricultural practices [32]. For instance, differences in maize landraces have been measured that considered phenological, morphological, or physiological traits related to yield [33–35].

While LA in plants has been found in a number of studies [36], relatively less is known on local adaptation due to farming. Farmer-mediated selection does not always go along with the more frequently reported environmental selection, although fertility, fitness, and yield are usually interdependent [37]. Adaptive divergence in quantitative traits is negatively correlated with the rate of population mixing [38], which suggests that LA should be also common in recently introduced landraces (i.e., a relatively reduced number of generations experiencing specific selective regimes). On the other hand, gene flow between populations, which is generally seen as a disruptive force for LA in the absence of strong adaptive selection [39], is more frequent for landraces within some centers of origin [40].

Establishing how populations respond to environmental conditions is not an easy challenge in agriculture as well asevolutionary biology [41]. The positive features of landraces are often linked to specific growing techniques and environmental conditions that cannot be easily reproduced in experimental stations, such as low-fertility soil [42]. For instance, PSTOL1 enhances grain yield in phosphorus-deficient soil [11]. In maize, LA is often associated with altitude, with highland landraces poorly performing in lowland areas, and vice versa [43,44]. For these reasons, reciprocal transplants to reveal underlying factors of LA may not be straightforward, especially for crop landraces adjusted to low-input farming or originating from marginal lands, which are of great interest as a source of adaptive traits useful to increase stress resistance [15]. Moreover, LA may also lead to an adaptability trade-off, and some landraces can adapt to a wide range of environments, whereas others can only adapt to a few environments [6,45].

To grasp adaptive diversity for current and future challenges, it is necessary to identify locations where agronomic and/or historical climatic conditions match predicted changes. The comparison of isolated landraces in similar yet geographically isolated environments would also provide the possibility to understand whether convergent solutions to a specific stress are established at the phenotypic and/or genetic level [46]. At least in natural populations, the probability of gene reuse in parallel or convergent phenotypic evolution was considered high [47].

The study of the phenotypic adaptation also requires establishing whether the observed phenotypic differences between populations are primarily genetically based or the result of phenotypic plasticity (i.e., a plastic response to the environment that does not require genetic changes) [48,49]. Adaptive phenotypic plasticity is typically associated with the magnitude of response of quantitative traits in relation to the environment [50]. However, yield stability across a range of conditions, which is a feature of many landraces [6], can be also a plastic response that is not always supported by a "distinct" phenotypic trait. Quantitative Trait Loci (QTLs), gene expression levels, and epigenetic mechanisms (i.e., those related to priming and acclimation) are likely to contribute to stress adaptation in plants [50,51]. For instance, DNA methylation has also been associated with drought resistance in upland rice, and to adaptation to higher altitude in maize landraces [52,53]. The adaptation of plant species and communities to global climate change is an important trait in breeding for a more climate-resilient agriculture, and is frequently associated with plasticity [54]. A meta-analysis indicated that evidence on the evolutionary adaptation to climate change is still relatively scarce [55].

3. Genomic Scans of Local Adaptation in Landraces

Landraces adaptive differentiation has been found in major crops, such as wheat, maize, rice, barley, and sorghum [56–60]. Uncovering the genetic basis of LA will shed light on the evolutionary forces acting on crops, and on the mechanisms underlying environmental adaptation, stress response, and yield. It is not fully clarified whether selection acts primarily on existing genetic variation within a crop (including introgression from modern varieties), or on new mutations specifically present in landraces' populations [61,62]. In Arabidopsis, local climatic adaptation was associated with environmental-specific selection on existing variants, as well as hard-selective sweep (e.g., a rapid increase of the frequency of new beneficial mutations) [63,64]. In addition, it has also been shown that barley landraces have a mosaic ancestry, with multiple genomic segments from local wild populations that can contribute to adaptive variation [65].

The development of more cost-effective techniques to investigate genetic diversity at the large genomic scale makes it possible to identify sequence variations associated, and hopefully responsible, for superior crop performance in low-input and more sustainable farming systems. Briefly, two strategies are usually employed [66], and both are based on a comparative approach. The first relies on the identification of loci that display significantly different genetic differentiation among populations under the assumption that selection pressure differed [67–69]. This strategy can be applied irrespective of hypotheses about the causative role of the environment. Given the likely presence of some environmental constraint for the landrace, attention is usually given to outlier loci that are subject to positive selection, although positive directional selection may not necessarily increase intraspecific variability. A second strategy aims to correlate environments and genotypes under the assumption that a selective pressure creates associations between allelic frequencies (at the selected loci) and environmental variables. In essence, by analyzing allelic frequencies, it is tested whether a sequence variant, haplotype, or allele is significantly associated with a specific environment or environmental factor (if identified), while controlling for neutral genetic structure [58,59]. If the phenotype of interest and/or the genetic basis of the trait are known (e.g., stress resistance, metabolite production, etc.), the analysis of LA in crop landraces can be also carried out focusing on specific candidate regions or by genome-wide scans (e.g., linkage mapping or genome-wide association studies (GWAS), respectively) [59]. Association mapping in crop landraces (e.g., barley, common bean, soybean, durum, and common wheat) can reveal previously undescribed candidate regions associated with agronomic traits, including biotic stress resistance [70–75]. QTL mapping classically requires structured populations (e.g., recombinant populations deriving from phenotypically divergent inbred lines). The development of introgression lines (ILs) from rice landraces led to the identification of QTLs for yield components and the isolation, by map-based cloning, of an allele (*NAL1*) that increases grain productivity [76]. Scientific and technological advances have also enabled the exploitation of panels of unrelated cultivars or genotypes for QTL studies (i.e., non-candidate-driven association mapping approaches, such as GWAS). For instance, the molecular and phenotypic analysis of 723 wheat landraces revealed markers associated with previously unidentified QTLs relative to different traits [77].

Each method has its pros and cons, as discussed in the literature [66,78,79]. Essentially, neutral and demographic processes can generate correlations between the environment and the genotype that are difficult to distinguish from those arising from LA. Unfortunately, the availability of genomic data does not solve this problem.

For any genomic scan of LA, the type of DNA marker is invariantly restricted to single-nucleotide polymorphisms (SNPs) because of the development of reasonably affordable high-throughput sequencing instruments. However, SNPs are not necessarily superior to other DNA markers, such as simple sequence repeats (SSRs) (also known as microsatellites or short tandem repeats, STRs), because the latter are more suitable to detect recent demographic events and private alleles due to their higher mutation rate and multiallelic nature, respectively. SNPs analysis based on next-generation sequencing (NGS) technologies, however, offers a possibility to investigate a number of polymorphisms that is currently unmatched by any other approach, and strongly increases the possibility to identify adaptive loci [78]. The number of DNA polymorphisms under investigation is important, because it will affect the estimation of the population structure and the generation of the null distribution for statistical hypothesis testing. On the other hand, linkage and more generally, the non-independence of loci, are possible confounding sources that are likely to be more significant when analyzing a very large number of polymorphisms [66].

The genomic analysis of a landrace requires more extended sampling compared to a genetically uniform cultivated plant variety. Therefore, methods that ascertain sequence variations in a fraction of the genome are usually employed. Nonetheless, SNP genotyping chips and other reduced representation methods (for a list of methods, see [80]) are considered not fully adequate for identifying the genomic signatures of LA, because of their reduced genomic sampling power [66,78]. Moreover, SNP arrays are not very effective at capturing rare and previously undescribed variants in diverse genetic resources, and may suffer from ascertainment bias deriving from the SNP discovery process [80,81].

The availability of a reference genome for many crops [82] strongly facilitates the genomic analysis of landraces by resequencing [59,83]. Nonetheless, polymorphisms at the single nucleotide level cannot be considered sufficient to account for the whole LA, and it is necessary to analyze other more computationally demanding structural variations (SV), such as in/del, copy-number variation (CNV), and insertion of transposable elements (TEs). Adaptation to high boron concentration in wheat landraces is associated with multiple genomic changes, such as tetraploid introgression, gene duplication, and variation in gene structure and expression [84]. Different genes conferring resistance to stress (e.g., flooding and metal toxicity), firstly isolated in landraces, display gene CNV [9]. In maize landraces, more than half of the SNPs associated with altitude were within large structural variants (inversions, centromeres, and pericentromeric regions) [85]. A loss-of-function retrotransposon insertion led to adaptation to cultivation at high latitudes in a photoperiod-insensitive soybean landrace [86]. The *de novo* detection of SVs requires a deeper sequence coverage compared with the low-fold approaches usually employed in resequencing [87,88]. To overcome some of these limitations, a metagenome-like assembly strategy based on a low-coverage population sequencing data was employed for the construction of the dispensable rice genome as a more cost- and labor-effective strategy [89]. The recent availability of long-read sequencing technologies (also known as third-generation sequencing) can greatly improve the analysis of genome structure, not only for chromosome scaffolding and haplotype phasing, but also for the identification of long (e.g., >50 bp) structural variants [90].

The concept of "pan-genome" as the sum of the "core genome" (containing genes/sequences present in all strains) and the "dispensable genome" was first developed in microbiology, and later applied also to plants science [91,92]. At the leastin some organisms, the "dispensable genome" significantly contributed to adaptation [91,93,94]. In maize and soybean, a substantial proportion of variation may lay in the "dispensable genome" [93,95]. It has been also suggested that the "dispensable genome" may have a role in the environmental adaptation in soybean [95].

For non-model and orphan species (i.e., those in which there has been little "omics" research), it is likely that genome-wide sequence analysis will be performed using a reduced representation method. Currently, genotyping by sequencing (GbS) represents one of the most affordable methodologies for SNP analysis in large populations. This approach is popular especially for GWA studies in crops because it can be also employed on plant species with complex and large genomes, including polyploids. A GbS-based survey of nucleotide diversity in soybean landraces revealed selective sweeps around starch metabolism genes; GWAS also provided insights into the origin and spread of haplotypes linked to agro-climatic traits [96].

When at least a reference transcriptome is available, exome sequencing, also known as whole-exome sequencing, may represent an affordable option for analyzing a well-characterized adaptive trait or very large plant populations [97], because of its reduced running cost when probes are already available. For instance, an investigation of the barley genomic variability related to environmental conditions was carried out starting from the exome sequencing data of more than 250 georeferenced landraces and wild accessions [97].

The study of RNA molecules by the so-called transcriptomics technologies is also an opportunity to uncover genetic variants, with the added possibility of providing information on the molecular basis of adaptation because differences in transcript abundance are a component of phenotypic variation, especially in response to the environment. In maize, gene expression analysis by microarray has underlined that drought tolerant landraces more rapidly respond to stress compared with susceptible landraces [98]. RNA-Seq, also known as whole transcriptome shotgun sequencing (WTSS), is at present the most widely employed methodology for transcriptomics studies and it has overshadowed chip-based technologies. In a wheat landrace, RNA-Seq highlighted pathways and genes potentially related to resistance against *Fusarium* [99]. The analysis of transcribed or coding sequences also provides the possibility of detecting and coding landrace-specific allelic variations. Lastly, RNA-Seq can be also employed to detect allele-specific expressions in hybrids of cultivars and landraces, which can potentially contribute to adaptation [100]. However, RNA-based genomic scans cannot provide direct information on structural elements (e.g., regulatory sequences, as well as SVs), and may not have enough genomic resolution in very large genomes, unless linkage disequilibrium (LD) is high [78].

Finally, the integration of omics approaches (e.g., from proteomics and metabolomics to foodomics and nutrigenomics) can contribute to understanding the link between landrace-specific bioactive compounds, their importance, and DNA sequence variation. GWA mapping based on metabolomics data (mGWA) has been carried out in crops such as maize, tomato, and rice, which in some cases also exploit landraces [101–103].

4. Current Opportunities and Challenges

Conventional high-input agriculture faces diverse and complex challenges. It is necessary to promote the development and implementation of new agricultural techniques in order to mitigate the negative impact on soil conservation, water management, and biodiversity, as well as increase crop resilience to stress and adaptability to new areas. Agriculture will not overcome these challenges without modern (bio-) technology. Crop landraces represent a readily available resource to address these issues, because they are cultivated material already adapted to low-input agriculture, marginal lands, or stressful environments. For instance, maize landraces have an evolved adaptability to a wider range of environmental conditions than teosinte [104]. Identifying the loci involved in LA provides the possibility of not only defining the genomic basis of adaptation to specific conditions in crops, but also improving our understanding of some agriculturally important traits. Although genetic variability in landraces is considered lower than in wild relatives, different studies have underlined that it is higher than those of improved varieties [105,106]. More crucially, such variability should be readily available not only for germplasm improvement, but also for breeding. Different genes isolated from landraces have been successfully used for breeding programs in major crops [9].

The classic approach to detect LA is to verify whether the phenotypic divergence in candidate traits between populations cannot be explained by drift alone. Current advances in sequencing technologies encourage a genomic characterization of landraces. The genomic scan for LA represents today the first, most affordable step towards the exploitation of the positive features of crop landraces, mainly because of the resources needed for the phenotypic characterization of ample populations in different environments. Advances in high-throughput plant phenotyping facilities give reasons to believe that in the near future landraces characterization will be accelerated [107–109].

Irrespective of the approach and methodology employed, it is common that genomic scans for adaptation provide hundreds of candidate loci. Their identification is the foundation for understanding the physiological basis of adaptation. Therefore, a limiting factor towards the genomic basis of LA in landraces is the functional validation of these loci, an effort that should include the comparison of the trait of interest in near isogenic material in specific agricultural conditions. The correlation between crop response in field and experimental conditions is a longstanding issue in agriculture, and it may be even more relevant for the validation of adaptive genetic variations [75]. The functional validation of candidate loci requires resources that often exceed the ones needed for the genomic analysis of LA, making necessary the selection of a limited number of variants by using additional computational methods [110,111]. Recent developments in plant biotechnology, including genome editing, give reason to believe that the validation of specific variants could be more easily achieved compared with more classic approaches, such as mutagenesis, genetic transformation, or the screening of natural and artificial populations [112]. However, the trait of interest may be highly polygenic, and LA may be the result of a number of relatively modest changes in allelic frequency that underlie or contribute to (unknown) phenotypic traits, as it is likely to occur for the natural variation of plant metabolites [113].

The identification of genetic loci and sequences responsible for LA will unlock the landraces' diversity for precision breeding and plant science. Adaptation to rapidly changing climate conditions and to low-input sustainable agriculture will also require new varieties with, for instance, modified planting time or increased resilience, as well as the (assisted) migration of crops. Landraces are central for developing high-value plant varieties better suited to local conditions, especially for cropping systems that evolve towards a reduced use of off-farm inputs. For all of these reasons, understanding the genomic basis of LA in landraces has the potential to alleviate the environmental impact of agriculture in the near future.

Acknowledgments: This work was supported by the "Salvaguardia della biodiversità agroalimentare in Campania" (SALVE) project, Programma di Sviluppo Rurale per la Campania 2007–2013, misura 214 az. f2.

Author Contributions: Giandomenico Corrado and Rosa Rao wrote and reviewed the article.

Conflicts of Interest: The authors declare no conflict of interest.

References

1. Rands, M.R.; Adams, W.M.; Bennun, L.; Butchart, S.H.; Clements, A.; Coomes, D.; Entwistle, A.; Hodge, I.; Kapos, V.; Scharlemann, J.P. Biodiversity conservation: Challenges beyond 2010. *Science* **2010**, *329*, 1298–1303. [CrossRef] [PubMed]
2. Ripple, W.J.; Chapron, G.; López-Bao, J.V.; Durant, S.M.; Macdonald, D.W.; Lindsey, P.A.; Bennett, E.L.; Beschta, R.L.; Bruskotter, J.T.; Campos-Arceiz, A. Saving the world's terrestrial megafauna. *BioScience* **2016**, *66*, 807–812. [CrossRef] [PubMed]
3. Food and Agriculture Organization (FAO). Background Paper 1. In *Agricultural Biodiversity*; Multifunctional character of agriculture and land conference; FAO: Maastricht, NL, USA, 1999; pp. 1–42.
4. Frison, E.A.; Cherfas, J.; Hodgkin, T. Agricultural biodiversity is essential for a sustainable improvement in food and nutrition security. *Sustainability* **2011**, *3*, 238–253. [CrossRef]
5. Lane, A.; Jarvis, A. Changes in climate will modify the geography of crop suitability: Agricultural biodiversity can help with adaptation. *SAT eJournal* **2007**, *4*, 1–12.
6. Zeven, A.C. Landraces: A review of definitions and classifications. *Euphytica* **1998**, *104*, 127–139. [CrossRef]
7. Villa, T.C.C.; Maxted, N.; Scholten, M.; Ford-Lloyd, B. Defining and identifying crop landraces. *Plant Genet. Resour.* **2005**, *3*, 373–384. [CrossRef]
8. Casañas, F.; Simó, J.; Casals, J.; Prohens, J. Toward an evolved concept of landrace. *Front. Plant Sci.* **2017**, *8*, 145. [CrossRef] [PubMed]
9. Mickelbart, M.V.; Hasegawa, P.M.; Bailey-Serres, J. Genetic mechanisms of abiotic stress tolerance that translate to crop yield stability. *Nat. Rev. Genet.* **2015**, *16*, 237. [CrossRef] [PubMed]
10. Chin, J.H.; Lu, X.; Haefele, S.M.; Gamuyao, R.; Ismail, A.; Wissuwa, M.; Heuer, S. Development and application of gene-based markers for the major rice QTL Phosphorus Uptake 1. *Theor. Appl. Genet.* **2010**, *120*, 1073–1086. [CrossRef] [PubMed]
11. Gamuyao, R.; Chin, J.H.; Pariasca-Tanaka, J.; Pesaresi, P.; Catausan, S.; Dalid, C.; Slamet-Loedin, I.; Tecson-Mendoza, E.M.; Wissuwa, M.; Heuer, S. The protein kinase pstol1 from traditional rice confers tolerance of phosphorus deficiency. *Nature* **2012**, *488*, 535. [CrossRef] [PubMed]

12. Gibson, R.W. A review of perceptual distinctiveness in landraces including an analysis of how its roles have been overlooked in plant breeding for low-input farming systems. *Econ. Bot.* **2009**, *63*, 242–255. [CrossRef]
13. Pícha, K.; Navrátil, J.; Švec, R. Preference to local food vs. Preference to "national" and regional food. *J. Food Prod. Mark.* **2017**, 1–21. [CrossRef]
14. Sims, R. Food, place and authenticity: Local food and the sustainable tourism experience. *J. Sustain. Tour.* **2009**, *17*, 321–336. [CrossRef]
15. Dwivedi, S.L.; Ceccarelli, S.; Blair, M.W.; Upadhyaya, H.D.; Are, A.K.; Ortiz, R. Landrace germplasm for improving yield and abiotic stress adaptation. *Trends Plant Sci.* **2016**, *21*, 31–42. [CrossRef] [PubMed]
16. Hammer, K.; Knüpffer, H.; Xhuveli, L.; Perrino, P. Estimating genetic erosion in landraces—Two case studies. *Genet. Resour. Crop Evol.* **1996**, *43*, 329–336. [CrossRef]
17. Ceccarelli, S. Landraces: Importance and use in breeding and environmentally friendly agronomic systems. In *Agrobiodiversity Conservation: Securing the Diversity of Crop Wild Relatives and Landraces*; CAB International: Wallingford, UK, 2012; pp. 103–117.
18. Cowling, W.A. Sustainable plant breeding. *Plant Breed.* **2013**, *132*, 1–9. [CrossRef]
19. Zohary, D. Unconscious selection and the evolution of domesticated plants. *Econ. Bot.* **2004**, *58*, 5–10. [CrossRef]
20. Bai, Y.; Lindhout, P. Domestication and breeding of tomatoes: What have we gained and what can we gain in the future? *Ann. Bot.* **2007**, *100*, 1085–1094. [CrossRef] [PubMed]
21. Davis, D.R.; Epp, M.D.; Riordan, H.D. Changes in USDA food composition data for 43 garden crops, 1950 to 1999. *J. Am. Coll. Nutr.* **2004**, *23*, 669–682. [CrossRef] [PubMed]
22. Powell, A.L.; Nguyen, C.V.; Hill, T.; Cheng, K.L.; Figueroa-Balderas, R.; Aktas, H.; Ashrafi, H.; Pons, C.; Fernández-Muñoz, R.; Vicente, A. Uniform ripening encodes a golden 2-like transcription factor regulating tomato fruit chloroplast development. *Science* **2012**, *336*, 1711–1715. [CrossRef] [PubMed]
23. Murphy, K.M.; Reeves, P.G.; Jones, S.S. Relationship between yield and mineral nutrient concentrations in historical and modern spring wheat cultivars. *Euphytica* **2008**, *163*, 381–390. [CrossRef]
24. Klee, H.J.; Tieman, D.M. Genetic challenges of flavor improvement in tomato. *Trends Genet.* **2013**, *29*, 257–262. [CrossRef] [PubMed]
25. Andreakis, N.; Giordano, I.; Pentangelo, A.; Fogliano, V.; Graziani, G.; Monti, L.M.; Rao, R. DNA fingerprinting and quality traits of Corbarino cherry-like tomato landraces. *J. Agric. Food Chem.* **2004**, *52*, 3366–3371. [CrossRef] [PubMed]
26. Baldina, S.; Picarella, M.E.; Troise, A.D.; Pucci, A.; Ruggieri, V.; Ferracane, R.; Barone, A.; Fogliano, V.; Mazzucato, A. Metabolite profiling of Italian tomato landraces with different fruit types. *Front. Plant Sci.* **2016**, *7*, 664. [CrossRef] [PubMed]
27. Scarano, D.; Rao, R.; Masi, P.; Corrado, G. SSR fingerprint reveals mislabeling in commercial processed tomato products. *Food Control* **2015**, *51*, 397–401. [CrossRef]
28. Patto, V.; Alves, N.; Almeida, C.S.; Mendes, P.; Satovic, Z. Is the bread making technological ability of portuguese traditional maize landraces associated with their genetic diversity? *Maydica* **2009**, *54*, 297–311.
29. Kawecki, T.J.; Ebert, D. Conceptual issues in local adaptation. *Ecol. Lett.* **2004**, *7*, 1225–1241. [CrossRef]
30. Aitken, S.N.; Whitlock, M.C. Assisted gene flow to facilitate local adaptation to climate change. *Annu. Rev. Ecol. Evol. Syst.* **2013**, *44*, 367–388. [CrossRef]
31. Brachi, B.; Meyer, C.G.; Villoutreix, R.; Platt, A.; Morton, T.C.; Roux, F.; Bergelson, J. Coselected genes determine adaptive variation in herbivore resistance throughout the native range of *Arabidopsis thaliana*. *Proc. Natl. Acad. Sci. USA* **2015**, *112*, 4032–4037. [CrossRef] [PubMed]
32. Joshi, J.; Schmid, B.; Caldeira, M.; Dimitrakopoulos, P.; Good, J.; Harris, R.; Hector, A.; Huss-Danell, K.; Jumpponen, A.; Minns, A. Local adaptation enhances performance of common plant species. *Ecol. Lett.* **2001**, *4*, 536–544. [CrossRef]
33. Stehli, A.; Soldati, A.; Stamp, P. Vegetative performance of tropical highland maize (*Zea. mays* L.) in the field. *J. Agron. Crop Sci.* **1999**, *183*, 193–198. [CrossRef]
34. Khan, Z.; Khalil, S.; Nigar, S.; Khalil, I.; Haq, I.; Ahmad, I.; Ali, A.; Khan, M. Phenology and yield of sweet corn landraces influenced by planting dates. *Sarhad. J. Agric.* **2009**, *25*, 153–157.
35. Ellis, R.; Summerfield, R.; Edmeades, G.; Roberts, E. Photoperiod, temperature, and the interval from sowing to tassel initiation in diverse cultivars of maize. *Crop Sci.* **1992**, *32*, 1225–1232. [CrossRef]

36. Leimu, R.; Fischer, M. A meta-analysis of local adaptation in plants. *PLoS ONE* **2008**, *3*, e4010. [CrossRef] [PubMed]

37. Mercer, K.L.; Perales, H.R. Evolutionary response of landraces to climate change in centers of crop diversity. *Evol. Appl.* **2010**, *3*, 480–493. [CrossRef] [PubMed]

38. Hendry, A.P.; Day, T.; Taylor, E.B. Population mixing and the adaptive divergence of quantitative traits in discrete populations: A theoretical framework for empirical tests. *Evolution* **2001**, *55*, 459–466. [CrossRef]

39. Tigano, A.; Friesen, V.L. Genomics of local adaptation with gene flow. *Mol. Ecol.* **2016**, *25*, 2144–2164. [CrossRef] [PubMed]

40. Van Heerwaarden, J.; Van Eeuwijk, F.; Ross-Ibarra, J. Genetic diversity in a crop metapopulation. *Heredity* **2010**, *104*, 28. [CrossRef] [PubMed]

41. Savolainen, O.; Lascoux, M.; Merilä, J. Ecological genomics of local adaptation. *Nat. Rev. Genet.* **2013**, *14*, 807–820. [CrossRef] [PubMed]

42. Ceccarelli, S. Adaptation to low/high input cultivation. *Euphytica* **1996**, *92*, 203–214. [CrossRef]

43. Lafitte, H.; Edmeades, G. Temperature effects on radiation use and biomass partitioning in diverse tropical maize cultivars. *Field Crops Res.* **1997**, *49*, 231–247. [CrossRef]

44. Mercer, K.; Martínez-Vásquez, Á.; Perales, H.R. Asymmetrical local adaptation of maize landraces along an altitudinal gradient. *Evol. Appl.* **2008**, *1*, 489–500. [CrossRef] [PubMed]

45. Hereford, J. A quantitative survey of local adaptation and fitness trade-offs. *Am. Nat.* **2009**, *173*, 579–588. [CrossRef] [PubMed]

46. Bennici, A. The convergent evolution in plants. *Riv. Biol.* **2002**, *96*, 485–489.

47. Conte, G.L.; Arnegard, M.E.; Peichel, C.L.; Schluter, D. The probability of genetic parallelism and convergence in natural populations. *Proc. R. Soc. B* **2012**, *279*, 5039–5047. [CrossRef] [PubMed]

48. Kingsolver, J.G.; Pfennig, D.W.; Servedio, M.R. Migration, local adaptation and the evolution of plasticity. *Trends Ecol. Evol.* **2002**, *17*, 540–541. [CrossRef]

49. Pajoro, A.; Verhage, L.; Immink, R.G. Plasticity versus adaptation of ambient–temperature flowering response. *Trends Plant Sci.* **2016**, *21*, 6–8. [CrossRef] [PubMed]

50. Des Marais, D.L.; Hernandez, K.M.; Juenger, T.E. Genotype-by-environment interaction and plasticity: Exploring genomic responses of plants to the abiotic environment. *Annu. Rev. Ecol. Evol. Syst.* **2013**, *44*, 5–29. [CrossRef]

51. Mirouze, M.; Paszkowski, J. Epigenetic contribution to stress adaptation in plants. *Curr. Opin. Plant Biol.* **2011**, *14*, 267–274. [CrossRef] [PubMed]

52. Xia, H.; Huang, W.; Xiong, J.; Tao, T.; Zheng, X.; Wei, H.; Yue, Y.; Chen, L.; Luo, L. Adaptive epigenetic differentiation between upland and lowland rice ecotypes revealed by methylation-sensitive amplified polymorphism. *PLoS ONE* **2016**, *11*, e0157810. [CrossRef] [PubMed]

53. Rius, S.P.; Emiliani, J.; Casati, P. P1 epigenetic regulation in leaves of high altitude maize landraces: Effect of UV-b radiation. *Front. Plant Sci.* **2016**, *7*. [CrossRef] [PubMed]

54. Valladares, F.; Gianoli, E.; Gómez, J.M. Ecological limits to plant phenotypic plasticity. *New Phytol.* **2007**, *176*, 749–763. [CrossRef] [PubMed]

55. Merilä, J.; Hendry, A.P. Climate change, adaptation, and phenotypic plasticity: The problem and the evidence. *Evol. Appl.* **2014**, *7*, 1–14. [CrossRef] [PubMed]

56. Iwaki, K.; Haruna, S.; Niwa, T.; Kato, K. Adaptation and ecological differentiation in wheat with special reference to geographical variation of growth habit and Vrn genotype. *Plant Breed.* **2001**, *120*, 107–114. [CrossRef]

57. Westengen, O.T.; Berg, P.R.; Kent, M.P.; Brysting, A.K. Spatial structure and climatic adaptation in african maize revealed by surveying SNP diversity in relation to global breeding and landrace panels. *PLoS ONE* **2012**, *7*, e47832. [CrossRef] [PubMed]

58. Lasky, J.R.; Upadhyaya, H.D.; Ramu, P.; Deshpande, S.; Hash, C.T.; Bonnette, J.; Juenger, T.E.; Hyma, K.; Acharya, C.; Mitchell, S.E. Genome-environment associations in sorghum landraces predict adaptive traits. *Sci. Adv.* **2015**, *1*, e1400218. [CrossRef] [PubMed]

59. Huang, X.; Sang, T.; Zhao, Q.; Feng, Q.; Zhao, Y.; Li, C.; Zhu, C.; Lu, T.; Zhang, Z.; Li, M. Genome-wide association studies of 14 agronomic traits in rice landraces. *Nat. Genet.* **2010**, *42*, 961–967. [CrossRef] [PubMed]

60. Pswarayi, A.; Van Eeuwijk, F.; Ceccarelli, S.; Grando, S.; Comadran, J.; Russell, J.; Pecchioni, N.; Tondelli, A.; Akar, T.; Al-Yassin, A. Changes in allele frequencies in landraces, old and modern barley cultivars of marker loci close to QTL for grain yield under high and low input conditions. *Euphytica* **2008**, *163*, 435–447. [CrossRef]

61. Bitocchi, E.; Nanni, L.; Rossi, M.; Rau, D.; Bellucci, E.; Giardini, A.; Buonamici, A.; Vendramin, G.G.; Papa, R. Introgression from modern hybrid varieties into landrace populations of maize (*Zea. mays* ssp. *mays* L.) in central italy. *Mol. Ecol.* **2009**, *18*, 603–621.

62. Massawe, F.; Dickinson, M.; Roberts, J.; Azam-Ali, S. Genetic diversity in bambara groundnut (*Vigna subterranea* (L.) Verdc) landraces revealed by aflp markers. *Genome* **2002**, *45*, 1175–1180. [CrossRef] [PubMed]

63. Fournier-Level, A.; Korte, A.; Cooper, M.D.; Nordborg, M.; Schmitt, J.; Wilczek, A.M. A map of local adaptation in *Arabidopsis thaliana*. *Science* **2011**, *334*, 86–89. [CrossRef] [PubMed]

64. Hancock, A.M.; Brachi, B.; Faure, N.; Horton, M.W.; Jarymowycz, L.B.; Sperone, F.G.; Toomajian, C.; Roux, F.; Bergelson, J. Adaptation to climate across the *Arabidopsis thaliana* genome. *Science* **2011**, *334*, 83–86. [CrossRef] [PubMed]

65. Poets, A.M.; Fang, Z.; Clegg, M.T.; Morrell, P.L. Barley landraces are characterized by geographically heterogeneous genomic origins. *Genome Biol.* **2015**, *16*, 173. [CrossRef] [PubMed]

66. Hoban, S.; Kelley, J.L.; Lotterhos, K.E.; Antolin, M.F.; Bradburd, G.; Lowry, D.B.; Poss, M.L.; Reed, L.K.; Storfer, A.; Whitlock, M.C. Finding the genomic basis of local adaptation: Pitfalls, practical solutions, and future directions. *Am. Nat.* **2016**, *188*, 379–397. [CrossRef] [PubMed]

67. Cavanagh, C.R.; Chao, S.; Wang, S.; Huang, B.E.; Stephen, S.; Kiani, S.; Forrest, K.; Saintenac, C.; Brown-Guedira, G.L.; Akhunova, A. Genome-wide comparative diversity uncovers multiple targets of selection for improvement in hexaploid wheat landraces and cultivars. *Proc. Natl. Acad. Sci. USA* **2013**, *110*, 8057–8062. [CrossRef] [PubMed]

68. Corrado, G.; Piffanelli, P.; Caramante, M.; Coppola, M.; Rao, R. SNP genotyping reveals genetic diversity between cultivated landraces and contemporary varieties of tomato. *BMC Genom.* **2013**, *14*, 835. [CrossRef] [PubMed]

69. Xia, H.; Zheng, X.; Chen, L.; Gao, H.; Yang, H.; Long, P.; Rong, J.; Lu, B.; Li, J.; Luo, L. Genetic differentiation revealed by selective loci of drought-responding EST-SSRs between upland and lowland rice in China. *PLoS ONE* **2014**, *9*, e106352. [CrossRef] [PubMed]

70. Miklas, P.N.; Coyne, D.P.; Grafton, K.F.; Mutlu, N.; Reiser, J.; Lindgren, D.T.; Singh, S.P. A major QTL for common bacterial blight resistance derives from the common bean great northern landrace cultivar Montana No. 5. *Euphytica* **2003**, *131*, 137–146. [CrossRef]

71. Liu, B.; Abe, J. QTL mapping for photoperiod insensitivity of a Japanese soybean landrace Sakamotowase. *J. Hered.* **2009**, *101*, 251–256. [CrossRef] [PubMed]

72. Mengistu, D.K.; Kidane, Y.G.; Catellani, M.; Frascaroli, E.; Fadda, C.; Pè, M.E.; Dell'Acqua, M. High-density molecular characterization and association mapping in ethiopian durum wheat landraces reveals high diversity and potential for wheat breeding. *Plant Biotechnol. J.* **2016**, *14*, 1800–1812. [CrossRef] [PubMed]

73. Muleta, K.T.; Rouse, M.N.; Rynearson, S.; Chen, X.; Buta, B.G.; Pumphrey, M.O. Characterization of molecular diversity and genome-wide mapping of loci associated with resistance to stripe rust and stem rust in Ethiopian bread wheat accessions. *BMC Plant Biol.* **2017**, *17*, 134. [CrossRef] [PubMed]

74. Sehgal, D.; Dreisigacker, S.; Belen, S.; Küçüközdemir, Ü.; Mert, Z.; Özer, E.; Morgounov, A. Mining centuries old in situ conserved Turkish wheat landraces for grain yield and stripe rust resistance genes. *Front. Genet.* **2016**, *7*. [CrossRef] [PubMed]

75. Mamo, B.E.; Barber, B.L.; Steffenson, B.J. Genome-wide association mapping of zinc and iron concentration in barley landraces from Ethiopia and Eritrea. *J. Cereal Sci.* **2014**, *60*, 497–506. [CrossRef]

76. Fujita, D.; Trijatmiko, K.R.; Tagle, A.G.; Sapasap, M.V.; Koide, Y.; Sasaki, K.; Tsakirpaloglou, N.; Gannaban, R.B.; Nishimura, T.; Yanagihara, S. Nal1 allele from a rice landrace greatly increases yield in modern indica cultivars. *Proc. Natl. Acad. Sci. USA* **2013**, *110*, 20431–20436. [CrossRef] [PubMed]

77. Liu, Y.; Lin, Y.; Gao, S.; Li, Z.; Ma, J.; Deng, M.; Chen, G.; Wei, Y.; Zheng, Y. A genome-wide association study of 23 agronomic traits in Chinese wheat landraces. *Plant J.* **2017**. [CrossRef] [PubMed]

78. Tiffin, P.; Ross-Ibarra, J. Advances and limits of using population genetics to understand local adaptation. *Trends Ecol. Evol.* **2014**, *29*, 673–680. [CrossRef] [PubMed]

79. Bergelson, J.; Roux, F. Towards identifying genes underlying ecologically relevant traits in *Arabidopsis thaliana*. *Nat. Rev. Genet.* **2010**, *11*, 867. [CrossRef] [PubMed]
80. Rasheed, A.; Hao, Y.; Xia, X.; Khan, A.; Xu, Y.; Varshney, R.K.; He, Z. Crop breeding chips and genotyping platforms: Progress, challenges, and perspectives. *Mol. Plant* **2017**, *10*, 1047–1064. [CrossRef] [PubMed]
81. Lachance, J.; Tishkoff, S.A. SNP ascertainment bias in population genetic analyses: Why it is important, and how to correct it. *Bioessays* **2013**, *35*, 780–786. [CrossRef] [PubMed]
82. Wendel, J.F.; Jackson, S.A.; Meyers, B.C.; Wing, R.A. Evolution of plant genome architecture. *Genome Biol.* **2016**, *17*, 37. [CrossRef] [PubMed]
83. Lai, J.; Li, R.; Xu, X.; Jin, W.; Xu, M.; Zhao, H.; Xiang, Z.; Song, W.; Ying, K.; Zhang, M. Genome-wide patterns of genetic variation among elite maize inbred lines. *Nat. Genet.* **2010**, *42*, 1027–1030. [CrossRef] [PubMed]
84. Pallotta, M.; Schnurbusch, T.; Hayes, J.; Hay, A.; Baumann, U.; Paull, J.; Langridge, P.; Sutton, T. Molecular basis of adaptation to high soil boron in wheat landraces and elite cultivars. *Nature* **2014**, *514*, 88. [CrossRef] [PubMed]
85. Navarro, J.A.R.; Willcox, M.; Burgueño, J.; Romay, C.; Swarts, K.; Trachsel, S.; Preciado, E.; Terron, A.; Delgado, H.V.; Vidal, V. A study of allelic diversity underlying flowering-time adaptation in maize landraces. *Nat. Genet.* **2017**, *49*, 476–480. [CrossRef] [PubMed]
86. Kanazawa, A.; Liu, B.; Kong, F.; Arase, S.; Abe, J. Adaptive evolution involving gene duplication and insertion of a novel Ty1/copia-like retrotransposon in soybean. *J. Mol. Evol.* **2009**, *69*, 164–175. [CrossRef] [PubMed]
87. Francia, E.; Pecchioni, N.; Policriti, A.; Scalabrin, S. CNV and structural variation in plants: Prospects of NGS approaches. In *Advances in the Understanding of Biological Sciences Using Next Generation Sequencing (NGS) Approaches*; Springer International Publishing: Gewerbestrasse, Switzerland, 2015; pp. 211–232.
88. Ye, K.; Hall, G.; Ning, Z. Structural variation detection from next generation sequencing. *Next Gener. Seq. Appl.* **2016**, *1*, 007. [CrossRef]
89. Yao, W.; Li, G.; Zhao, H.; Wang, G.; Lian, X.; Xie, W. Exploring the rice dispensable genome using a metagenome-like assembly strategy. *Genome Biol.* **2015**, *16*, 187. [CrossRef] [PubMed]
90. Jiao, W.-B.; Schneeberger, K. The impact of third generation genomic technologies on plant genome assembly. *Curr. Opin. Plant Biol.* **2017**, *36*, 64–70. [CrossRef] [PubMed]
91. Medini, D.; Donati, C.; Tettelin, H.; Masignani, V.; Rappuoli, R. The microbial pan-genome. *Curr. Opin. Genet. Dev.* **2005**, *15*, 589–594. [CrossRef] [PubMed]
92. Morgante, M.; De Paoli, E.; Radovic, S. Transposable elements and the plant pan-genomes. *Curr. Opin. Plant Biol.* **2007**, *10*, 149–155. [CrossRef] [PubMed]
93. Hirsch, C.N.; Foerster, J.M.; Johnson, J.M.; Sekhon, R.S.; Muttoni, G.; Vaillancourt, B.; Peñagaricano, F.; Lindquist, E.; Pedraza, M.A.; Barry, K. Insights into the maize pan-genome and pan-transcriptome. *Plant Cell* **2014**, *26*, 121–135. [CrossRef] [PubMed]
94. Marroni, F.; Pinosio, S.; Morgante, M. Structural variation and genome complexity: Is dispensable really dispensable? *Curr. Opin. Plant Biol.* **2014**, *18*, 31–36. [CrossRef] [PubMed]
95. Li, Y.-H.; Zhou, G.; Ma, J.; Jiang, W.; Jin, L.-G.; Zhang, Z.; Guo, Y.; Zhang, J.; Sui, Y.; Zheng, L. De novo assembly of soybean wild relatives for pan-genome analysis of diversity and agronomic traits. *Nat. Biotechnol.* **2014**, *32*, 1045–1052. [CrossRef] [PubMed]
96. Morris, G.P.; Ramu, P.; Deshpande, S.P.; Hash, C.T.; Shah, T.; Upadhyaya, H.D.; Riera-Lizarazu, O.; Brown, P.J.; Acharya, C.B.; Mitchell, S.E. Population genomic and genome-wide association studies of agroclimatic traits in sorghum. *Proc. Natl. Acad. Sci. USA* **2013**, *110*, 453–458. [CrossRef] [PubMed]
97. Russell, J.; Mascher, M.; Dawson, I.K.; Kyriakidis, S.; Calixto, C.; Freund, F.; Bayer, M.; Milne, I.; Marshall-Griffiths, T.; Heinen, S. Exome sequencing of geographically diverse barley landraces and wild relatives gives insights into environmental adaptation. *Nat. Genet.* **2016**, *48*, 1024–1030. [CrossRef] [PubMed]
98. Hayano-Kanashiro, C.; Calderón-Vázquez, C.; Ibarra-Laclette, E.; Herrera-Estrella, L.; Simpson, J. Analysis of gene expression and physiological responses in three Mexican maize landraces under drought stress and recovery irrigation. *PLoS ONE* **2009**, *4*, e7531. [CrossRef] [PubMed]
99. Xiao, J.; Jin, X.; Jia, X.; Wang, H.; Cao, A.; Zhao, W.; Pei, H.; Xue, Z.; He, L.; Chen, Q. Transcriptome-based discovery of pathways and genes related to resistance against Fusarium head blight in wheat landrace Wangshuibai. *BMC Genom.* **2013**, *14*, 197. [CrossRef] [PubMed]

100. Aguilar-Rangel, M.R.; Montes, R.A.C.; González-Segovia, E.; Ross-Ibarra, J.; Simpson, J.K.; Sawers, R.J. Allele specific expression analysis identifies regulatory variation associated with stress-related genes in the mexican highland maize landrace Palomero Toluqueño. *PeerJ* **2017**, *5*, e3737. [CrossRef] [PubMed]

101. Riedelsheimer, C.; Lisec, J.; Czedik-Eysenberg, A.; Sulpice, R.; Flis, A.; Grieder, C.; Altmann, T.; Stitt, M.; Willmitzer, L.; Melchinger, A.E. Genome-wide association mapping of leaf metabolic profiles for dissecting complex traits in maize. *Proc. Natl. Acad. Sci. USA* **2012**, *109*, 8872–8877. [CrossRef] [PubMed]

102. Chen, W.; Gao, Y.; Xie, W.; Gong, L.; Lu, K.; Wang, W.; Li, Y.; Liu, X.; Zhang, H.; Dong, H. Genome-wide association analyses provide genetic and biochemical insights into natural variation in rice metabolism. *Nat. Genet.* **2014**, *46*, 714–721. [CrossRef] [PubMed]

103. Sauvage, C.; Segura, V.; Bauchet, G.; Stevens, R.; Do, P.T.; Nikoloski, Z.; Fernie, A.R.; Causse, M. Genome-wide association in tomato reveals 44 candidate loci for fruit metabolic traits. *Plant Physiol.* **2014**, *165*, 1120–1132. [CrossRef] [PubMed]

104. Ruiz Corral, J.A.; Durán Puga, N.; Sánchez González, J.D.J.; Ron Parra, J.; González Eguiarte, D.R.; Holland, J.; Medina García, G. Climatic adaptation and ecological descriptors of 42 Mexican maize races. *Crop Sci.* **2008**, *48*, 1502–1512. [CrossRef]

105. Tang, S.; Knapp, S.J. Microsatellites uncover extraordinary diversity in native American land races and wild populations of cultivated sunflower. *TAG Theor. Appl. Genet.* **2003**, *106*, 990–1003. [CrossRef] [PubMed]

106. Warburton, M.; Reif, J.; Frisch, M.; Bohn, M.; Bedoya, C.; Xia, X.; Crossa, J.; Franco, J.; Hoisington, D.; Pixley, K. Genetic diversity in CIMMYT nontemperate maize germplasm: Landraces, open pollinated varieties, and inbred lines. *Crop Sci.* **2008**, *48*, 617–624. [CrossRef]

107. Shakoor, N.; Lee, S.; Mockler, T.C. High throughput phenotyping to accelerate crop breeding and monitoring of diseases in the field. *Curr. Opin. Plant Biol.* **2017**, *38*, 184–192. [CrossRef] [PubMed]

108. Tanger, P.; Klassen, S.; Mojica, J.P.; Lovell, J.T.; Moyers, B.T.; Baraoidan, M.; Naredo, M.E.B.; McNally, K.L.; Poland, J.; Bush, D.R. Field-based high throughput phenotyping rapidly identifies genomic regions controlling yield components in rice. *Sci. Rep.* **2017**, *7*, 42839. [CrossRef] [PubMed]

109. Yang, W.; Guo, Z.; Huang, C.; Duan, L.; Chen, G.; Jiang, N.; Fang, W.; Feng, H.; Xie, W.; Lian, X. Combining high-throughput phenotyping and genome-wide association studies to reveal natural genetic variation in rice. *Nat. Commun.* **2014**, *5*, 5087. [CrossRef] [PubMed]

110. Joost, S.; Vuilleumier, S.; Jensen, J.D.; Schoville, S.; Leempoel, K.; Stucki, S.; Widmer, I.; Melodelima, C.; Rolland, J.; Manel, S. Uncovering the genetic basis of adaptive change: On the intersection of landscape genomics and theoretical population genetics. *Mol. Ecol.* **2013**, *22*, 3659–3665. [CrossRef] [PubMed]

111. Rellstab, C.; Gugerli, F.; Eckert, A.J.; Hancock, A.M.; Holderegger, R. A practical guide to environmental association analysis in landscape genomics. *Mol. Ecol.* **2015**, *24*, 4348–4370. [CrossRef] [PubMed]

112. Cardi, T. Cisgenesis and genome editing: Combining concepts and efforts for a smarter use of genetic resources in crop breeding. *Plant Breed.* **2016**, *135*, 139–147. [CrossRef]

113. Rowe, H.C.; Hansen, B.G.; Halkier, B.A.; Kliebenstein, D.J. Biochemical networks and epistasis shape the *Arabidopsis thaliana* metabolome. *Plant Cell* **2008**, *20*, 1199–1216. [CrossRef] [PubMed]

diversity

MDPI

Review

NGS-Based Genotyping, High-Throughput Phenotyping and Genome-Wide Association Studies Laid the Foundations for Next-Generation Breeding in Horticultural Crops

Nunzio D'Agostino * and Pasquale Tripodi

CREA Research Centre for Vegetable and Ornamental Crops, 84098 Pontecagnano Faiano, Italy;
pasquale.tripodi@crea.gov.it
* Correspondence: nunzio.dagostino@crea.gov.it; Tel.: +39-089-386-243

Received: 26 July 2017; Accepted: 13 September 2017; Published: 15 September 2017

Abstract: Demographic trends and changes to climate require a more efficient use of plant genetic resources in breeding programs. Indeed, the release of high-yielding varieties has resulted in crop genetic erosion and loss of diversity. This has produced an increased susceptibility to severe stresses and a reduction of several food quality parameters. Next generation sequencing (NGS) technologies are being increasingly used to explore "gene space" and to provide high-resolution profiling of nucleotide variation within germplasm collections. On the other hand, advances in high-throughput phenotyping are bridging the genotype-to-phenotype gap in crop selection. The combination of allelic and phenotypic data points via genome-wide association studies is facilitating the discovery of genetic *loci* that are associated with key agronomic traits. In this review, we provide a brief overview on the latest NGS-based and phenotyping technologies and on their role to unlocking the genetic potential of vegetable crops; then, we discuss the paradigm shift that is underway in horticultural crop breeding.

Keywords: plant genetic resources; vegetable crops; horticulture; single nucleotide polymorphisms; genetic diversity; phenotyping; genome-wide association studies; genomic selection

---

1. The Use of Plant Genetic Resources in Vegetable Crop Improvement

The main challenges for vegetable crop improvement are linked to the sustainable development of agriculture, food security, evolution of dietary styles, the growing consumers' demand for food quality, the spread of non-communicable diseases, and, finally, the under-nutrition due to deficiencies in vitamins and minerals ("hidden hunger"). The design and the development of breeding programs can provide effective responses to all of these challenges. One of the recognized hubs in breeding activities is the exploitation of crop genetic diversity aiming at the beneficial allele-hunting process.

The first part of this review will focus on the importance of biodiversity in horticultural crops, on the main initiatives for germplasm conservation and on the assessment of potentiality and constraints for the use of biodiversity in breeding programs.

1.1. Erosion of Genetic Diversity in Crops

Loss of genetic diversity in crop species, also referred to as genetic erosion, is a step-by-step process that has begun with human population growth and the expansion of human activities whose effects have had serious consequences in ancient, traditional, and modern agriculture.

Early farmers gradually abandoned their nomadic hunter-gatherer habits in favor of semi-sedentary/sedentary agriculture as the primary mode of supplying plant food resources [1].

This process, which is dated back to 12,000 years ago, is known as plant domestication and involved a broad spectrum of transitions, which has led to an increased adaptation of plants to cultivation and utilization by humans [2]. In vegetable crops, the "domestication syndrome" led to combinations of different traits, such as changes in plant architecture and reproductive strategy, the increase of fruit and seed size, and the loss of secondary metabolites [3].

Although the phases of domestication and diversification have not been homogeneous in all areas of the World, in the early stages they have caused the selection of naturally occurring variants in the wild ancestors of crops (i.e., crop wild relatives; CWR) to perpetuate only those species suitable to survive in various agro-ecological habitats (pre-domestication). Afterwards, continuous rounds of selection have completely re-shaped wild into domesticated species [1]. Such a process, known as selective breeding, lasted until early 1900's generating a multitude of landraces (LR) with an intermediate level of variability and differentiation that have been mainly selected for specific adaptation to local environments and for desirable quality traits.

An exciting example on the history of domestication and genetic improvement concerns the cultivated tomato (*Solanum lycopersicum* L.). *S. pimpinellifolium*, the small-fruited species considered to be its wild ancestor, has been subjected to main domestication cycles, firstly in Peru and Ecuador, and later in Mexico [4]. The resulting domesticated cultivars have been brought to Europe and then spread all over the World, where an intensive selection has been carried out in the last two centuries leading to cultivars with an increased fruit size (~100 times larger than its ancestor) [5], but less flavorful than heirloom varieties [6].

The advent of modern agriculture and the impact of the "Green Revolution", from the second half of the 20th century onwards, have introduced high-yielding and phenotypically uniform varieties that are better adapted to industrial agriculture (Figure 1). However, despite the gains breeders have seen during the past years, most of the natural variability has been lost as farmers abandoned traditional landraces in favor of hybrids that are generally more marketable and valuable. This has led to a sharp reduction of genetic diversity, which made crop substantially devoid of differences in quality traits [6]. Although cultivars carrying resistances are continuously released, a reduction of genetic diversity and the extensive use of few crop varieties can lead to critical consequences in the emergence of new pests and diseases [7].

It has been estimated that only 200 out of more than 275,000 species of flowering plants have been domesticated [8,9], and a relatively small number of these accounts for 95% of food supply [10]. Among horticultural crops, species belonging to the Solanaceae family along with those from Poaceae, Fabaceae, and Rosaceae explain more than 60% of the World calories [11]. Food production based on such restricted number of species represents a risk that should be not left out in the next decades with the human population is expected to reach 9.6 billion by 2050 [12], the occurrence of extreme weather events, the reduced availability of natural resources and the related increase in food demand [13]. If on one hand, the number of crop commodities contributing to food supply is expected to grow [14], on the other hand, minor crops cannot completely face the challenges of the new millennium. For the latter reason, CWR and LR, being tailored to extreme conditions and not having been particularly conditioned by the selection process, might be exploited for breeding purposes with innovative genomic and phenomic approaches since they are valuable repositories of traits. Indeed, the thorough exploration of plant genetic resources (PGR) and the preservation and use of CWR and LR are primary targets for the future progress of breeding programs.

1.2. Strategies for Collection and Conservation of Plant Genetic Resources

The first concerns on the effects of genetic erosion date back to the half of the 20th century, when Vavilov called attention in crop diversity conservation, motivating the institution of gene banks [15,16]. Since then, strategies of in-situ and ex-situ management have been proposed [17]. In-situ conservation is generally applied in the natural habitat where crop is cultivated and offers the possibility to maintain the ecosystem health as well as the dynamic evolution of CWR diversity in relation to parallel environmental changes [18]. However, these advantages do not match the easiness to define a comprehensive network of protected areas for all species [17]. Ex situ strategies better fit with the possibility to conserve genetically representative populations in designed areas other than natural habitats.

Seeds are the most common form of conservation due to the relatively low cost of management. Other methods include parts of plants in tissue culture or cryo-preserved and mature individuals in field collections (i.e., arboretum). To date, more than 1750 gene banks conserve ~7 million of accessions worldwide in the form of seeds [19].

Although preserving PGR is essential to ensure the development of new crop varieties, the lack of adequate data on the accessions stored in gene banks is the main issue that hinders their effective use [20]. The large size of these collections complicates their characterization, making the identification of novel/beneficial alleles expensive. An approach that enhances the exploitation of the huge variability present in gene banks is the development of core collections (CC), which are represented by a reduced set of accessions capturing most of the variability with minimum repetitiveness [21]. While the definition of CC was mainly based on ecological and geographical information, the advent of novel genomics technologies allowed the efficiency of the selection and utilization of large germplasm collections stored in gene banks to be improved. Since then, several studies have been published, and different types of collections have been proposed [22].

By exploiting the information stored in different databases, a CC can be assembled fairly easily on an as-needed basis depending on target trait(s). We report as examples the tool developed by the Centre for Genetic Resource (CGN; The Netherlands), that allows a maximum diversity subset of accessions to be selected on-line based on user defined selection criteria (https://goo.gl/XhEuwR); the PowerCore program [23], and the Signal Processing Tool method recently developed by Borrayo and Takeya [24].

An additional approach, which relies on the combination of environmental data with specific plant characteristics, is based on the Focused Identification of Germplasm Strategy (FIGS). FIGS allows for the identification of a core of accessions with a higher probability of containing specific "target" traits subjected to the selection pressure of the environment from which they were originally sampled [25]. The method, which maximizes the possibility to capture specific adaptive traits by means of high-throughput geographic information system (GIS) technologies, has been mainly used to identify sources of resistance to abiotic and biotic stresses in cereals, but has not yet successfully applied in horticultural crops [26]. Despite the potentiality of CC, debate on their effectiveness and criteria of selection to adopt are still underway [27].

Collection, conservation, and utilization of PGR in agriculture are therefore major issues involving several international bodies and substantial investments. The International Treaty on Plant Genetic Resources for Food and Agriculture (FAO) and the Convention on Biological Diversity (CBD) have defined strategies to avoid loss of plant genetic diversity estimated to be 25–35% [28]. Established efforts have included actions to improve the effectiveness in-situ conservation, secure safety ex-situ storage, strengthening efforts of public and private breeders in PGR characterization and utilization [28]. More recently, global research alliances [29] and transnational projects [30,31] have been stipulated with the aim to unlock the potential of vegetable crop diversity by means of innovative genomics and phenomic approaches, and make it available for researchers and breeders.

1.3. Importance of Plant Genetic Resources and Biodiversity in Breeding Programs

In the last 35 years, several breeding programs involving the use of CWR have been developed for a wide range of crops. The pioneering studies conducted since the early nineties evidenced how wild relatives can contribute with their beneficial traits to crop improvement, highlighting the key role of genome mapping for the efficient use of genetic diversity [32]. Furthermore, advances in the phylogeny and taxonomy of plant species as well as the development of novel molecular technologies allowed the exploitation of genetic variability even in more distantly related *taxa*. The transfer of novel/beneficial alleles from wild to cultivated species is not always easy to pursue, and it requires that reproductive barriers between different gene pools be overcome (Figure 1). Gene pools [33,34], referred to as the portion of genetic diversity available for breeding, are defined based on the cross-ability between the crop itself and the primarily non-domesticated species. Nevertheless, most of the potential resources to be used in breeding programs are in the primary gene pool, where, as general rule, gene transfer is easy and immediate. Conventional techniques that rely on the cross between crops and their close wild relatives are still in the norm [17], even if novel plant breeding techniques (NPBT) provide opportunities to: (i) transfer genes from genotypes of the same or sexually compatible species (cisgenesis and intragenesis); (ii) induce mutations in target genes; or, (iii) investigate species in distantly related *taxa* as a useful and extensive reservoir of alleles [35].

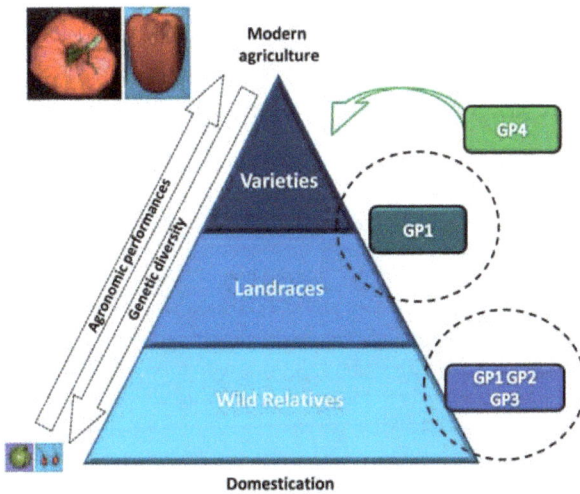

Figure 1. Global scenario from domestication to modern agriculture. Wild species of tomato and pepper (*Solanum habrochaites* and *Capsicum chacoense* at the base of the triangle) are characterized by wide genetic variability, which can be used to improve modern varieties (*Solanum lycopersicum* and *Capsicum annuum*, at the top of the triangle). An intermediate step of domestication is represented by landraces which have a broadening genetic variation linked to adaptation to local environments. Transfer of alleles can be possible within gene pools (GP) [33]. Four different GP levels include: (i) species with easy crossing ability resulting in fruitful hybrids and fertile off-springs (primary gene pool, GP1); (ii) less closely related species that generates weak or sterile hybrids and are characterized by difficulty in obtaining advanced generations (secondary gene pool, GP2); (iii) species requiring sophisticated techniques for gene transfer such as embryo rescue, somatic fusion, grafting, and bridge species (tertiary gene pool, GP3); and, (iv) distantly-related species belonging to different families or kingdoms for which gene transfer is not possible sexually but through direct gene transfer by means of genetic engineering (fourth gene pool, GP4).

Studies on the use of NPBT are being increasingly published. A non-exhaustive but extensive list of genome editing approaches applied to vegetable crops is reported in a recently published review we co-authored [36].

Despite their potentiality, CWR have not been well exploited due to their phenotype that is unsuitable for modern agriculture and to the poor value of economically important features, such as yield. Even more, the quantitative inheritance of relevant agronomic traits is often in linkage with undesirable characteristics (i.e., linkage drag) [15], which makes the use of wild species complicated since it would require efficient selection procedures [37].

The dissection of wild germplasm and the identification of genes underlying quantitative traits have been a central target over the past 35 years. Introgression of hundreds of genes from wild to cultivated species has been possible with the advent of DNA sequencing technologies, which have facilitated the establishment of numerous experimental mapping populations and related linkage maps in many horticultural species [38]. Bi-parental crosses offered the chance to fix alleles from exotic materials in advanced generations (inbred backcross lines, IBLs; recombinant inbred lines, RILs) leading to the identification of several quantitative trait *loci* (QTL) [39]. This approach, still largely used, has the main limitation that is due to the fact that only allelic diversity between the two parents is investigated. In addition, a lack of recombination due to continuous self-fertilization cycles reduces mapping resolution.

Advances in genomics have enabled the implementation of several genome-wide association studies (GWAS) [40] in order to investigate the larger variation present in CC. As it will be discussed in a greater extent at a later stage of this review, GWAS leads to better precision mapping through the identification of single nucleotide polymorphisms (SNPs) strictly related to traits of interest. Main constraints affect this method, such as the lacking of gene flow from wild relatives and several other drawbacks, as reviewed by Korte and Farlow [41].

To address these limitations, multi-parent advanced generation inter-cross (MAGIC) populations [42] are being developed to increase mapping resolution through multiple generations of recombination and provide a high statistical power afforded by a linkage-based design [43]. Furthermore, the combination of genomes from more founders allows for a larger allelic diversity to be explored, generating new phenotypes that constitute a highly valuable pre-breeding resource [44]. MAGIC, hence, represents an intermediate population, which overcomes the main constraints of bi-parental and GWAS. However, the statistical complexity of the analysis and the time required for its development make MAGIC utilization still challenging. As for horticultural species, MAGIC populations have been developed only in tomato to investigate the genetic basis of fruit weight [44].

Genome-wide introgression lines (ILs) represent a further option that increases both mapping resolution and statistical power for minor QTL detection. When compared with the populations mentioned above, each IL includes only small genomic regions from CWR [37] as the results of marker-assisted selection at early stages. Despite the large efforts required for their development, ILs represent a valid source for both genetic studies and breeding purposes, having the advantage of wiping out the linkage drag effect by transferring only the *loci* of interest [45].

A comprehensive use of exotic germplasm is still far away. High-throughput technologies in the field of genomics and phenomics will allow speediness and accuracy in germplasm characterization to be enhanced and will play a central role in the coming decades, leading to a unique opportunity for next generation precision breeding.

2. NGS-Based Genotyping for Genetic Diversity Evaluation

In the pre-genomic era, the assessment of genetic diversity has been traditionally carried out via morphological and cytogenetic characterization, or through the analysis of isozymes. Since '90s onwards, the techniques based on different types of DNA molecular markers have been preferentially used to measure the level of genetic variation [46]. Marker types and genotyping techniques have evolved over time; indeed time consuming, too costly, cumbersome and/or challenging techniques have been replaced by simpler, less expensive, and more efficient alternatives. Because of their extraordinary abundance in the genome and their usually bi-allelic nature, single nucleotide polymorphisms have quickly become the markers of choice to dissect the genetic variability of PGR [47].

The release into the public domain of complete, near-complete or partial genome sequence of the most important vegetable crops [36], the development of SNP detection assays [48,49] and high-density genotyping arrays [50–52] caused a shift from small-/medium-scale to large-scale SNP genotyping.

Both of the above methods allow for thousands of SNPs to be discovered and ample genetic variability across germplasm collections to be captured; however, their design relies on a priori knowledge of the sequence space of the species under investigation and usually cannot be easily modified to fit in with custom experimental designs. This has encouraged the development of novel but still high-throughput and time- and cost-saving SNP discovery methods based on next-generation sequencing (NGS) technologies. Several strategies, methods and protocols for NGS-based genotyping have been developed so far [53,54]. We will go through some of these with a special focus on genotype-by-sequencing (GBS).

NGS, coupled with the availability of high-quality reference genomes of horticultural crops, have expedited the re-sequencing of many individuals to identify a large number of SNPs and investigate within- and between-species sequence variation [5,55–58]. Although the above cited re-sequencing projects were fruitful from a scientific standpoint, the re-sequencing was characterized by low depth of coverage for several individuals. Indeed, within these studies, the average sequence depth varied from 5-fold to 36-fold coverage. Obviously, the greater the depth of coverage is, than the greater the reliability of SNP calling. As general rule, a minimum of 10–20× coverage depth is indicated for reliable variant calling. Unfortunately, because of cost issues this standard is not always applicable.

An alternative strategy to whole genome re-sequencing is to generate a reduced representation of the genome by using on-array- or liquid-based hybridization methods to enrich and capture target genomic regions and possibly identify promising alleles having a potential application in crop breeding programs [59–61].

Unlike sequence capture and targeted re-sequencing, restriction enzyme-based enrichment techniques [62], while not allowing specific target sequences in the genome to be investigated, are the methods of choice for SNP discovery and genotyping.

These methods include three common steps: (i) DNA digestion with restriction enzymes (REs); (ii) ligation with sequencing platform-specific adapters; and, (iii) PCR amplification to increase the DNA yield into sequencing libraries. Conversely, they differ for the size selection step (that is necessary to filter out DNA fragments of desired size) that can be carried out at any point in the protocol workflow or it may be entirely dismissed. Obviously, different protocols result in different data outcomes. Even if it is desirable to dealing with dense distribution of SNP markers and uniform sequence coverage across samples, all of these methods have limitations due to inconsistency in the number of: (i) reads per sample; (ii) reads per polymorphic site; and, (iii) sequenced sites per sample. The combination of all these items can result in a huge number of low quality or missing data. Jiang, et al. [63] suggested a number of improvements to be made in individual steps of the wet-lab workflow to minimize biases in NGS library construction and to increase the degree of reliability and robustness of downstream sequence data.

The relevance of isolating, enriching, and sequencing of specific genomic *loci* for SNP discovery, was initially proved with restriction site-associated DNA sequencing (RAD-Seq) [64]. However, in a short while, RAD-Seq was paralleled and replaced by GBS [65]. At present, GBS is the most favorite technology for high-throughput SNP discovery and it is generating remarkable knowledge on the nature and extent of genetic diversity within germplasm collections [66,67]. GBS was originally applied in maize to identify SNP markers in a RIL population [65]. Since then, several studies have been reported in literature for large-scale SNP discovery and genotyping of RILs [68,69], bi-parental mapping population segregating for important agronomical traits [70,71]; ethylmethane sulfonate- (EMS) induced mutant populations [72,73]; and unrelated individuals in small or large size populations [74,75]. All of these efforts, most of which were carried out on vegetable crop species, aimed to have available a high number of SNP data points for concomitant or future genome-wide association studies.

Usually, ApeKI is the restriction endonuclease used to produce restriction fragments among individuals in a population. It is a methylation sensitive enzyme that shows preferential cleavage for lower or single copy regions of the genome that are generally the richest in genes. In 2012, Poland, et al. [76] modified the original protocol based on ApeKI, by using two restriction enzymes (PstI/MspI) to achieve a higher SNP density. Of course rare-cutter, frequent cutter, or methylation insensitive RE(s) can be used alone or in combination to obtain the desired number of SNP markers.

Intuitively, one major limitation of GBS is its random access to the genome. Such randomness can be adjusted by appropriately selecting the combination of REs to be used. Moreover, the choice of RE(s) is crucial since it affects DNA fragment size distribution as well as the number of fragments in the GBS library. These two parameters, in turn, influence sequencing depth and ultimately the number of SNPs identified [77]. GBS has been used in a large list of vegetable crops showing that this method is efficient for large-scale, low cost genotyping despite all of the limitations mentioned above.

As a rule, GBS tags are aligned to a reference genome to identify SNPs from aligned tags. Several reference-based SNP calling pipelines are available so far [78]. Although a reference genome can facilitate GBS data analysis, several reference-independent SNP calling pipeline (e.g., Stacks and TASSEL-UNEAK) have been developed [79,80] and successfully applied for PGR diversity studies [81,82].

The starting list of SNPs is then subjected to filtering (call rate; minimum depth of coverage; and minor allele frequency, etc.) and, in some cases, it can also be pruned by removing all of the SNPs that are in high linkage disequilibrium (LD). These operations generate high quality SNP datasets that are fed into the population structure analysis software. Such software are able to categorize individuals into ethnically similar cluster based on allele frequency estimates [83,84], or not [85]. Recently, we proved the strength of combining a parametric (STRUCTURE) with a non-parametric method (AWclust) in defining the genetic structure of a population of *Capsicum annuum* accessions [74].

More recently, next-generation marker genotyping platforms have evolved to address specific issues. Yang, et al. [86] developed a semi-automated primer design pipeline to convert GBS-derived SNPs into amplicon sequencing (AmpSeq) markers. This has become necessary because GBS alone is unworkable for highly heterozygous species, for which a large number of missing data and the under-calling of heterozygous sites are very common. The AmpSeq strategy starts from the design of primer pairs by using GBS tags as templates; then, the resulting amplicons are used for genotyping via NGS. In this way, it is possible to circumvent GBS-related issues and to develop reliable SNP-based markers for the high-throughput screening of heterozygous crops.

A further technique, named rAMPSeq, has been established by Buckler, et al. [87] to develop repetitive sequence-based markers for robust genotyping. Although the authors are well aware that the design of rAMPSeq sacrifice several strengths of the GBS method, they feel that the method they have proposed can be revolutionary for breeding and conservation biology.

All considered, the key role of NGS on SNP markers identification, genetic diversity assessment, and population structure analysis in horticultural crops is unquestionable, regardless of the genotyping strategies available. With its unprecedented throughput and scalability, NGS is enabling investigations on genetic diversity at a level never before possible.

3. Advanced Phenomics in Plant Breeding

Phenomics is becoming increasingly important in genetic studies and precision agriculture since it allows for the accurate characterization of multiple traits in crops and a better understanding of phenotypic changes due to underlying heritable genetic variation.

As reported in the previous section, the tremendous advancement of cutting-edge sequencing technologies allowed the identification of thousands of SNPs at affordable costs; at the same time, methods to assess plant traits progressed more slowly, generating what is known as the "phenotyping bottleneck" [88]. Biochemical and metabolomic phenotyping of quality traits and especially abiotic and biotic stress evaluation are the main cause of the "phenotyping bottleneck". Further on in the text, we will provide several examples on how imaging methods are being used to collect phenotypic data points for complex traits in vegetable crops.

The need for increasing the ability to investigate a large amount of traits in a non-destructive manner and with an acceptable accuracy has become a major target in plant breeding. The labor-intensive and costly nature of phenotyping, motivated, in the past decade, the research community to develop automated technologies for high-throughput plant phenotyping (HTPP). This has guaranteed massive progress in the dissection of a wide range of qualitative, agronomical, morphological, and physiological traits, as well in the investigation of traits related to biotic and abiotic stresses.

These technologies, which rely on automated non-invasive sensing methods, are able to capture plant features on the basis of image analysis. At present, several systems are available, including conventional RGB/CIR cameras, spectroscopy (multispectral and hyperspectral remote sensing), thermal infrared systems, fluorescence and tridimensional (3D) imaging, and magnetic resonance imagers (MRI) [89,90].

Features and potentialities of these tools are widely documented [91]; herein, we briefly list which implications (benefits and constraints) emerged from their use on horticultural crop phenotyping.

RGB/CIR cameras are extensively used to analyze traits such as plant canopy and biomass, which can be estimated by combining the imagery falling into red, blue, and green light (RGB), and color infrared (CIR). Derived indexes can be further calculated as the "leaf area index" (LAI) or "normalized difference vegetation index" (NDVI). Multispectral (MS) and hyperspectral (HS) analysis are more addressed to those traits linked to crop physiological status (e.g., nutrient and water content, photosynthetic efficiency, etc.) using a set of images which covers the entire range of radiation from visible (VIS; 400–765 nm) to near-infrared (NIR; 765 to 3200 nm). Thermal imaging is better suited to evaluate (i) the stage of infections by pathogens or if diseases are spreading through the crops and (ii) responses to abiotic stresses (e.g., drought and heat tolerance). It is based on measurements of the variation in foliage surface temperature, which is related to differences in stomatal conductance due to various stresses that alter water balance and transpiration [92]. Plant metabolic status can be assessed by means of fluorescence, capturing excitation, and emission spectra during the absorption of radiation in shorter wavelengths by chlorophyll [90]. Vegetation canopy structure and topographic maps can be estimated by 3D imaging, while MRI detects nuclear resonance signals from isotopes to generate images of the internal structures of the plant. These devices can be used alone or in combination to deliver different data outcomes.

These tools have been firstly applied in cereals, Arabidopsis, and industrial crops [90]. Studies in horticultural crops are currently being conducted. Chlorophyll fluorescence imaging has been used in tomato transgenic plants and young industrial chicory to investigate drought tolerance [93] and cold stress resistance [94], respectively. RGB cameras and 3D imaging have been applied in recombinant inbreds of pepper to inspect canopy structure and plant architecture [95]. MRI have been applied in bean to resolve root structure under resource competition [96].

While analyses in controlled-environments provided major breakthroughs for targeted applications, such as the possibility to investigate roots and seeds in a non-destructive manner by means of MRI-3D combination, analyses in open-field are much more laborious and troublesome due to the heterogeneous nature of soils and the inability to control external factors (e.g., climate).

When evaluating different classes of traits, their nature must be taken into account and how they interact with the external environment. As an example, fruit morphometric traits, can be easily and accurately assessed in a repetitive manner on plants cultivated in greenhouse as well as roots in hydroponics, aeroponics or pots. Contrariwise, yield potential as well as the detrimental effects of abiotic stresses, are better evaluated via field-based phenotyping (FBP) experiments [97]. FBP is generally applied to estimate the genotype x environment (GxE) interaction in a large number of individuals in wide collections or extensive mapping populations [98]. Platforms carrying multiple sensors on wheeled vehicles can be used to measure plant traits on extensive surfaces and investigate GxE interaction. A successful example is provided by the Field Scan Analyzer (LemnaTec, Aachen, Germany), which is a fully automated system designed to capture deep phenotyping data (growth and physiology traits throughout the crop cycle) from crops growing in field environments. Other options are provided by drones equipped with multispectral imaging systems that are able to collect thousands of images in a relatively small amount of time. These technologies are having a positive impact in the reliable and accurate estimate of phenotypic data points as well as in the development of high-yielding, stress tolerant and disease resistant plant varieties.

Although knowledge on semi- and fully-automatic phenotyping systems is rapidly spreading within scientific community, their accessibility still represents a major constraint due to high costs and difficulties in data management. Indeed, one of the main challenges of HTPP and FBP is the processing and analysis of millions of captured images. The development of software for managing and analyzing large and complex datasets is underway to better integrate the different sources of data underlying various plant developmental processes [98].

4. Linking Genotype to Phenotype

In recent times, the selection process carried out by breeders has largely benefited from the improvement of genotyping and phenotyping techniques, but primarily it has taken advantage of the link between phenotypic and genotypic data sets. Uncovering genotype-phenotype relationships is a key requirement in establishing a basic understanding of complex traits and to elucidate how genetic differences can determine individual differences in agronomic performances.

Technical advancements in high-throughput genotyping, in the accuracy and mechanization of phenotyping, as well as the development of efficient statistical methods and bioinformatic tools laid the foundation for genome-wide association studies. This approach, introduced for the first time in human genetics a dozen of years ago [99], is becoming the most popular method to statistically associate genomic *loci* (also referred to as QTN; quantitative trait nucleotides) to simple or complex traits of interest [100]. Even more, the results from GWAS have also been employed to understanding the genetic bases of the domestication and artificial selection processes and to point out how breeding has influenced the genetic variability of modern crops [101]. Below are some examples from recent literature related to the application of GWAS in vegetable crops to reveal useful alleles in genes encoding for agronomic traits.

A genome-wide association study based on a SNP catalogue from whole-genome re-sequencing of 398 modern, heirloom, and wild tomato accessions, and a targeted metabolome quantification of sugars, acids, and volatiles permitted identification of candidate genetic *loci* capable of altering 21 of the chemicals contributing to consumer liking as well as to overall flavor intensity of tomatoes [6]. A similar study founded on polymorphic simple sequence repeat (SSR) markers and volatile quantification in a collection of 174 diverse tomato accessions was performed by Zhang, et al. [102]. The authors identified via GWAS already known as well as novel *loci* that could be important in controlling the volatile metabolism in tomato.

Genome-wide association studies based on SNPs generated by GBS have proved effective for the identification of new alleles affecting morpho-agronomic and fruit quality-related traits in horticultural crops. As an example, we report the work by Nimmakayala, et al. [103], which had the purpose of identifying SNP markers associated with capsaicinoid content and fruit weight traits in pepper (*Capsicum annuum* L.). A similar approach was used by Pavan, et al. [104], who performed preliminary GWAS and identified SNP(s)/trait associations for flowering time and seed-related traits in a collection of 72 accessions of *Cucumis melo* from Apulia. Additional GWAS for fruit firmness were performed in a larger collection of melon from Asia and Western hemisphere [105].

A further genome-wide association study by Cericola, et al. [106], focused on fruit-related traits but based on SNPs from RAD-tag sequencing, was performed in eggplant (*Solanum melongena* L.). A panel of 191 accessions, including breeding lines, old varieties, and landraces was genotyped and scored for anthocyanin pigmentation and fruit color. As result, different already known QTLs were validated and novel marker/trait associations were observed.

While the number of GWAS for quality traits is increasing rapidly, there are few papers describing the association analysis aimed at the identification of alleles for resistance or tolerance to abiotic or biotic stresses in horticultural crops.

To the best of our knowledge, no studies have been published so far aiming at the detection of significant genotype–phenotype associations for abiotic stress (e.g., drought; heat, flooding, salinity, cold).

As for biotic stresses, we did not find any work on fruit or leafy vegetables. We can report only a few examples in legumes. Hart and Griffiths [107] used GBS to genotype a set of RILs in common bean (*Phaseolus vulgaris* L.) and GWAS to identify SNPs associated with the resistance to *bean yellow mosaic virus*. The GBS approach was also used by Saxena, et al. [108] for SNP identification and genotyping of three different mapping populations (2 RILs and one F_2) segregating for sterility mosaic disease (SMD) in pigeonpea (*Cajanus cajan* (L.) Millspaugh). The authors identified a total of 10 QTLs, including three major QTLs governing SMD resistance in pigeonpea.

GWAS are based on three cornerstones: (i) a high-density marker catalogue from genome-wide assessment of the diversity among individuals in a population; (ii) knowledge on population structure and allele frequency spectrum; and, (iii) robust phenotypic data for each individual within the population (Figure 2).

In addition to a large panel of markers (preferentially thousands of SNPs), GWAS require a large sample size (i.e., from hundreds to thousands of individuals) enclosing the maximum variability for the traits under study to achieve adequate statistical power. Population sampling is one of the key steps and should always include accessions collected from different locations and possibly hierarchically structured. Indeed, population structure represents a considerable source of confounding in GWAS [41]; for that reason, a suitable characterization of population structure is advisable to prevent false positive (type I error) or negative (type II error) SNP(s)-trait associations.

As for phenotypic records, it is more advisable to dealing with quantitative data (integer or real-valued numbers) rather than binary data (presence/absence). As it is easily perceived, the former provides more statistical power than the latter.

Statistical analysis for capturing significant genotype–phenotype associations is performed by examining each SNP independently (multiple testing). In general, only those markers that meet the Bonferroni corrected p-value threshold ($\alpha = 0.05$) can be considered significantly associated with phenotypic observations. Alternative approaches to Bonferroni correction have been proposed for establishing significance in GWAS: among these, the use of false discovery rate (FDR) or permutation testing [109].

Association analysis can be performed via general linear model (GLM) or mixed linear model (MLM). It has been demonstrated that the statistical power of MLM is higher than that of GLM and better suited to the quantitative nature of the traits under study, markedly reducing type I errors [110]. This is possible because MLM is able to handle the population structure and the covariance between individuals due to genetic relatedness (i.e., kinship matrix). In other words, MLM takes into account the confounding effect of the genetic background.

Even though GWAS is by now a recognized and powerful tool to investigate genotype-phenotype relationship, it still suffers of some limitations: (i) rare alleles cannot be detected; (ii) alleles having small effects on the phenotype are difficult to be identified; and (iii) the most significant SNPs are not always the true causative/predictive factors for a given trait.

It is very likely that genomic selection (GS) strategies will replace GWAS in the next future [111]. Albeit, comparable to GWAS, GS relies on high-throughput genotyping and phenotyping of unrelated individuals in a population (i.e., training population; TP) and allows markers in strong LD with causal variants to be identified [112], it has singular features. Results from TP are used to predict genomic estimated breeding values (GEBV) on a different set of individuals (i.e., breeding population; BP), which has been previously genotyped with dense marker coverage across genome, but whose phenotype is unknown. In this way, time and expense in phenotyping diminish considerably since phenotyping is needed only in the initial phase to enhance the accuracy of the prediction model. In crops GS has been mainly applied in cereals [113,114], while emerging applications in vegetables are reported only for tomato [115,116]. Both works on tomato assessed the potential of GS approach to improve fruit quality traits. Duangjit, et al. [115] used a panel of 163 tomato accessions that were genotyped using the SolCAP Infinium assay (Illumina) and phenotyped for a total of 35 metabolic traits. Training and breeding populations were assembled including an increasing/decreasing number of accessions from the original population.

Yamamoto, et al. [116] used 96 big-fruited F_1 tomato varieties as a TP (that was phenotyped for soluble solids content and total fruit weight) and evaluated the predictability of the GS models in their progeny populations.

These pioneering studies demonstrated that GS models could predict phenotypes paving the way for a successful implementation of GS in vegetable crops.

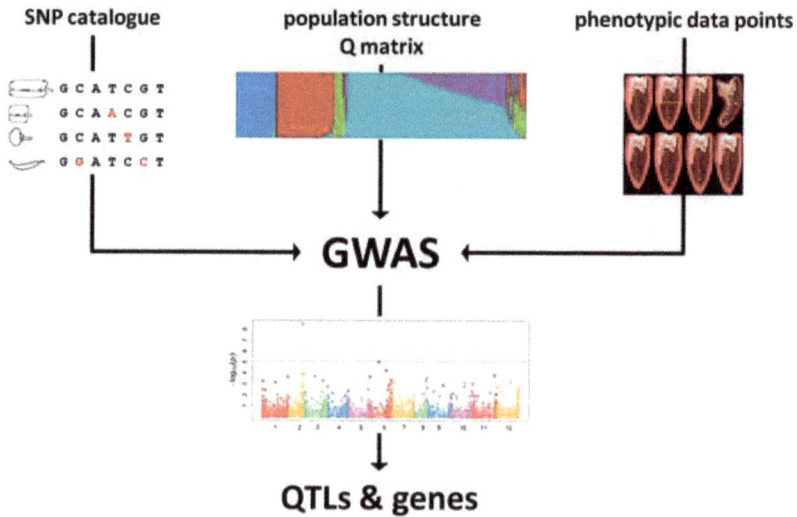

Figure 2. Genome-wide association studies (GWAS) are based on three pillars: (i) a high-density single nucleotide polymorphisms (SNP) catalogue derived from diversity assessment of individuals in a germplasm collection; (ii) knowledge on population structure and allele frequency spectrum (Q matrix); and, (iii) phenotypic data points for each individual within the population. Generally results from GWAS are displayed as Manhattan plots showing genome-wide *p*-values of SNP(s)-trait associations. GWAS allow the causal gene(s) or QTL(s) associated with the trait(s) of interest to be identified.

5. Outlook

Future research on horticultural crops should aim at advances in shelf life extension and, at the same time, at improving the quality and enhance the health benefits derived from their consumption (e.g., increase the levels of bioactive compounds). Breeders should also consider taste and flavor, while breeding for products with improved nutritional attributes. Additional urgent needs concern the mitigation of abiotic and biotic stress factors to counter climate change related adverse effects on vegetable crop yield, as well as the development of "low input vegetable crops" because of increasing needs to achieve sustainable practices. We think the research community has developed over the last decade a complete toolbox, full of effective items, to face forthcoming challenges in horticulture.

This review provides useful and contemporary information at one place, and supports the notion that efficient management and exploitation of PGR is essential for recovering the repertoire of alleles that has been left behind by the artificial selection process.

With the widespread availability of NGS technologies SNP/allele discovery become feasible and affordable even for species with no genome sequence. Sequence-based information is increasingly used to assess genetic variability within and between closely-related species and restriction enzyme-based enrichment is making diversity studies on PGR more informative than before. Indeed, genetic diversity assessment of horticultural crop collections is a prerequisite for breeding purposes. It first allows guiding decision making for their conservation. Secondly, it provides novel options (i.e., new/beneficial alleles) to breeders for the greater exploitation of horticultural germplasm aiming at the development of new hybrids and varieties.

On the other hand, advances in phenotyping technologies are essential for capitalizing on the developments in NGS-based genotyping; are accelerating the discovery of trait-allele associations and are improving our ability to uncover genotype-phenotype relationship (Figure 3).

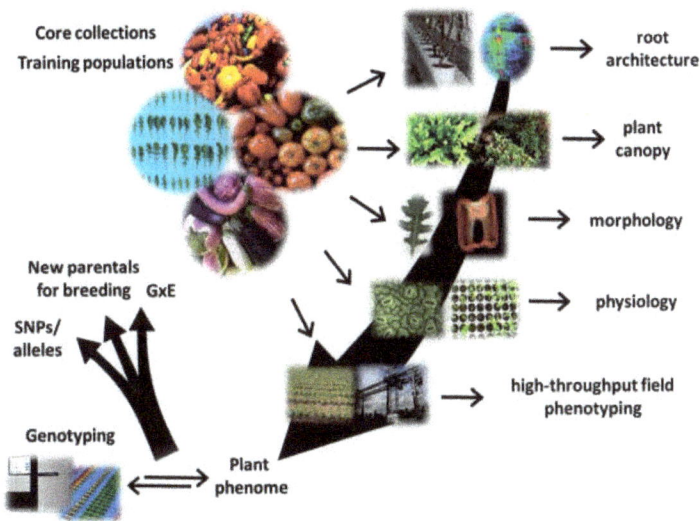

Figure 3. Large-scale phenotyping and its impact on plant breeding. Core collection (CC) and/or training populations (TP) developed in various fruit and leafy vegetable crops (in the figure from the top clockwise: pepper, tomato, eggplant, and rocket salad) can be deeply assessed through innovative phenotyping tools for different categories of traits. The integration with genotyping data lead to the identification of: (i) SNPs and alleles associated with target traits; (ii) accessions which can be used as parentals for future breeding programs; and, (iii) the basis of the genotype per environment (GxE) interaction.

The combination of genotyping with accurate phenotyping is defining the current practices in the breeding of vegetable crops. Nonetheless, we have described how the association between genetic variations and a phenotypic change is nothing but simple. Since GWAS enter the scientific scenario, our ability to link genotypes to phenotypes is markedly improved. Thanks to their predictive power, GWAS have experienced rapid growth in the agri-genomics field and are allowing us to assemble some of the pieces of the puzzle that depict how genotypes articulate with phenotypes. Although the puzzle is partially assembled and with many pieces missing, it is already unraveling the genetic bases of complex agronomical traits and it is making available different alleles to be used to cope with forthcoming challenges.

From the review of recent literature it is clear that GWAS in horticultural crops are more common for traits related to plant morphology and quality, while SNP(s)/trait associations for abiotic and biotic stresses are quite neglected. This is explained by the fact that fruit and leafy vegetables are a major source of biologically active compounds with human health-promoting properties.

Likely, it is less time consuming to establish small or large size populations of unrelated individuals with contrasting phenotypes than develop mapping population segregating for traits related to abiotic/biotic stresses. Furthermore, phenotyping horticultural crops for abiotic/biotic stresses is still not within everyone's means and requires *ad-hoc* facilities and field trials across a broad range of environments. Nevertheless, we expect a change of course in the near future and that GWAS under abiotic/biotic stresses will be undertaken in vegetable crops as in the case for staple crops.

We also believe that it is only a matter of time before GWAS will routinely applied in vegetable crops to identify the so-called non-phenotypic variations, such as eQTL (expression QTL) and/or methylation QTL (meQTL) [117,118].

While the identification of novel and valuable alleles via GWAS is a well-established and quite reliable process, their transfer to elite vegetable crops is still limited. As mentioned above, the increasing spread of NPBT will ease targeted genome editing in vegetable crops facilitating the combination of new sets of alleles.

Finally, we expect genomic selection to cause a paradigm shift in horticultural crop breeding by greatly expediting genomics-driven crop design [119].

The few published papers on GS application in horticultural crops, we briefly illustrated, show how GS is still in its infancy and that its great potential is far from being fully exploited. Indeed, GS could lead to a new scenario within commercial plant breeding programs, facilitating the selection of potential parentals among those with the highest GEBV value.

Further efforts in the development of novel statistical models able to capture more faithfully genotype-phenotype relationships as well as software and databases implementation will, however, required for higher precision mapping.

GWAS, GS, and complex mapping populations (i.e., MAGIC) represent nowadays the main pillars for next generation precision breeding. Although less emphasis to research programs in plant breeding is given by universities and research institutes [120], the work of the scientific community in the next decades will be crucial in the development of appropriate populations with the aim to better utilize genetic diversity within vegetable crops.

Acknowledgments: This work has received funding from the EU Horizon 2020 research and innovation programme under grant agreement No. 677379 (G2PSOL, Linking genetic resources, genomes and phenotypes of Solanaceous crops).

Author Contributions: Nunzio D'Agostino provided the outlines of the review and the key ideas. Pasquale Tripodi contributed to the concept and layout of the manuscript. Nunzio D'Agostino and Pasquale Tripodi contributed to the writing and editing of the review.

Conflicts of Interest: The authors declare no conflict of interest.

Abbreviations

AmpSeq	amplicon sequencing
CBD	convention on biological diversity
CC	core collections
CIR	color infrared
CWR	crop wild relatives
EMS	ethylmethane sulfonate
eQTL	expression QTL
FAO	food and agriculture organization of the United Nations
FBP	field-based phenotyping
FDR	false discovery rate
FIGS	focused identification of germplasm strategy
GBS	genotype-by-sequencing
GEBV	genomic estimated breeding values
GIS	geographic information system
GLM	general linear model
GP	gene pools
GS	genomic selection
GWAS	genome-wide association studies
HS	hyperspectral

HTPP	high-throughput plant phenotyping
IBLs	inbred backcross lines
ILs	introgression lines
LAI	leaf area index
LD	linkage disequilibrium
LR	landraces
MAGIC	multi-parent advanced generation inter-cross
meQTL	methylation QTL
MLM	mixed linear model
MRI	magnetic resonance imagers
MS	multispectral
NDVI	normalized difference vegetation index
NGS	next generation sequencing
NPBT	novel plant breeding techniques
PCR	polymerase chain reaction
PGR	plant genetic resources
QTN	quantitative trait nucleotides
QTL	quantitative trait loci
RAD-Seq	restriction site-associated DNA sequencing
REs	restriction enzymes
RILs	recombinant inbred lines
SMD	sterility mosaic disease
SNPs	single nucleotide polymorphisms
SSR	simple sequence repeat
TP	training population

References

1. Meyer, R.S.; Purugganan, M.D. Evolution of crop species: Genetics of domestication and diversification. *Nat. Rev. Genet.* **2013**, *14*, 840–852. [CrossRef] [PubMed]
2. Gepts, P. Crop domestication as a long-term selection experiment. In *Plant Breeding Reviews*; John Wiley & Sons, Inc.: Oxford, UK, 2010; pp. 1–44.
3. Meyer, R.S.; DuVal, A.E.; Jensen, H.R. Patterns and processes in crop domestication: An historical review and quantitative analysis of 203 global food crops. *New Phytol.* **2012**, *196*, 29–48. [CrossRef] [PubMed]
4. Flint-Garcia, S.A. Genetics and consequences of crop domestication. *J. Agric. Food Chem.* **2013**, *61*, 8267–8276. [CrossRef] [PubMed]
5. Lin, T.; Zhu, G.; Zhang, J.; Xu, X.; Yu, Q.; Zheng, Z.; Zhang, Z.; Lun, Y.; Li, S.; Wang, X.; et al. Genomic analyses provide insights into the history of tomato breeding. *Nat. Genet.* **2014**, *46*, 1220–1226. [CrossRef] [PubMed]
6. Tieman, D.; Zhu, G.; Resende, M.F.; Lin, T.; Nguyen, C.; Bies, D.; Rambla, J.L.; Beltran, K.S.O.; Taylor, M.; Zhang, B.; et al. A chemical genetic roadmap to improved tomato flavor. *Science* **2017**, *355*, 391–394. [CrossRef] [PubMed]
7. Massawe, F.; Mayes, S.; Cheng, A. Crop diversity: An unexploited treasure trove for food security. *Trends Plant. Sci.* **2016**, *21*, 365–368. [CrossRef] [PubMed]
8. Diamond, J. Evolution, consequences and future of plant and animal domestication. *Nature* **2002**, *418*, 700–707. [CrossRef] [PubMed]
9. Gornall, R.J. Recombination systems and plant domestication. *Biol. J. Linn. Soc.* **1983**, *20*, 375–383. [CrossRef]
10. Brozynska, M.; Furtado, A.; Henry, R.J. Genomics of crop wild relatives: Expanding the gene pool for crop improvement. *Plant Biotechnol. J.* **2016**, *14*, 1070–1085. [CrossRef] [PubMed]
11. Roberts, M.J.; Schlenker, W. World supply and demand of food commodity calories. *Am. J. Agric. Econ.* **2009**, *91*, 1235–1242. [CrossRef]
12. World Population Ageing. Available online: http://www.un.org/en/development/desa/population/publications/pdf/ageing/WorldPopulationAgeing2013 (accessed on 24 July 2017).

13. Godfray, H.C.J.; Beddington, J.R.; Crute, I.R.; Haddad, L.; Lawrence, D.; Muir, J.F.; Pretty, J.; Robinson, S.; Thomas, S.M.; Toulmin, C. Food security: The challenge of feeding 9 billion people. *Science* **2010**, *327*, 812–818. [CrossRef] [PubMed]

14. Khoury, C.K.; Bjorkman, A.D.; Dempewolf, H.; Ramirez-Villegas, J.; Guarino, L.; Jarvis, A.; Rieseberg, L.H.; Struik, P.C. Increasing homogeneity in global food supplies and the implications for food security. *Proc. Natl. Acad. Sci. USA* **2014**, *111*, 4001–4006. [CrossRef] [PubMed]

15. Tanksley, S.D.; McCouch, S.R. Seed banks and molecular maps: Unlocking genetic potential from the wild. *Science* **1997**, *277*, 1063–1066. [CrossRef] [PubMed]

16. Dyer, G.A.; López-Feldman, A.; Yúnez-Naude, A.; Taylor, J.E. Genetic erosion in maize's center of origin. *Proc. Natl. Acad. Sci. USA* **2014**, *111*, 14094–14099. [CrossRef] [PubMed]

17. Maxted, N.; Kell, S.P. *Establishment of a Global Network for the in Situ Conservation of Crop Wild Relatives: Status and Needs*; FAO Commission on Genetic Resources for Food and Agriculture: Rome, Italy, 2009; p. 266.

18. Maxted, N.; Ford-Lloyd, B.V.; Hawkes, J.G. Complementary conservation strategies. In *Plant Genetic Conservation, The In Situ Approach*; Maxted, N., Ford-Lloyd, B.V., Hawkes, J.G., Eds.; Springer: Dordrecht, The Netherlands, 1997; pp. 15–39.

19. Gruber, K. Agrobiodiversity: The living library. *Nature* **2017**, *544*, S8–S10. [CrossRef] [PubMed]

20. Koo, B.; Wright, B.D. The optimal timing of evaluation of genebank accessions and the effects of biotechnology. *Am. J. Agric. Econ.* **2000**, *82*, 797–811. [CrossRef]

21. Brown, A.H.D. Core collections: A practical approach to genetic resources management. *Genome* **1989**, *31*, 818–824. [CrossRef]

22. McKhann, H.I.; Camilleri, C.; Bérard, A.; Bataillon, T.; David, J.L.; Reboud, X.; Le Corre, V.; Caloustian, C.; Gut, I.G.; Brunel, D. Nested core collections maximizing genetic diversity in arabidopsis thaliana. *Plant. J.* **2004**, *38*, 193–202. [CrossRef] [PubMed]

23. Kim, K.-W.; Chung, H.-K.; Cho, G.-T.; Ma, K.-H.; Chandrabalan, D.; Gwag, J.-G.; Kim, T.-S.; Cho, E.-G.; Park, Y.-J. Powercore: A program applying the advanced m strategy with a heuristic search for establishing core sets. *Bioinformatics* **2007**, *23*, 2155–2162. [CrossRef] [PubMed]

24. Borrayo, E.; Takeya, M. Signal-Processing Tools for Core-Collection Selection from Genetic-Resource Collections. Available online: https://f1000research.com/articles/4-97/v1 (accessed on 14 September 2017).

25. Khazaei, H.; Street, K.; Bari, A.; Mackay, M.; Stoddard, F.L. The figs (focused identification of germplasm strategy) approach identifies traits related to drought adaptation in vicia faba genetic resources. *PLoS ONE* **2013**, *8*, e63107. [CrossRef] [PubMed]

26. Street, K.; Bari, A.; Mackay, M.; Amri, A. How the focused identification of germplasm strategy (figs) is used to mine plant genetic resources collections for adaptive traits. In *Enhancing Crop Genepool Use: Capturing Wild Relative and Landrace Diversity for Crop Improvement*; Maxted, N., Dulloo, M.E., Ford-Lloyd, B.V., Eds.; CABI: Boston, MA, USA, 2016; p. 54.

27. Odong, T.L.; Jansen, J.; van Eeuwijk, F.A.; van Hintum, T.J.L. Quality of core collections for effective utilisation of genetic resources review, discussion and interpretation. *Theor. Appl. Genet.* **2013**, *126*, 289–305. [CrossRef] [PubMed]

28. Fao: Second Global Plan of Action for Plant Genetic Resources. Available online: http://www.fao.org/agriculture/crops/thematic-sitemap/theme/seeds-pgr/gpa/en/ (accessed on 24 July 2017).

29. Divseek. Available online: http://www.divseek.org (accessed on 21 July 2017).

30. G2p-sol: Linking Genetic Resources, Genomes and Phenotypes of Solanaceous Crops. Available online: http://www.g2p-sol.eu/ (accessed on 21 January 2017).

31. Traditom: Traditional Tomato Varieties and Cultural Practices. Available online: http://traditom.eu/ (accessed on 21 July 2017).

32. Hajjar, R.; Hodgkin, T. The use of wild relatives in crop improvement: A survey of developments over the last 20 years. *Euphytica* **2007**, *156*, 1–13. [CrossRef]

33. Harlan, J.R.; de Wet, J.M.J. Toward a rational classification of cultivated plants. *Taxon* **1971**, *20*, 509–517. [CrossRef]

34. Gepts, P.; Papa, R. Possible effects of (trans)gene flow from crops on the genetic diversity from landraces and wild relatives. *Environ. Biosaf. Res.* **2003**, *2*, 89–103. [CrossRef]

35. Cardi, T. Cisgenesis and genome editing: Combining concepts and efforts for a smarter use of genetic resources in crop breeding. *Plant Breed.* **2016**, *135*, 139–147. [CrossRef]

36. Cardi, T.; D'Agostino, N.; Tripodi, P. Genetic transformation and genomic resources for next-generation precise genome engineering in vegetable crops. *Front. Plant. Sci* **2017**, *8*, 241. [CrossRef] [PubMed]

37. Zamir, D. Improving plant breeding with exotic genetic libraries. *Nat. Rev. Genet.* **2001**, *2*, 983–989. [CrossRef] [PubMed]

38. Prohens, J.; Gramazio, P.; Plazas, M.; Dempewolf, H.; Kilian, B.; Díez, M.J.; Fita, A.; Herraiz, F.J.; Rodríguez-Burruezo, A.; Soler, S.; et al. Introgressiomics: A new approach for using crop wild relatives in breeding for adaptation to climate change. *Euphytica* **2017**, *213*, 158. [CrossRef]

39. Holland, J.B. Genetic architecture of complex traits in plants. *Curr. Opin. Plant Biol.* **2007**, *10*, 156–161. [CrossRef] [PubMed]

40. Zhu, C.; Gore, M.; Buckler, E.S.; Yu, J. Status and prospects of association mapping in plants. *Plant Genome* **2008**, *1*, 5–20. [CrossRef]

41. Korte, A.; Farlow, A. The advantages and limitations of trait analysis with gwas: A review. *Plant Methods* **2013**, *9*, 29. [CrossRef] [PubMed]

42. Huang, B.E.; Verbyla, K.L.; Verbyla, A.P.; Raghavan, C.; Singh, V.K.; Gaur, P.; Leung, H.; Varshney, R.K.; Cavanagh, C.R. Magic populations in crops: Current status and future prospects. *Theor. Appl. Genet.* **2015**, *128*, 999–1017. [CrossRef] [PubMed]

43. King, E.G.; Merkes, C.M.; McNeil, C.L.; Hoofer, S.R.; Sen, S.; Broman, K.W.; Long, A.D.; Macdonald, S.J. Genetic dissection of a model complex trait using the drosophila synthetic population resource. *Genome Res.* **2012**, *22*, 1558–1566. [CrossRef] [PubMed]

44. Pascual, L.; Desplat, N.; Huang, B.E.; Desgroux, A.; Bruguier, L.; Bouchet, J.-P.; Le, Q.H.; Chauchard, B.; Verschave, P.; Causse, M. Potential of a tomato magic population to decipher the genetic control of quantitative traits and detect causal variants in the resequencing era. *Plant Biotechnol. J.* **2015**, *13*, 565–577. [CrossRef] [PubMed]

45. Alseekh, S.; Ofner, I.; Pleban, T.; Tripodi, P.; Di Dato, F.; Cammareri, M.; Mohammad, A.; Grandillo, S.; Fernie, A.R.; Zamir, D. Resolution by recombination: Breaking up solanum pennellii introgressions. *Trends Plant Sci.* **2013**, *18*, 536–538. [CrossRef] [PubMed]

46. Henry, R.J. *Plant Genotyping: The DNA Fingerprinting of Plants*; CABI Publishing: Oxford, UK, 2001.

47. Mammadov, J.; Aggarwal, R.; Buyyarapu, R.; Kumpatla, S. Snp markers and their impact on plant breeding. *Int. J. Plant Genom.* **2012**, *2012*, 728398. [CrossRef] [PubMed]

48. Smith, S.M.; Maughan, P.J. Snp genotyping using kaspar assays. In *Plant genotyping: Methods and Protocols*; Batley, J., Ed.; Springer: New York, NY, USA, 2015; pp. 243–256.

49. Shen, G.-Q.; Abdullah, K.G.; Wang, Q.K. The taqman method for snp genotyping. In *Single Nucleotide Polymorphisms: Methods and Protocols*; Komar, A.A., Ed.; Humana Press: Totowa, NJ, USA, 2009; pp. 293–306.

50. LaFramboise, T. Single nucleotide polymorphism arrays: A decade of biological, computational and technological advances. *Nucleic Acids Res.* **2009**, *37*, 4181–4193. [CrossRef] [PubMed]

51. Sim, S.C.; Durstewitz, G.; Plieske, J.; Wieseke, R.; Ganal, M.W.; Van Deynze, A.; Hamilton, J.P.; Buell, C.R.; Causse, M.; Wijeratne, S.; et al. Development of a large snp genotyping array and generation of high-density genetic maps in tomato. *PLoS ONE* **2012**, *7*, e40563. [CrossRef] [PubMed]

52. Jaccoud, D.; Peng, K.; Feinstein, D.; Kilian, A. Diversity arrays: A solid state technology for sequence information independent genotyping. *Nucleic Acids Res.* **2001**, *29*, E25. [CrossRef] [PubMed]

53. Davey, J.W.; Hohenlohe, P.A.; Etter, P.D.; Boone, J.Q.; Catchen, J.M.; Blaxter, M.L. Genome-wide genetic marker discovery and genotyping using next-generation sequencing. *Nat. Rev. Genet.* **2011**, *12*, 499–510. [CrossRef] [PubMed]

54. Ray, S.; Satya, P. Next generation sequencing technologies for next generation plant breeding. *Front Plant Sci.* **2014**, *5*, 367. [CrossRef] [PubMed]

55. Aflitos, S.; Schijlen, E.; Jong, H.; Ridder, D.; Smit, S.; Finkers, R.; Wang, J.; Zhang, G.; Li, N.; Mao, L.; et al. Exploring genetic variation in the tomato (solanum section lycopersicon) clade by whole-genome sequencing. *Plant. J.* **2014**, *80*, 136–148. [PubMed]

56. Schmutz, J.; McClean, P.E.; Mamidi, S.; Wu, G.A.; Cannon, S.B.; Grimwood, J.; Jenkins, J.; Shu, S.; Song, Q.; Chavarro, C.; et al. A reference genome for common bean and genome-wide analysis of dual domestications. *Nat. Genet.* **2014**, *46*, 707–713. [CrossRef] [PubMed]

57. Guo, S.; Zhang, J.; Sun, H.; Salse, J.; Lucas, W.J.; Zhang, H.; Zheng, Y.; Mao, L.; Ren, Y.; Wang, Z.; et al. The draft genome of watermelon (*Citrullus lanatus*) and resequencing of 20 diverse accessions. *Nat. Genet.* **2013**, *45*, 51–58. [CrossRef] [PubMed]

58. Causse, M.; Desplat, N.; Pascual, L.; Le Paslier, M.C.; Sauvage, C.; Bauchet, G.; Berard, A.; Bounon, R.; Tchoumakov, M.; Brunel, D.; et al. Whole genome resequencing in tomato reveals variation associated with introgression and breeding events. *BMC Genom.* **2013**, *14*, 791. [CrossRef] [PubMed]

59. Terracciano, I.; Cantarella, C.; D'Agostino, N. Hybridization-based enrichment and next generation sequencing to explore genetic diversity in plants. In *Dynamics of Mathematical Models in Biology: Bringing Mathematics to Life*; Rogato, A., Zazzu, V., Guarracino, M., Eds.; Springer International Publishing: Cham, Switzerland, 2016; pp. 117–136.

60. Terracciano, I.; Cantarella, C.; Fasano, C.; Cardi, T.; Mennella, G.; D'Agostino, N. Liquid-phase sequence capture and targeted re-sequencing revealed novel polymorphisms in tomato genes belonging to the mep carotenoid pathway. *Sci. Rep.* **2017**, *7*, 5616. [CrossRef] [PubMed]

61. Ruggieri, V.; Anzar, I.; Paytuvi, A.; Calafiore, R.; Cigliano, R.A.; Sanseverino, W.; Barone, A. Exploiting the great potential of sequence capture data by a new tool, super-cap. *DNA Res.* **2017**, *24*, 81–91. [CrossRef] [PubMed]

62. Cronn, R.; Knaus, B.J.; Liston, A.; Maughan, P.J.; Parks, M.; Syring, J.V.; Udall, J. Targeted enrichment strategies for next-generation plant biology. *Am. J. Bot.* **2012**, *99*, 291–311. [CrossRef] [PubMed]

63. Jiang, Z.; Wang, H.; Michal, J.J.; Zhou, X.; Liu, B.; Woods, L.C.; Fuchs, R.A. Genome wide sampling sequencing for snp genotyping: Methods, challenges and future development. *Int. J. Biol. Sci.* **2016**, *12*, 100–108. [CrossRef] [PubMed]

64. Baird, N.A.; Etter, P.D.; Atwood, T.S.; Currey, M.C.; Shiver, A.L.; Lewis, Z.A.; Selker, E.U.; Cresko, W.A.; Johnson, E.A. Rapid snp discovery and genetic mapping using sequenced rad markers. *PLoS ONE* **2008**, *3*, e3376. [CrossRef] [PubMed]

65. Elshire, R.J.; Glaubitz, J.C.; Sun, Q.; Poland, J.A.; Kawamoto, K.; Buckler, E.S.; Mitchell, S.E. A robust, simple genotyping-by-sequencing (gbs) approach for high diversity species. *PLoS ONE* **2011**, *6*, e19379. [CrossRef] [PubMed]

66. Voss-Fels, K.; Snowdon, R.J. Understanding and utilizing crop genome diversity via high-resolution genotyping. *Plant. Biotechnol. J.* **2016**, *14*, 1086–1094. [CrossRef] [PubMed]

67. Taranto, F.; D'Agostino, N.; Tripodi, P. An overview of genotyping by sequencing in crop species and its application in pepper. In *Dynamics of Mathematical Models in Biology: Bringing Mathematics to Life*; Rogato, A., Zazzu, V., Guarracino, M., Eds.; Springer International Publishing: Cham, Switzerland, 2016; pp. 101–116.

68. Boutet, G.; Alves Carvalho, S.; Falque, M.; Peterlongo, P.; Lhuillier, E.; Bouchez, O.; Lavaud, C.; Pilet-Nayel, M.-L.; Rivière, N.; Baranger, A. Snp discovery and genetic mapping using genotyping by sequencing of whole genome genomic DNA from a pea ril population. *BMC Genom.* **2016**, *17*, 121. [CrossRef] [PubMed]

69. Verma, S.; Gupta, S.; Bandhiwal, N.; Kumar, T.; Bharadwaj, C.; Bhatia, S. High-density linkage map construction and mapping of seed trait qtls in chickpea (*Cicer arietinum* L.) using genotyping-by-sequencing (gbs). *Sci. Rep.* **2015**, *5*, 17512. [CrossRef] [PubMed]

70. Ren, Y.; McGregor, C.; Zhang, Y.; Gong, G.; Zhang, H.; Guo, S.; Sun, H.; Cai, W.; Zhang, J.; Xu, Y. An integrated genetic map based on four mapping populations and quantitative trait loci associated with economically important traits in watermelon (*Citrullus lanatus*). *BMC Plant. Biol.* **2014**, *14*, 33. [CrossRef] [PubMed]

71. Celik, I.; Gurbuz, N.; Uncu, A.T.; Frary, A.; Doganlar, S. Genome-wide snp discovery and qtl mapping for fruit quality traits in inbred backcross lines (ibls) of solanum pimpinellifolium using genotyping by sequencing. *BMC Genom.* **2017**, *18*, 1. [CrossRef] [PubMed]

72. Sidhu, G.; Mohan, A.; Zheng, P.; Dhaliwal, A.K.; Main, D.; Gill, K.S. Sequencing-based high throughput mutation detection in bread wheat. *BMC Genom.* **2015**, *16*, 962. [CrossRef] [PubMed]

73. Mishra, A.; Singh, A.; Sharma, M.; Kumar, P.; Roy, J. Development of ems-induced mutation population for amylose and resistant starch variation in bread wheat (*Triticum aestivum*) and identification of candidate genes responsible for amylose variation. *BMC Plant Biol.* **2016**, *16*, 217. [CrossRef] [PubMed]

74. Taranto, F.; D'Agostino, N.; Greco, B.; Cardi, T.; Tripodi, P. Genome-wide snp discovery and population structure analysis in pepper (*Capsicum annuum*) using genotyping by sequencing. *BMC Genom.* **2016**, *17*, 943. [CrossRef] [PubMed]

75. Pavan, S.; Lotti, C.; Marcotrigiano, A.R.; Mazzeo, R.; Bardaro, N.; Bracuto, V.; Ricciardi, F.; Taranto, F.; D'Agostino, N.; Schiavulli, A.; et al. A distinct genetic cluster in cultivated chickpea as revealed by genome-wide marker discovery and genotyping. *Plant. Genome* **2017**, *10*. [CrossRef] [PubMed]

76. Poland, J.A.; Brown, P.J.; Sorrells, M.E.; Jannink, J.L. Development of high-density genetic maps for barley and wheat using a novel two-enzyme genotyping-by-sequencing approach. *PLoS ONE* **2012**, *7*, e32253. [CrossRef] [PubMed]

77. Schröder, S. Optimization of genotyping by sequencing (gbs) data in common bean (*Phaseolus vulgaris* L.). *Mol. Breed.* **2016**, *36*, 6. [CrossRef]

78. Torkamaneh, D.; Laroche, J.; Belzile, F. Genome-wide snp calling from genotyping by sequencing (gbs) data: A comparison of seven pipelines and two sequencing technologies. *PLoS ONE* **2016**, *11*, e0161333. [CrossRef] [PubMed]

79. Catchen, J.; Hohenlohe, P.A.; Bassham, S.; Amores, A.; Cresko, W.A. Stacks: An analysis tool set for population genomics. *Mol. Ecol.* **2013**, *22*, 3124–3140. [CrossRef] [PubMed]

80. Lu, F.; Lipka, A.E.; Glaubitz, J.; Elshire, R.; Cherney, J.H.; Casler, M.D.; Buckler, E.S.; Costich, D.E. Switchgrass genomic diversity, ploidy, and evolution: Novel insights from a network-based snp discovery protocol. *PLoS Genet.* **2013**, *9*, e1003215. [CrossRef] [PubMed]

81. Russell, J.; Hackett, C.; Hedley, P.; Liu, H.; Milne, L.; Bayer, M.; Marshall, D.; Jorgensen, L.; Gordon, S.; Brennan, R. The use of genotyping by sequencing in blackcurrant (*Ribes nigrum*): Developing high-resolution linkage maps in species without reference genome sequences. *Mol. Breed.* **2014**, *33*, 835–849. [CrossRef]

82. Huang, Y.F.; Poland, J.A.; Wight, C.P.; Jackson, E.W.; Tinker, N.A. Using genotyping-by-sequencing (gbs) for genomic discovery in cultivated oat. *PLoS ONE* **2014**, *9*, e102448. [CrossRef] [PubMed]

83. Pritchard, J.K.; Stephens, M.; Donnelly, P. Inference of population structure using multilocus genotype data. *Genetics* **2000**, *155*, 945–959. [PubMed]

84. Alexander, D.H.; Novembre, J.; Lange, K. Fast model-based estimation of ancestry in unrelated individuals. *Genome Res.* **2009**, *19*, 1655–1664. [CrossRef] [PubMed]

85. Awclust. Available online: https://sourceforge.net/projects/awclust/ (accessed on 20 July 2017).

86. Yang, S.; Fresnedo-Ramirez, J.; Wang, M.; Cote, L.; Schweitzer, P.; Barba, P.; Takacs, E.M.; Clark, M.; Luby, J.; Manns, D.C.; et al. A next-generation marker genotyping platform (ampseq) in heterozygous crops: A case study for marker-assisted selection in grapevine. *Hortic. Res.* **2016**, *3*, 16002. [CrossRef] [PubMed]

87. Buckler, E.S.; Ilut, D.C.; Wang, X.; Kretzschmar, T.; Gore, M.A.; Mitchell, S.E. Rampseq: Using repetitive sequences for robust genotyping. *bioRxiv* **2016**. [CrossRef]

88. Furbank, R.T.; Tester, M. Phenomics—Technologies to relieve the phenotyping bottleneck. *Trends Plant Sci.* **2011**, *16*, 635–644. [CrossRef] [PubMed]

89. Araus, J.L.; Cairns, J.E. Field high-throughput phenotyping: The new crop breeding frontier. *Trends Plant Sci.* **2014**, *19*, 52–61. [CrossRef] [PubMed]

90. Li, L.; Zhang, Q.; Huang, D. A review of imaging techniques for plant phenotyping. *Sensors* **2014**, *14*, 20078–20111. [CrossRef] [PubMed]

91. Fritsche-Neto, R.; Borém, A. *Phenomics: How Next-Generation Phenotyping Is Revolutionizing Plant Breeding*; Springer: Dordrecht, Switzerland, 2015; pp. 1–142.

92. Liew, O.; Chong, P.; Li, B.; Asundi, A. Signature optical cues: Emerging technologies for monitoring plant health. *Sensors* **2008**, *8*, 3205–3239. [CrossRef] [PubMed]

93. Mishra, K.B.; Iannacone, R.; Petrozza, A.; Mishra, A.; Armentano, N.; La Vecchia, G.; Trtílek, M.; Cellini, F.; Nedbal, L. Engineered drought tolerance in tomato plants is reflected in chlorophyll fluorescence emission. *Plant Sci.* **2012**, *182*, 79–86. [CrossRef] [PubMed]

94. Lootens, P.; Devacht, S.; Baert, J.; van Waes, J.; van Bockstaele, E.; Roldán-Ruiz, I. Evaluation of cold stress of young industrial chicory (*Cichorium intybus* L.) by chlorophyll a fluorescence imaging. Ii. Dark relaxation kinetics. *Photosynthetica* **2011**, *49*, 185–194. [CrossRef]

95. Van der Heijden, G.; Song, Y.; Horgan, G.; Polder, G.; Dieleman, A.; Bink, M.; Palloix, A.; van Eeuwijk, F.; Glasbey, C. Spicy: Towards automated phenotyping of large pepper plants in the greenhouse. *Funct. Plant Biol.* **2012**, *39*, 870–877. [CrossRef]

96. Rascher, U.; Blossfeld, S.; Fiorani, F.; Jahnke, S.; Jansen, M.; Kuhn, A.J.; Matsubara, S.; Märtin, L.L.A.; Merchant, A.; Metzner, R.; et al. Non-invasive approaches for phenotyping of enhanced performance traits in bean. *Funct. Plant Biol.* **2011**, *38*, 968–983. [CrossRef]

97. Pauli, D.; Chapman, S.C.; Bart, R.; Topp, C.N.; Lawrence-Dill, C.J.; Poland, J.; Gore, M.A. The quest for understanding phenotypic variation via integrated approaches in the field environment. *Plant Physiol.* **2016**, *172*, 622–634. [CrossRef] [PubMed]

98. White, J.W.; Andrade-Sanchez, P.; Gore, M.A.; Bronson, K.F.; Coffelt, T.A.; Conley, M.M.; Feldmann, K.A.; French, A.N.; Heun, J.T.; Hunsaker, D.J.; et al. Field-based phenomics for plant genetics research. *Field Crop. Res.* **2012**, *133*, 101–112. [CrossRef]

99. Hirschhorn, J.N.; Daly, M.J. Genome-wide association studies for common diseases and complex traits. *Nat. Rev. Genet.* **2005**, *6*, 95–108. [CrossRef] [PubMed]

100. Brachi, B.; Morris, G.P.; Borevitz, J.O. Genome-wide association studies in plants: The missing heritability is in the field. *Genome Biol.* **2011**, *12*, 232. [CrossRef] [PubMed]

101. Huang, X.; Han, B. Natural variations and genome-wide association studies in crop plants. *Annu. Rev. Plant Biol.* **2014**, *65*, 531–551. [CrossRef] [PubMed]

102. Zhang, J.; Zhao, J.; Xu, Y.; Liang, J.; Chang, P.; Yan, F.; Li, M.; Liang, Y.; Zou, Z. Genome-wide association mapping for tomato volatiles positively contributing to tomato flavor. *Front. Plant Sci.* **2015**, *6*, 1042. [CrossRef] [PubMed]

103. Nimmakayala, P.; Abburi, V.L.; Saminathan, T.; Alaparthi, S.B.; Almeida, A.; Davenport, B.; Nadimi, M.; Davidson, J.; Tonapi, K.; Yadav, L.; et al. Genome-wide diversity and association mapping for capsaicinoids and fruit weight in capsicum annuum L. *Sci. Rep.* **2016**, *6*, 38081. [CrossRef] [PubMed]

104. Pavan, S.; Marcotrigiano, A.R.; Ciani, E.; Mazzeo, R.; Zonno, V.; Ruggieri, V.; Lotti, C.; Ricciardi, L. Genotyping-by-sequencing of a melon (*Cucumis melo* L.) germplasm collection from a secondary center of diversity highlights patterns of genetic variation and genomic features of different gene pools. *BMC Genom.* **2017**, *18*, 59. [CrossRef] [PubMed]

105. Nimmakayala, P.; Tomason, Y.R.; Abburi, V.L.; Alvarado, A.; Saminathan, T.; Vajja, V.G.; Salazar, G.; Panicker, G.K.; Levi, A.; Wechter, W.P.; et al. Genome-wide differentiation of various melon horticultural groups for use in gwas for fruit firmness and construction of a high resolution genetic map. *Front. Plant Sci.* **2016**, *7*, 1437. [CrossRef] [PubMed]

106. Cericola, F.; Portis, E.; Lanteri, S.; Toppino, L.; Barchi, L.; Acciarri, N.; Pulcini, L.; Sala, T.; Rotino, G.L. Linkage disequilibrium and genome-wide association analysis for anthocyanin pigmentation and fruit color in eggplant. *BMC Genom.* **2014**, *15*, 896. [CrossRef] [PubMed]

107. Hart, J.P.; Griffiths, P.D. Genotyping-by-sequencing enabled mapping and marker development for the by-2 potyvirus resistance allele in common bean. *Plant Genome* **2015**, *8*. [CrossRef]

108. Saxena, R.K.; Kale, S.M.; Kumar, V.; Parupali, S.; Joshi, S.; Singh, V.; Garg, V.; Das, R.R.; Sharma, M.; Yamini, K.N.; et al. Genotyping-by-sequencing of three mapping populations for identification of candidate genomic regions for resistance to sterility mosaic disease in pigeonpea. *Sci. Rep.* **2017**, *7*, 1813. [CrossRef] [PubMed]

109. Bush, W.S.; Moore, J.H. Chapter 11: Genome-wide association studies. *PLoS Comput. Biol.* **2012**, *8*, e1002822. [CrossRef] [PubMed]

110. Zhang, Z.; Ersoz, E.; Lai, C.Q.; Todhunter, R.J.; Tiwari, H.K.; Gore, M.A.; Bradbury, P.J.; Yu, J.; Arnett, D.K.; Ordovas, J.M.; et al. Mixed linear model approach adapted for genome-wide association studies. *Nat. Genet.* **2010**, *42*, 355–360. [CrossRef] [PubMed]

111. Bhat, J.A.; Ali, S.; Salgotra, R.K.; Mir, Z.A.; Dutta, S.; Jadon, V.; Tyagi, A.; Mushtaq, M.; Jain, N.; Singh, P.K.; et al. Genomic selection in the era of next generation sequencing for complex traits in plant breeding. *Front. Genet.* **2016**, *7*, 221. [CrossRef] [PubMed]

112. Hamblin, M.T.; Buckler, E.S.; Jannink, J.-L. Population genetics of genomics-based crop improvement methods. *Trends Genet.* **2011**, *27*, 98–106. [CrossRef] [PubMed]

113. Xu, Y.; Xie, C.; Wan, J.; He, Z.; Prasanna, B.M. Marker-assisted selection in cereals: Platforms, strategies and examples. In *Cereal Genomics II*; Gupta, P.K., Varshney, R.K., Eds.; Springer: Dordrecht, The Netherlands, 2013; pp. 375–411.

114. He, S.; Schulthess, A.W.; Mirdita, V.; Zhao, Y.; Korzun, V.; Bothe, R.; Ebmeyer, E.; Reif, J.C.; Jiang, Y. Genomic selection in a commercial winter wheat population. *Theor. Appl. Genet.* **2016**, *129*, 641–651. [CrossRef] [PubMed]

115. Duangjit, J.; Causse, M.; Sauvage, C. Efficiency of genomic selection for tomato fruit quality. *Mol. Breed.* **2016**, *36*, 29. [CrossRef]

116. Yamamoto, E.; Matsunaga, H.; Onogi, A.; Ohyama, A.; Miyatake, K.; Yamaguchi, H.; Nunome, T.; Iwata, H.; Fukuoka, H. Efficiency of genomic selection for breeding population design and phenotype prediction in tomato. *Heredity* **2017**, *118*, 202–209. [CrossRef] [PubMed]

117. Druka, A.; Potokina, E.; Luo, Z.; Jiang, N.; Chen, X.; Kearsey, M.; Waugh, R. Expression quantitative trait loci analysis in plants. *Plant Biotechnol. J.* **2010**, *8*, 10–27. [CrossRef] [PubMed]

118. Vidalis, A.; Živković, D.; Wardenaar, R.; Roquis, D.; Tellier, A.; Johannes, F. Methylome evolution in plants. *Genome Biol.* **2016**, *17*, 264. [CrossRef] [PubMed]

119. Desta, Z.A.; Ortiz, R. Genomic selection: Genome-wide prediction in plant improvement. *Trends Plant Sci* **2014**, *19*, 592–601. [CrossRef] [PubMed]

120. Cheng, Z.-M.; Werner, D.J. Will the Traditional Horticultural Breeding and Genetics Research Be Fairly Valued in Academia. Available online: https://www.nature.com/articles/hortres201553 (accessed on 14 September 2017).

diversity

MDPI

Review

International Instruments for Conservation and Sustainable Use of Plant Genetic Resources for Food and Agriculture: An Historical Appraisal

Andrea Sonnino

Biotechnologies and Agroindustry Division, Casaccia Research Centre, ENEA, Via Anguillarese 301, 00060 Rome, Italy; andrea.sonnino@enea.it

Received: 22 September 2017; Accepted: 24 October 2017; Published: 1 November 2017

Abstract: This paper critically reviews the evolution of concepts and principles that inspired the adoption and enforcement of international instruments related to the conservation, exchange and sustainable use of plant genetic resources for food and agriculture, including agreements, governance and programs. The review spans from the pioneering attempts to regulate this matter, to the negotiations that led to the current regulatory framework, covering the creation of the Panel of Experts on Plant Exploration and Introduction of Food and Agriculture Organization (FAO) in 1965, the establishment of the International Board for Plant Genetic Resources (IBPGR) in 1974 and the FAO Commission on Plant Genetic Resources for Food and Agriculture in 1983, the adoption of the International Undertaking in 1983 and, more recently (2001), the International Treaty for Plant Genetic Resources for Food and Agriculture. The conceptual contribution, offered by Prof. Scarascia Mugnozza and other visionary scholars, to the establishment of these international instruments, is highlighted.

Keywords: governance; treaty; farmers' rights; gene banks

1. Introduction

Genetic resources for food and agriculture, including plants, animals, micro-organisms and invertebrates, play an essential role in feeding mankind. They ensure the strategic reservoir necessary to maintain efficient food production systems and to adapt them for new socio-economic and environmental conditions. They contribute to the delivery of vital ecosystem services, such as pollination and pest control. It is therefore strategically crucial to preserve diversity through sustainable use, including diversity both among and within species.

The exchange of genetic resources started at the dawn of agriculture. One of the earliest documented germplasm collecting trips was sent by Queen Hatshepsut of Egypt, to Punt, to collect incense trees for the royal gardens (about 1500 BC) [1]. Her successor, Tuthmose III, organized a mission to Syria to collect seeds, plants and trees [2]. There has been so much exchange of genetic resources throughout the history of agriculture that nowadays more than two thirds of all food is produced using genetic resources grown outside their areas of origin, generating a high interdependence among countries in regard to their food supply and production systems [3].

The importance of the conservation of genetic resources was recognized as early as 795 AD, when Charlemagne gave instructions by means of the decree, *Capitulare de villis*, to plant in each farm, 77 different plant food species [4]. Keeping control of genetic resources was considered crucial by China to preserve the silk production monopoly, but ultimately, they lost it in 550 AD, when two Nestorian monks smuggled silkworm eggs and mulberry seeds into the Byzantine Empire for Emperor Justinian I [5]. Another early attempt to regulate the use of plant genetic resources was made in 1414 AD, by the Venetian Senate, which reported that massive exploitation of cypress in Crete was putting it in danger of extinction, and therefore forbade exports of this strategic resource used for naval constructions [6]. Brazil banned any export of rubber trees or seeds to maintain a monopoly on the production of rubber, but the monopoly was broken in 1876, when Henry Wickham smuggled a large number of seeds and initiated rubber production in British colonies, such as Ceylon and Malaysia. Less than fifty years later, the Brazilian rubber industry was annihilated by competition from Southeast Asia [7].

Even if the history of deliberate or accidental international movement of genetic resources is as old as agriculture, it is only in the last 50 years that the scientific developments of germplasm exploration and the economic interests accrued by germplasm exploitation have prompted the exigency of organizing and regulating the conservation and use of genetic resources on a global level [8].

This paper critically reviews the evolution of concepts and principles that inspired the adoption and enforcement of international instruments related to the conservation, exchange and sustainable use of plant genetic resources (PGRs) for food and agriculture, including agreements, conventions, treaties, governance and programs. The most important international events relevant to the establishment and evolution of international instruments, related to the conservation and sustainable utilization of PGRs, are reported in Table 1, in chronological order. Concepts and principles underpinning the international dialogue and, as a consequence, the international instruments related to the conservation and sustainable utilization of the PGRs, are discussed, following a thematic order, disregarding the chronological sequence, in order to ensure an organic analysis. The intellectual contribution, offered by Prof. GianTommaso Scarascia Mugnozza and other visionary scholars, to the establishment of these international instruments is highlighted, along with the discussion of the conceptual evolution about PGRs [9]. More details on the history of the international regulation of PGR conservation and exchange are available in [8,10–12].

2. Evolution of Underpinning Concepts and Principles

2.1. Genetic Erosion

The rapid expansion of plant breeding during the second half of the 20th century brought the introduction of a big number of improved varieties, which progressively replaced old landraces, especially in developing countries. Not all the genes present in the farmers' varieties are also contained in modern varieties. In addition, the release of commercial varieties into traditional farming systems often leads to a reduction in the number of varieties cultivated in a given area. This phenomenon was noticed and denounced as a peril by Harlan and Martini [13], and Frankel [14], among others. The term 'genetic erosion' was coined at the *Technical Conference Plant Exploration, Utilization and Conservation of Plant Genetic Resources* of FAO/IBP (International Biological Programme), held in Rome in 1967 [15], to describe this loss of individual genes and combinations of genes. At this conference, the genetic erosion issue was first systematically addressed, urging for a coordinated global programme of collection and conservation of PGRs. Erna Bennet from FAO and Otto Frankel from Australia were the major contributors to these concepts.

After the outbreak of corn-leaf blight in the United States of America (USA) and coffee rust in Brazil, the loss of genetic variability was identified as a cause of crop vulnerability to pests, diseases and other adverse factors [16]. Scarascia Mugnozza subsequently pointed out that genetic erosion is not only a problem of environmental significance, but it has also socio-economic, political and ethical relevance [17].

In 1991, the FAO Conference agreed that a report on the State of the World's PGRs should be prepared. The report was submitted at the *Fourth International Technical Conference for the Conservation and Utilization of Plant Genetic Resources*, in 1996 [18]. It has been an important global instrument to assess genetic erosion and to inform decision-making related to PGR management. The State of the World's PGRs identified the replacement of local varieties by modern varieties as the main cause of genetic erosion. Other causes of genetic erosion and loss of biodiversity included the emergence of new pests, weeds and diseases, habitat degradation, urbanization, land clearing through deforestation, and bush fires.

A *Second Report on the State of the World's Plant Genetic Resources* identified the most significant changes that occurred between 1996 and 2009 [19].

With the adoption of the Programme of Work on Climate Change and Genetic Resources for Food and Agriculture, in 2013 and the approval of the Voluntary Guidelines to Support the Integration of Genetic Diversity into National Climate Change Adaptation Planning, in 2015, attention was paid to climate change as major driver of genetic erosion. The importance of PGRs for coping with climate change was therein underlined.

2.2. What to Conserve

To counter genetic erosion, a number of exploration and collection missions were undertaken, especially after World War II. The problem arose about what to collect and conserve. At the beginning, germplasm collection expeditions adopted a 'mission-oriented approach', focusing on targets specific to single plant breeding projects [8]. The collected material was mainly used in small, purpose-specific collections, mostly located at plant breeding stations. It was objected that this approach, while responding to immediate individual or organizational needs, had little effect on genetic erosion. The opposite approach, called, by Frankel [20], a 'generalist approach', was directed towards collecting as much as possible, especially in plant centres of origin, and to conserve the collected materials in centralized ex situ gene banks, which serve networks of plant breeding programs. This approach was facilitated by the adoption of long-term cold storage facilities, developed in the 1960s.

The debate about the object of PGR conservation was, in the 1960s and 1970s, limited to the ambit of food and agriculture and dominated by the consideration of their economic and strategic values. The rise in concern about environmental degradation, especially during the process of preparation of the United Nations (UN) Conference on Environment and Development (UNCED) and the Convention on Biological diversity (CBD), prompted an international discussion on the conservation of biological diversity, on a large scale [21]. This debate led to the establishment of two parallel arenas: (i) an agricultural arena, based in the FAO and Consultative Group on International Agricultural Research (CGIAR), and focused on the use and economic value of genetic resources, and (ii) an environmental arena, grounded by the United Nations Environment Programme (UNEP) and conservationist organizations, and devoted to more general aspects, including those related to non-use and non-value of diversity [8]. These two political arenas often entered into conflict with each other, especially for allocation of financial resources, but also benefitted from exchange of ideas and concepts.

2.3. How to Conserve and Use PGRs

During the pioneering years, PGR-related activities were focused on the collection and ex situ conservation of germplasm. Although the value of ex situ collections remains undoubted by the scientific community, several scholars, including Erna Bennet and Robert Allard [15], expressed concern that the storage of seeds in gene banks does not allow natural evolution to happen, to create new diversity, and thus proposed to adopt in situ conservation strategies in [22,23]. Initially this proposal did not find unanimous support. The first formal acceptance of in situ conservation was given in 1973, by the FAO/IBP *Technical Conference on Genetic Resources* [8]. The concept was reinforced by the recommendations of the 1981 FAO/IBP *Technical Conference on Genetic Resources*. The 1996 Leipzig Declaration appropriated both the ex situ and in situ approaches, considering them as not mutually exclusive, but complementary components of conservation programmes.

The CBD explicitly promotes in situ conservation, through the establishment of protected areas and natural parks [11]. In addition, on-farm conservation is often adopted to grow, utilize and conserve landraces, native varieties and other local materials, within their original landscapes and traditional farming systems.

The *community-based conservation* approach, as proposed by M.S. Swaminathan, involves programs for local variety and landrace conservation in indigenous communities and/or rural families, who receive compensation and access to the stored PGRs, in exchange for their services [24]. The "Scarascia Mugnozza Community Genetic Resources Centre", established in Chennai, India with Italian funding [9], allows for the exchange of genetic resources between farmers of the same community and therefore the widening of the genetic basis of crops [25]. The Potato Park in Peru, where potato genetic resources are conserved by local communities, along with their cultural heritage, is another good example of application of this approach [26].

Integration of ex situ, in situ, on-farm and community-based approaches in PGR conservation strategies was advocated by Scarascia Mugnozza when he addressed the FAO Conference in 1995 [17].

2.4. The Contribution of Biotechnologies to PGR Management

The contribution of biotechnologies to decision-making in PGR management and mitigation of genetic erosion was recognized by several internationally-agreed documents [27]. Agenda 21, adopted by UNCED in 1992 (Table 1), explicitly recognizes the potential of biotechnology. Article 19 of the CBD also stated the need for further development of internationally-agreed principles on risk assessment and management of all aspects of biotechnology and established an Open-Ended Ad Hoc Working Group on Biosafety to develop a draft protocol on biosafety, specifically focusing on transboundary movements of living modified organisms (LMO), resulting from modern biotechnology, that may have adverse effects on the conservation and sustainable use of biological diversity. The Cartagena Protocol on Biosafety defines a living modified organism (LMO) as "any living organism that possesses a novel combination of genetic material, obtained through the use of modern biotechnology". In general, use of the term LMO is considered to be functionally the same as the term, genetically modified organism (GMO). After a long and complicated negotiation, in 2001, the CBD Conference of the Parties adopted the Cartagena Protocol on Biosafety (entered into force in 2003). The aim of the Protocol is to protect biological diversity from the potential environmental risks posed by LMOs. It established that an advance informed agreement (AIA) ensures that countries are provided with the information necessary to make informed decisions before agreeing to imports. The protocol references to a precautionary approach, as defined in Principle 15 of the Rio Declaration on Environment and Development. The protocol also establishes a Biosafety Clearing-House, to facilitate the exchange of information on LMOs.

The Commission on Genetic Resources for Food and Agriculture (CGRFA) has examined the developments in biotechnology, as they relate to conservation and use of PGRs, several times, starting from 1989, when it requested FAO to draft a Code of Conduct for Biotechnology. Preliminary drafts of this proposed legal instrument were examined by CGRFA, in 1991, 1993, and 1995. After several postponements, the possible adoption of a Code of Conduct was ultimately dropped, in consideration of the existence of other possibly overlapping international mechanisms. The drafts and the accompanying documents were nonetheless sent to the CBD secretariat and contributed to the negotiation of a biosafety protocol. CGRFA examined the advancements of biotechnologies in 1999 [28] and in 2011 [29], appreciating the contributions that biotechnologies were giving to both conservation and use of PGRs, calling for consideration and management of inherent risks, and urging for capacity development initiatives in developing countries. In particular, the decisive contributions offered by biotechnological approaches, especially from genomics and transcriptomics, to the measurement of plant genetic variation, to the protection of diversity, and to gene bank management, were underlined. A more comprehensive review of the biotechnologies, applied to the Management of Genetic Resources for Food and Agriculture, is presented in [30].

The current status of biotechnology development and application in naturally regenerated forests and in planted forests was examined by the International Symposium on Forest Biotechnology for Smallholders, held in Foz do Iguaçu, Brazil, in 2015 [31].

CGRFA paid more focused attention to the potential impacts of Genetic Use Restriction Technologies (GURTs) on agrobiodiversity, in 2002 [32]; and on genomics and genetic resources for food and agriculture, in 2007 [33]. The background paper on the effects of (trans)gene flow on the conservation and sustainable use of genetic resources, discussed by CGRFA in 2007, analysed the delicate issue of the preservation of the genetic identity of PGR specimens conserved in ex situ collections and the possible perils posed by their proximity with Genetically Modified Organisms (GMOs) [34].

Already in 1995, Scarascia Mugnozza had noted the intimate connection of biotechnology to biodiversity and warned FAO member countries about the danger of developing countries remaining excluded from the benefits of biotechnology application [17]. This concept was further reinforced by the International Technical FAO Conference on *Agricultural biotechnologies in developing countries*, held in Guadalajara, Mexico, in 2010 [35], where the possible benefits of biotechnologies for small farmers, producers and consumers were underlined. The FAO International Symposium on *The role of agricultural biotechnologies in sustainable food systems and nutrition*, held in 2016, reaffirmed that biotechnologies can be used, following agroecological principles, to enhance productivity, while ensuring sustainability, and to conserve genetic resources and use, and indigenous knowledge [36].

2.5. Who Owns PGRs and How the Owners Should Be Compensated by the Users

The 22nd Session of the FAO Conference, held in Rome in 1983, marked a breakthrough for the international regulation of PGRs, in that it adopted the first international, even if not legally binding, instrument to regulate conservation and use of PGRs, the *International Undertaking on Plant Genetic Resources*. It stated that "plant genetic resources should be considered as a common heritage of mankind and be available without restrictions for plant breeding, scientific and development purposes to all countries and institutions concerned" [37]. The common heritage principle was implicitly used as the principle governing the diffusion of PGRs from centers of domestication into new cultivation areas, since the discovery of the Americas and before [38]. The notion that while tropical countries hold the greatest deal of diversity, advanced countries, which hold the technical capacity to collect and use PGRs, benefit the most from their use [39] and that private companies were appropriating and exploiting developing countries' natural resources, without compensating them or even requesting their permission [12], sparked wide dissatisfaction among scientists and politicians of the third world. This unease was given a conceptual framework, with the publication of the book, *The Seeds of the Earth: a Private or Public Resource*, by Pat Mooney [40]. In recognition of this issue, third world representatives maintained that the Undertaking covered, in addition to landraces, wild species, and similar PGRs, "special genetic stocks (including elite and current breeders' lines)," which should also be made "available without restriction". This measure was opposed by private seed companies, supported by a number of developed countries that refused to sign the undertaking.

The 25th Session of the FAO Conference (1989) tried to conciliate the rights of developing and developed countries by adopting Resolution 4/89 (*Agreed Interpretation of the Undertaking*) and Resolution 5/89 [8]. The first document recognised that plant breeders' rights were not inconsistent with the international undertaking; the latter introduced the concept of farmers' rights, described as "rights arising from the past, present and future contributions of farmers in conserving, improving, and making available plant genetic resources, particularly those in the centers of origin/diversity." The resolution also stated that farmers' rights "are vested in the international community, as the trustee for present and future generations of farmers, for the purpose of ensuring full benefits to farmers, and supporting the continuation of their contributions."

The Keystone Dialogues, held in 1988, 1990 and 1991, in Keystone, Madras (now Chennai) and Oslo respectively, and chaired by Prof. Mankombu Sambasivan Swaminathan, were instrumental in framing the issue of farmers' rights and providing a basis for their international recognition [41], as well as the recommendations which arose from two meetings of the Crucible Group [42]. The PGR ownership question was considered again by the 1991 FAO Conference. Resolution 3/91 stated that "the concept of mankind's heritage, as applied in the Undertaking on Plant Genetic Resources, is subject to the sovereignty of the states over their genetic resources". The CBD, in 1992, abandoned definitively, the concept of common heritage of mankind, when it stipulated that the nations hold sovereign rights on natural resources within their boundaries and that the authority to regulate access to PGRs rests with national governments, which are nevertheless requested to facilitate the access [12]. Resolution 3 of the Nairobi Final Act, adopted by countries in the framework of the CBD, included the famers' rights question among the outstanding, unsettled matters. The notion that famers contributed, and will keep contributing to, the conservation and development of PGRs, and that a financial mechanism to compensate this contribution should be established, was also asserted by Scarascia Mugnozza in front of the FAO Conference, in 1995 [17].

The official endorsement of farmers' rights and national sovereignty over PGRs arrived with the approval of the 2001 International Treaty on Plant Genetic Resources for Food and Agriculture (entered into force in 2004). Article 9 of the Treaty, recognizing the enormous contribution that farmers have made to the development of the world's wealth of PGRs, calls for ensuring that they share the benefits derived from the use of these resources. The responsibility for implementing farmers' rights is given to governments. The measures that could be taken to protect, promote and realize these rights are:

(a) Protection of traditional knowledge, relevant to PGRs;
(b) Ensuring the right to equitably participate in sharing benefits arising from the utilization of PGRs;
(c) Ensuring the right to participate in making decisions, on a national level, on matters related to the conservation and sustainable use of PGRs; and
(d) Safeguarding the farmers' rights to save, use, exchange and sell farm-saved seed/propagating materials, subject to national law and as appropriate.

The Treaty (Article 10) establishes the Multilateral System of Access and Benefit Sharing of PGRs. Access to PGRs is subject to the issue of a standard Material Transfer Agreement (MTA). Those who access genetic materials through the Multilateral System agree to pay to a financial mechanism, established by the Treaty, of an equitable share of the benefits arising from their commercial use (Article 13). With these provisions, the Treaty shifted the famers' rights issue from direct economic compensation, which would have been difficult to implement and would have had negative effects on access [8], to a social dimension. The benefit-sharing mechanism, settled by the Treaty, in fact recognizes the moral obligation towards the famers, who developed and maintained the PGRs, and generates resources for conservation and development of PGRs at large. The long negotiation that led to the adoption of the Treaty benefitted greatly from the contributions by José Esquinas-Alcázar from FAO [10].

2.6. Ownership and Legal Status of Ex Situ Collections Held by International Agricultural Research Centres (IARCs)

Since their establishment, the International Agricultural Research Centres (IARCs) of the CGIAR collected, conserved in their gene banks and utilized, hundreds of thousands of samples of PGRs, that formed the basis for the development of improved varieties, including those utilized in the green revolution [11]. The adoption of the international undertaking, and particularly Article 7, called for the development of an 'international network of collections in gene banks under the auspices or jurisdiction of FAO'. In compliance with the undertaking, the IARCs signed an agreement with FAO, placing their collection within this network [10]. Through this agreement, the IARCs were recognized as trustees of collections, renounced to claim ownership over the PGRs they conserved, and committed themselves to providing the international community with the designated germplasm.

While the ownership of national collections was usually vested in countries where the gene-banks were located, the ownership and legal status of ex situ collections, held by IARCs, was still unclear [8], especially in light of the provisions of the CBD. The convention, in fact, offers signing parties rights over their natural resources, even if the sovereign rights can be exerted only over PGRs of proved origin and acquired, after the entry into force of the CBD itself. Most the PGRs held by IARCS remain therefore excluded from this provision.

Article 15 of the International Treaty on Plant Genetic Resources for Food and Agriculture calls upon the IARCs to place their germplasm collections under the purview of the treaty. This was stipulated by agreements with the Governing Body of the Treaty, signed by IARCs and a number of other international centres [10]. IARCs began using the standard Material Transfer Agreement of the Treaty whenever transferring germplasm, using it not only for plants covered by the Treaty [43].

2.7. Governance

During the 1960s and the 1970s, FAO played the role of a practically unique international forum on PGRs. It was soon evident that PGR collection and conservation had not only technical, but also financial implications. The CGIAR, following the recommendations of the ad hoc Working Group of the Technical Advisory Committee (TAC) at its meeting, in Beltsville, in 1972, took a number of important decisions related to CGIAR base collections (see Table 1) and made a notable offer to cover the costs of a global conservation network [8]. This move brought aboutthe establishment of the *International Board for Plant Genetic Resources* (IBPGR), with the task of coordinating an international plant genetic resources programme, including emergency collecting missions, and building and expanding a network of national, regional and international gene banks [44]. To conciliate the technical and financial aspects of the PGR programme with the political facets of the PGR issue, the board was placed in FAO premises, and FAO acted as the secretariat. The story of collaboration between IBGR and FAO, even if quite conflictive, was at the end of the day fruitful, in that it complemented mutual functions [8]. In the last part of the 1980s, IBPGR acquired increased autonomy. In 1991, five countries signed the initial agreement that transformed IBPGR into the *International Plant Genetic Resources Institute* (IPGRI), with the CGIAR Centre being totally autonomous from the FAO [44] Forty-three additional countries joined the agreement at this second time [45]. The new institute is hosted by Italy [9]. Since 2006, the institute has used the name, Bioversity.

In 1983, the FAO Conference decided to establish the *Commission on Plant Genetic Resources for Food and Agriculture (CPGRFA)*. The commission provides the only permanent forum for governments to discuss and negotiate matters relevant to genetic resources, including the sustainable use and conservation of PGRs and the fair and equitable sharing of benefits derived from their use. In 1995, the FAO Conference broadened the commission's mandate to cover all components of biodiversity, relevent to food and agriculture and modified its name into *Commission on Genetic Resources for Food and Agriculture (CGRFA)*. With the entrance into force of the international treaty, all the matters related to its implementation became ruled by a Governing Body.

2.8. Securing the International PGR Collections

The history of collection, conservation and use of PGRs was constantly accompanied by underfunding problems, with the increasing volume of germplasm to be conserved conflicting with shrinking resources [46]. In addition, the financial endowment of gene banks is imperilled by political instabilities and fluctuations of stakeholders' commitments, not to mention wars, social unrest, natural disasters and other risks. The Global Crop Diversity Trust (now renamed Crop Trust) was established in 2004, as an independent organization under international law, that holds the permanent, self-sustaining Crop Diversity Endowment Fund. Each year, a portion of the fund's value, generated from investment income earned, is used to ensure the conservation and maintenance of

PGRs held in gene banks. The purpose is to subtract international collections from political instabilities and ensure them a long-term, predictable funding source. Unfortunately, the endowment fund's target size of the Crop Trust has not been reached yet [46].

In 2008, the Svalbard Global Seed Vault was established in the Svalbard Islands, Norway. It is a fail-safe seed storage facility, built to protect PGRs held in traditional seed banks around the world from natural or man-made disasters and thus prevent the loss of crop diversity. The duplicated seed samples, conserved by the seed vault, have already been used to retrieve germplasm collections damaged by war events in Syria [46].

3. Conclusions

It is difficult to deny that concepts and principles supporting the international movement on the conservation and sustainable use of PGRs have greatly evolved during the last 50–60 years. Notions generated by a tiny group of visionary pioneers are now shared by the public at large and have become common knowledge. The evolution of scientific and political thinking, accompanied by the increase in public awareness, has allowed for the achievement of important results that can be summarized as follows:

(a) Approximately 7.4 million germplasm accessions, representing more than 16,500 plant species are currently secured in 1750 gene banks worldwide [19];

(b) Conserved germplasm provides broad genetic diversity that is increasingly used for genetic studies and plant breeding programs, with undeniable benefits for present and future world food production [47];

(c) Newly developed technologies, including molecular biology-based technologies, are greatly contributing to the conservation, exploration and characterization of PGRs, opening unprecedented opportunities for their sustainable utilization [30,31].

(d) The adoption of the International Treaty on Plant Genetic Resources for Food and Agriculture has provided a regulatory framework to access, exchange and benefit sharing of PGRs [10].

Consideration of the results achieved should not distract attention from the outstanding problems, the most important of which are:

(a) Genetic erosion is far from being stopped, as attested to by the Second Report on the State of the World's Plant Genetic Resources [19];

(b) Payment for environmental services, such as the in situ conservation of PGRs, is not always recognized, for farmers and local communities,

(c) Conservation programmes are chronically underfunded [46],

(d) The impact of climate change on crop genetic diversity is not fully understood,

(e) Appropriate capacities and adequate infrastructures to explore and exploit biodiversity are still lacking in many developing countries [46].

Finally, the transition towards sustainable food and agriculture systems urges us to integrate the genetic resource conservation and utilization issue into a holistic view of sustainability. Only a systems approach, that studies not only the single components, but also considers the interplay among them, can, in fact, indicate sustainable strategies to face the multiple challenges before us.

Table 1. Events relevant to the establishment and evolution of international instruments related to the conservation and sustainable utilization of Plant Genetic Resources (PGRs).

Year	Event	Main Output(s) of Relevance for PGRs and [Reference]	Underpinning Principle(s)
1961	FAO (Food abd Agriculture Organization) *Technical Meeting on Plant Exploration and Introduction*, Rome, 10–20 July.	Report of the meeting.	Mission-driven approach: Conservation and use closely linked, tied to plant breeding, dominance of ex situ collections, mainly in developed countries.
1965	Establishment of the FAO *Panel of Experts on Plant Exploration and Introduction*.	Six meetings between 1967 and 1975.	Formulation of criteria for the conservation and use of PGRs.
1967	FAO/IBP *Technical Conference Plant Exploration, Utilization and Conservation of Plant Genetic Resources*, Rome, 18–26 September.	Publication [15].	Generalist approach: Rising concern about gene erosion of landraces and wild relatives. Large, long-term ex situ collections. In situ conservation as a complementary/alternative strategy.
1969	*Third Session of the FAO Panel of Experts on Plant Exploration and Introduction*, Rome, 25–28 March.	Report [48].	List of priority geographic areas for exploration and conservation of PGRs.
1971	Founding meeting of the *Consultative Group on International Agricultural Research* (CGIAR), Washington, DC, USA, 19 May 1971.	Resolution, establishment of the Technical Advisory Committee.	Linkage between agricultural research and development.
	Issue of the report of the Agricultural Board of the National Research Council.	[16]	Genetic uniformity as a cause of vulnerability to epidemics.
	UN Conference on Human Environment, Stockholm, 5–16 June.	Articles 39–45 of the *United Nations (UN) Declaration on the Human Environment*.	Genetic resources issue brought into the international agenda. Clear division of tasks between in situ (wild relatives) and ex situ (cultivated plants) conservation.
1972	TAC (Technical Advisory Committee) Ad Hoc Working Group, Beltsville, 20–25 March.	Report on *The Collection, Evaluation and Conservation of Plant Genetic Resources*.	Establishment of the World Network of Genetic Resources Centres. Establishment of a coordinating centre. Support to gene banks already existing in International Agricultural Research Centres (IARCs) of the CGIAR; Establishment of additional gene banks in other IARCs of the CGIAR.
1973	FAO/IBP *Technical Conference on Genetic Resources*, 12–16 March, Rome.	Plan of action [49].	Recommendation to establish in situ collections.

Table 1. *Cont.*

Year	Event	Main Output(s) of Relevance for PGRs and [Reference]	Underpinning Principle(s)
1974	Establishment of the *International Board for Plant Genetic Resources* (IBPGR).	[44]	Coordination of an international PGR programme. FAO acted as secretariat.
	FAO/IBP *Technical Conference on Genetic Resources*, 12–16 March, Rome.	Publication [50].	In situ collections best method to conserve variability of wild species.
1981	21st Session of the FAO Conference, Rome, 7–25 November.	Resolution 6/81.	Need for an international agreement to ensure the conservation, maintenance and free exchange of PGRs Requests FAO to prepare projects for two options: an international agreement and an international gene bank.
1983	22nd Session of the FAO Conference, Rome, 5–23 November.	Resolution 8/83: Adoption of the *International Undertaking on Plant Genetic Resources* Establishment of the *Commission on Plant Genetic Resources for Food and Agriculture (CGRFA) and of the Global System on Plant Genetic Resources* [51].	Shared principles, non-legally binding, on conservation and access to PGRs: PGRs are a common heritage of humankind Genetic stocks and breeding lines included Freedom of exchange through a network of gene banks Supervision through the Commission.
	25th Session of the FAO Conference, Rome, 11–29 November.	Resolution 4/89: Adoption of an agreed interpretation of the international undertaking. Resolution 5/89: Farmers' rights.	Plant breeders' rights are not inconsistent with the International Undertaking. Recognition of Farmers' Rights.
1989	3rd Regular Session of CGRFA, Rome, 17–21 April.	Call for the development of *The International Network of Ex Situ Collections under the Auspices of FAO.* Report	Lack of clarity regarding the legal situation of the ex situ collections. The Commission requested FAO to draft a code of conduct for biotechnology, as it affects conservation and use of genetic resources.
	Transformation of the IBPGR into the *International Plant Genetic Resources Institute* (IPGRI).		Independence of IPGRI.
1991	26th Session of the FAO Conference, Rome, 9–27 November.	Resolution 3/91.	Recognition of the sovereign rights of nations over their PGRs Agreement on the development of the 1st State of the World's Plant Genetic Resources and Global Plan of Action on PGRs.

Table 1. *Cont.*

Year	Event	Main Output(s) of Relevance for PGRs and [Reference]	Underpinning Principle(s)
1992	UN *Conference on Environment and Development* (UNCED) Rio de Janeiro, 3–14 June.	*Convention on Biological Diversity* (CBD) (entered into force on 29 December 1993).	Biodiversity vs. genetic resources Need for a protocol setting out appropriate procedures for safe transfer, handling and use of any living modified organism resulting from biotechnology that may have adverse effects on the conservation and sustainable use of biological diversity Establishment of the Open-ended Ad Hoc Working Group on Biosafety.
		Chapter 14 of Agenda 21.	Call for the strengthening of the FAO *Global System on Plant Genetic Resources*.
		Chapter 16 of Agenda 21.	Biotechnology can assist in the conservation of biological resources through, for example, ex situ techniques Need for further development of internationally-agreed principles on risk assessment and management of all aspects of biotechnology.
		Adoption of Resolution 3 of the Nairobi Final Act.	Recognises matters not addressed by the convention: Access to ex situ collections The questions surrounding farmers' rights Requests FAO forum to address these matters.
1994	1st Extraordinary Session of the CGRFA.	Start of negotiations for the revision of the international undertaking, 12 centres of the CGIAR sign an agreement with FAO, placing their collections under the auspices of FAO.	Centres agree to hold the designated germplasm "in trust for the benefit of the international community".
	Establishment of the "Scarascia Mugnozza Community Genetic Resources Centre", Chennai.	[25]	Community-based conservation of PGRs.

Table 1. *Cont.*

Year	Event	Main Output(s) of Relevance for PGRs and [Reference]	Underpinning Principle(s)
1995	28th Session of the FAO Conference, Rome, 20 October.	19th McDougall Memorial Lecture "The protection of biodiversity and the conservation and use of genetic resources for food and agriculture: potential and perspectives" by G.T. Scarascia Mugnozza [17].	Biodiversity loss is not only an important environmental problem, but also a socio-economic, political and ethical problem. Conservation and access to PGRs are essential interests of humankind and are strictly interconnected with food security issues, Exploration and ex situ conservation of PGRs are essential but must be integrated by in situ, on farm, a community-level conservation strategy. PGRs should be conserved and made available to scientists and farmers, but access should be regulated by international agreements. Farmers, especially those who work in centres of origin of cultivated plant species, contributed and will contribute, to the conservation and development of PGRs; a financial mechanism to compensate this contribution (*Farmers' rights*) should be established. There is also a need to develop equitable mechanisms for technological transfer from industrialized countries to developing countries.
	Science Academies Summit at the M.S. Swaminathan Research Foundation, 8–11 July, Madras.	Madras Declaration.	Appealed to scientists of the world for the maintenance and use of biodiversity of genetic resources important for food and agriculture.
1996	4th International Technical Conference on Plant Genetic Resources, Leipzig, 17–23 June.	Leipzig Declaration, *Global Plan for Conservation and of Sustainable Utilization of PGRs for Food and Agriculture* [52], *State of the World's PGRs* [18].	Appropriation of in situ and ex situ approaches
	World Food Summit, Rome, 13–17 November.	Solemn support to Leipzig Plan of Action [52].	Fair and equitable sharing of benefits arising for the use of genetic resources.
1999	1st Extraordinary meeting of the Conference of the Parties to the CBD, Cartagena, Colombia, 22 February.	Decision EM-1/1.	

Table 1. *Cont.*

Year	Event	Main Output(s) of Relevance for PGRs and [Reference]	Underpinning Principle(s)
2000	Resumed Session of the Conference of the Parties to the CBD, Montreal, Canada, 24–29 January.	Adoption of the Cartagena protocol on biosafety to the convention on biological diversity (decision EM-I/3).	Protect biological diversity from the potential risks posed by living modified organisms (LMOs) resulting from modern biotechnology Advance informed agreement (AIA) procedure, for transboundary movements of LMOs Reference to a precautionary approach, Biosafety Clearing-House to facilitate the exchange of information on LMOs.
2001	31st Session of the FAO Conference, Rome, 2–13 November.	Resolution 3/2001: adoption of the *International Treaty for Plant Genetic Resources (ITPGR)* (entered into force on 11 September 2004).	Legally binding, Recognition of farmers' rights Access to PGRs Fair and equitable sharing of the benefits derived from PGR use.
2002	6th Ordinary Meeting of the Conference of the Parties to the Convention on Biological Diversity, The Hague, Netherlands, 7–19 April.	Decision VI/26: *Strategic Plan for the Convention on Biological Diversity 2002–2010*.	Biodiversity is the living foundation for sustainable development The rate of loss is still accelerating The threats must be addressed, The convention is an essential instrument for achieving sustainable development.
	UN World Summit on Sustainable Development, Johannesburg, South Africa, 26 August–6 September.	Johannesburg declaration on sustainable development.	Emphasis on social and economic aspects of sustainable development.
2004	Establishment of the Global Crop Diversity Trust (now renamed Crop Trust).	Endowment fund, the income from which will be used to support the conservation of distinct and important crop diversity, in perpetuity, through existing institutions.	Ensuring an absolutely dependable source of funding for the conservation of PGRs.

Table 1. *Cont.*

Year	Event	Main Output(s) of Relevance for PGRs and [Reference]	Underpinning Principle(s)
2005	Publication of the *Millennium Ecosystem Assessment*.	Ecosystems and human well-being [53].	Over the past 50 years, humans have changed ecosystems more rapidly and extensively than in any comparable period of time, to meet rapidly growing demands for food, fresh water, timber, fiber, and fuel The changes made have contributed to substantial net gains in human well-being, but at growing environmental costs Degradation of ecosystem services could grow significantly worse during the first half of this century and is a barrier to achieving the Millennium Development Goals, Reversing the degradation of ecosystems, while meeting increasing demands for their services can be partially met, but involve significant changes in policies, institutions, and practices that are not currently under way.
		Standard Material Transfer Agreement (SMTA).	Regulation of the access to and use of PGRs. SMTA is the legal instrument through which the Multilateral System of Access and Benefit Sharing operates.
2006	1st meeting of the Governing Body of International Treaty for Plant Genetic Resources (ITPGR). Madrid, 12–16 June.	Relationship agreement between the governing body of the treaty and Global Crop Diversity Fund.	Recognition of the Crop Trust as an "essential element" of the treaty's funding strategy, in regard to ex situ conservation and availability of PGRs, and as an independent scientific organization in raising and disbursing funds.
		Agreements between the governing body and 12 international agricultural research centers (including the Centro Agronómico Tropical de Investigación y Enseñanza —CATIE).	Ex situ gene bank collections are put under the ITPGR (replaces agreement between IARCs and FAO).
2008	Establishment of the Svalbard Global Seed Vault, Svalbard, 26 February.		Long term conservation.
2009	12th Regular Session of the CGRFA, Rome, 19–23 October.	2nd report on the state of the world's plant genetic resources [19].	
	36th Session of the FAO Conference, Rome, 18–23 November.	Resolution 18/2009.	Special nature of PGRs in the context of negotiations of the International Regime on Access and Benefit-Sharing of the CBD.

Table 1. *Cont.*

Year	Event	Main Output(s) of Relevance for PGRs and [Reference]	Underpinning Principle(s)
2010	International Technical FAO Conference on *Agricultural biotechnologies in developing countries: Options and opportunities in crops, forestry, livestock, fisheries and agro-industry to face the challenges of food insecurity and climate change* (ABDC-10), Guadalajara; Mexico, 1–4 March 2010.	Report [35].	Agricultural biotechnologies are being applied to an increasing extent, But they have not been widely used in many developing countries, and have not sufficiently benefited smallholder farmers and producers and consumers.
2010	10th meeting of the Conference of the Parties to the Convention on Biological Diversity, Nagoya, Japan, 18–29 October.	Decision X/1: *Nagoya Protocol on Access to Genetic Resources and the Fair and Equitable Sharing of Benefits Arising from their Utilization* (ABS) *to the Convention on Biological Diversity* (entered into force on 12 October 2014).	Establishing more predictable conditions for access to genetic resources, helping to ensure benefit-sharing when genetic resources leave the country providing the genetic resources Also covers traditional knowledge, associated with genetic resources.
		Decision X/2: *II Strategic Plan for Biodiversity, including the Aichi Biodiversity Targets, for the 2011–2020 period.*	Recognition that the objectives of the 1st Strategic Plan were not achieved Establishment of twenty headline Aichi Biodiversity Targets for 2015 or 2020, organized under five strategic goals.
2011	13th Regular Session of the CGRFA, Rome, 18–22 July 2011.	Background paper CGRFA-13/11/3 *Status and trends of biotechnologies applied to the conservation and utilization of genetic resources for food and agriculture and matters relevant for their future development.*	Biotechnologies largely used for conservation and use of PGRs Many developing countries miss capacities.
	143rd Session of the FAO Council, Rome, 28 November–2 December.	Second global plan of action for the conservation and sustainable utilization of PGRs.	Need for a roadmap on climate change and genetic resources for food and agriculture.
2012	*United Nations Conference on Sustainable Development* (UNCSD), also known as Rio 2012, Rio+20, Rio de Janeiro, Brazil, 13–22 June.	Outcome document *The Future We Want.*	Necessity to promote, enhance and support more sustainable agriculture [. . .] that improves food security, eradicates hunger and is economically viable, while conserving land, water, plant and animal genetic resources, biodiversity and ecosystems and enhancing resilience to climate change and natural disasters.
2013	14th Regular Session of the CGRFA, Rome, 15–19 April.	Programme of work on climate change and genetic resources for food and agriculture.	Importance of genetic resources for food and agriculture for coping with climate change.

Table 1. *Cont.*

Year	Event	Main Output(s) of Relevance for PGRs and [Reference]	Underpinning Principle(s)
2015	International Symposium on Forest Biotechnology for Smallholders, Foz do Iguaçu, Paraná, Brazil, 19–22 May 2015.	Background paper [31].	Biotechnologies largely used for both planted and naturally regenerated forests.
	39th Session of the FAO Conference, Rome, 6–13 June.	Approval of the *Voluntary Guidelines to Support the Integration of Genetic Diversity into National Climate Change Adaptation Planning.*	Importance of genetic resources for food and agriculture for coping with climate change.
2016	FAO International Symposium on *The role of agricultural biotechnologies in sustainable food systems and nutrition*, Rome, 15–17 February.	Proceedings.	Biotechnologies can be used in production systems, based on agroecological principles, to enhance productivity while ensuring sustainability, conservation of genetic resources and use of indigenous knowledge.

Acknowledgments: The critical reading of the manuscript prior to its submission by Barbara Di Giovanni and Laura Padovani, ENEA, is gratefully acknowledged. The author is also very grateful to the anonymous peer reviewers, who offered very valuable comments.

Conflicts of Interest: The author declares no conflict of interest.

References

1. Harlan, J.R. *Crops and Man*; American Society of Agronomy: Madison, WI, USA, 1992.
2. Ryder, E.J. Perspectives on germplasm. *HortScience* **2003**, *38*, 922–927.
3. Khoury, C.K.; Achicanoy, H.A.; Bjorkman, A.D.; Navarro Racines, C.; Guarino, L.; Flores Palacios, X.; Struik, P.C. *Estimation of Countries' Interdependence in Plant Genetic Resources Provisioning National Food Supplies and Production Systems*; The International Treaty on Plant Genetic Resources for Food and Agriculture: Rome, Italy, 2015.
4. Avanzato, D.; Vassallo, I. *Following Almond Footprints (Amygdalis communis L.). Cultivation and Culture, Folk and History, Tradition and Uses*; ISHS: Leuven, Belgium, 2006.
5. Cherry, R.H. History of sericulture. *Bull. Entomol. Assoc. Am.* **1987**, *33*, 83–85. [CrossRef]
6. Baumann, H. *The Greek Plant Word: In Myth, Art and Literature*; Timber Press: Portland, Oregon, 1993.
7. Foltz, J.D. Valuation and ownership of genetic resources in agriculture. In *Public Policy in Food and Agriculture*; EOLSS Publishers: Paris, France, 2009; pp. 260–286.
8. Pistorius, R. *Scientists, Plants and Politics—A History of the Plant Genetic Resources Movement*; IPGRI: Rome, Italy, 1997.
9. Sonnino, A. Internazionalizzazione della ricerca e cooperazione scientifica internazionale—L'attualità dell'insegnamento di Gian Tommaso Scarascia Mugnozza. *Rendiconti dell'Accademia delle Scienze* **2015**, *XXXIX*, 203–216.
10. Esquinas-Alcázar, J.; Hilmi, A.; López Noriega, I. *A Brief History of the Negotiations on the International Treaty on Plant Genetic Resources for Food and Agriculture*; FAO: Rome, Italy, 2012.
11. Scarascia-Mugnozza, G.T.; Perrino, P. The history of ex situ conservation and use of plant genetic resources. In *Managing Plant Genetic Diversity*; Engels, V., Ramanatha, R., Brown, A., Jackson, M.T., Eds.; CABI Publishing: New York, NY, USA, 2002; pp. 1–22.
12. Sullivan, S.N. Plant Genetic Resources and the Law: Past, Present, and Future. *Plant Physiol.* **2004**, *135*, 10–15. [CrossRef] [PubMed]
13. Harlan, H.V.; Martini, M.L. Problems and results in barley breeding. *Yearb. Agric.* **1936**, *1936*, 303–346.
14. Frankel, O.H. The development and maintenance of superior genetic stocks. *Heredity* **1950**, *4*, 89–102. [CrossRef] [PubMed]
15. Frankel, O.H.; Bennet, E. *Genetic Resources in Plants—Their Exploration and Conservation*; Blackwell Scientific: Oxford, UK, 1970.
16. National Research Council. *Genetic Vulnerability of Major Crops*; National Academy of Sciences: Washington, DC, USA, 1972.
17. Scarascia Mugnozza, G.T. *The Protection of Biodiversity and the Conservation and Use of Genetic Resources for Food and Agriculture: Potential and Perspectives*; FAO: Rome, Italy, 1995.
18. FAO. *Report on the State of the World's Plant Genetic Resources*; FAO: Rome, Italy, 1998.
19. FAO. *The Second Report on the State of the World's Plant Genetic Resources*; FAO: Rome, Italy, 2010.
20. Frankel, O.H. Conservation of crop genetic resources and their wild relatives: On overview. In *Crop Genetic Resources for Today and Tomorrow*; Frankel, O.H., Hawkes, J.G., Eds.; Cambridge University Press: Cambridge, UK, 1975.
21. Padovani, L.M.; Carrabba, P.; Di Giovanni, B.; Mauro, F. *Biodiversità-Risorse per lo Sviluppo*; ENEA: Rome, Italy, 2009.
22. Bennet, E. *FAO/IBP Technical Conference on the Exploration, Utilization and Conservation of Plant Genetic Resources*; FAO: Rome, Italy, 1968.
23. Hawkes, J.G. International workshop on dynamic in situ conservation of wild relatives of major cultivated plants: Summary of final discussion and recommendations. *Israel. J. Bot.* **1991**, *40*, 529–536.
24. Swaminathan, M.S. The Past, Present and Future Contributions of Farmers to the Conservation and Development of Genetic Diversity. In *Managing Plant Genetic Diversity*; Engels, J.M.M., Rao, V.R., Brown, A., Jackson, M.T., Eds.; CABI Publishing: New York, NY, USA, 2002; pp. 23–32.

25. Bala Ravi, S.; Rani, M.G.; Swaminathan, S. Conservation of plant genetic resources at the Scarascia Mugnozza. In *Memorie di Scienze Fisiche e Naturali*; Aracne Editrice: Rome, Italy, 2010; pp. 47–58.

26. Argumedo, A. The Potato Park, Peru: Conserving agrobiodiversity in an Andean Indigenous Biocultural Heritage Area. In *Protected Landscapes and Agrobiodiversity Values*; Amend, T., Brown, J., Kothari, A., Phillips, A., Stolton, S., Eds.; Kasparek Verlag: Heidelberg, Germany, 2008; pp. 45–58.

27. Sonnino, A. Biodiversidad y biotecnologías: El eslabón estratégico. In *Biodiversidad, Biotecnologia y Derecho un Crisol Para la Sustentabilidad*; Ivone, V., Ed.; Aracne Editrice: Rome, Italy, 2011; pp. 299–320.

28. Spillane, C. *Recent Developments in Biotechnology as They Relate to Plant Genetic Resources for Food and Agriculture*; FAO: Rome, Italy, 1999.

29. Lidder, P.; Sonnino, A. *Biotechnologies for the Management of Genetic Resources for Food and Agriculture, Commission on Genetic Resources for Food and Agriculture*; FAO: Rome, Italy, 2011.

30. Lidder, P.; Sonnino, A. Biotechnologies for the Management of Genetic Resources for Food and Agriculture. *Adv. Genet.* **2012**, *78*, 1–168. [PubMed]

31. Sonnino, A. Current status of biotechnology development and application in forestry. In Proceedings of the International Symposium on Forest Biotechnology for Smallholders, Foz do Iguaçu, Paraná, Brazil, 19–22 May 2015; FAO: Rome, Italy, 2015.

32. Visser, B.; Eaton, D.; Louwaars, N.; van der Meer, I.; Beekwilder, J.; van Tongeren, F. *Potential Impacts of Genetic Use Restriction Technologies (GURTS) on Agrobiodiversity and Agricultural Production Systems*; FAO: Rome, Italy, 2002.

33. Fears, R. *Genomics and Genetic Resources for Food and Agriculture*; FAO: Rome, Italy, 2007.

34. Heinemann, J.A. *A Typology of the Effects of (Trans)gene Flow on the Conservation and Sustainable Use of Genetic Resources*; FAO: Rome, Italy, 2007.

35. FAO. *Biotechnologies for Agricultural Development—Proceedings of the FAO international Technical Conference on "Agricultural Biotechnologies in Developing Countries: Options and Opportunities in Crops, Forestry, Livestock, Fisheries and Agro-Industry to Face the Challenges of Food Insecurity and Climate Change" (ABDC-10)*; FAO: Rome, Italy, 2011.

36. FAO. *Proceedings of the FAO International Symposium on the Role of Agricultural Biotechnologies in Sustainable Food Systems and Nutrition*; FAO: Rome, Italy, 2016.

37. FAO. Report of the Conference of FAO. In Proceedings of the Twenty-Second Session, Rome, Italy, 5–23 November 1983.

38. Gepts, P. Who Owns Biodiversity, and How Should the Owners Be Compensated? *Plant Physiol.* **2004**, *134*, 1295–1307. [CrossRef] [PubMed]

39. Kloppenburg, J.R. *First the Seed: The Political Economy of Plant Biotechnology*; Cambridge University Press: Cambridge, UK, 1988.

40. Mooney, P.R. *The Seeds of the Earth: A Private or Public Resource*; Inter Pares: Ottawa, ON, Canada, 1979.

41. Keystone Center. Final Consensus Report: Global Initiative for the Security and Sustainable Use of Plant Genetic Resources. In Proceedings of the Oslo Plenary Session (Third Plenary Session), Oslo, Norway, 31 May–4 June 1991.

42. The Crucible Group. *People, Plants, and Patents: The Impact of Intellectual Property on Trade, Plant Biodiversity, and Rural Society*; International Development Research Centre: Ottawa, ON, Canada, 1994.

43. Ozgediz, S. *The CGIAR at 40: Institutional Evolution of the World's Premier Agricultural Research Network*; CGIAR: Washington, DC, USA, 2012.

44. IBPGR. *Programme and Structure of the International Board for Plant Genetic Resources*; International Board for Plant Genetic Resources: Rome, Italy, 1986.

45. IPGRI. *The Mulino at Maccarese*; International Plant Genetic Resources Institute: Rome, Italy, 2001.

46. Fu, Y.B. The Vulnerability of Plant Genetic Resources Conserved Ex Situ. *Crop Sci.* **2017**, *57*, 1–15. [CrossRef]

47. Dulloo, M.E.; Thormann, I.; Fiorino, E.; De Felice, S.; Rao, V.R.; Snook, L. Trends in Research using Plant Genetic Resources from Germplasm Collections: From 1996 to 2006. *Crop Sci.* **2013**, *53*, 1217–1227. [CrossRef]

48. FAO. *Report of the Third Session of the FAO Panel of Experts on Plant Exploration and Introduction*; FAO: Rome, Italy, 1969; pp. 25–28.

49. Frankel, O.H.; Hawkes, J.G. (Eds.) *Crop Genetic Resources for Today and Tomorrow*; Cambridge University Press: Cambridge, UK, 1975.

50. Williams, J.T.; Holden, J.H. (Eds.) *Crop Genetic Resources: Conservation & Evaluation*; Allen & Unwin: Winchester, MA, USA, 1984.

51. Sonnino, A. Salvaguardia delle risorse genetiche vegetali: Problemi attuali e nuove opportunità. *Biol. Ital.* **1998**, *7*, 19–22.

52. FAO. *Global Plan of Action for the Conservation and Sustainable Use of Plant Genetic Resources for Food and Agriculture*; FAO: Rome, Italy, 1996.

53. Millennium Ecosystem Assessment. *Ecosystems and Human Well-Being: Synthesis*; Island Press: Washington, DC, USA, 2005.

diversity

MDPI

Review

Twenty Years of Tomato Breeding at EPSO-UMH: Transfer Resistance from Wild Types to Local Landraces—From the First Molecular Markers to Genotyping by Sequencing (GBS)

Pedro Carbonell, Aranzazu Alonso, Adrián Grau, Juan Francisco Salinas, Santiago García-Martínez and Juan José Ruiz *

Department of Applied Biology, Miguel Hernández University, Carretera Beniel km 3,2, 03312 Orihuela, Alicante, Spain; pcarbonell@umh.es (P.C.); aalonso@umh.es (A.A.); agrau@umh.es (A.G.); salinas.kiter.92@gmail.com (J.F.S.); sgarcia@umh.es (S.G.-M.)
* Correspondence: juanj.ruiz@umh.es; Tel.: +34-096-674-9615

Received: 8 November 2017; Accepted: 25 February 2018; Published: 27 February 2018

Abstract: In 1998, the plant breeding team at the School of Engineering of Orihuela (EPSO), part of the Miguel Hernández University (UMH) in Elche, commenced a tomato breeding program. Marker-assisted selection and backcrossing were used to simultaneously introduce three genes (*Tm-2a*, *Ty-1*, and *Sw-5*) that confer resistance to relevant viruses, such as tomato mosaic virus (ToMV), tomato yellow curl virus (TYLCV), and tomato spotted wilt virus (TSWV), to traditional varieties of local tomatoes, specifically the "Muchamiel" and the "De la pera" types. After each backcross, cleaved amplified polymorphic sequence (CAPS) molecular markers were used to select the plants with the resistance genes of interest. A previously described marker was used for TSWV, and new markers were designed for ToMV, and TYLCV using available sequences in the National Center for Biotechnology Information (NCBI) database. In parallel to the breeding program, several molecular markers—Sequence Related Amplified Polymorphism (SRAP), Simple Sequence Repeats (SSRs), Amplified Fragment Length Polymorphisms (AFLPs), Single Nucleotide Polymorphisms (SNPs), and (GATA)$_4$ probes—were used to study genetic variability, and to identify a collection of Spanish and Italian traditional tomato varieties. The results showed a limited genetic variability among cultivated tomato varieties. The breeding lines Muchamiel UMH 1200, and De la pera 1203 (both with homozygous resistance to the three viruses) were the first new varieties that were obtained. They were included in the Register of Protected Plant Varieties in 2013. Lines without a resistance to TYLCV were also developed, and protected in 2017. We have begun to use SNP massive genotyping for studies of genetic association, and for selecting plants with the *Ty-1* gene with less linkage drag. Molecular markers have been extremely useful in identifying the different steps of the tomato breeding program at EPSO-UMH.

Keywords: Muchamiel; De la pera; ToMV; TYLCV; TSWV

1. Introduction

Over the years, traditional agriculture has produced an enormous range of local plant varieties that have been adapted to specific environmental conditions and for local uses and preferences. With no knowledge of genetics or statistics, farmers have created new varieties by saving the seeds from the best plants for the following year. These plants have been selected for their organoleptic qualities (including flavor, aroma and texture), their adaptation to the local environment (resistance to frost and drought, proper fruit set, etc.), and their suitability for different local uses (such as fresh consumption, canning, and drying). In southeastern Spain, there are a number of local tomato varieties, including the "Muchamiel" from Alicante, the "Tres cascos" from Elche, the "De la pera" from the Vega Baja del Segura region of Alicante, the "Valenciano" from Valencia, the "Flor de Baladre" from Murcia, and the "Morunos" from different regions. The Muchamiel and the De la pera, which have traditionally been cultivated in the province of Alicante, are particularly esteemed for their exceptional organoleptic quality. In local markets, they can be sold at a price that is six times higher than that of hybrid varieties [1].

Intensive farming, however, requires varieties that are consistently and highly productive in different systems and cycles, that are exceptionally uniform in size, shape and color, and that show a certain level of resistance to the different infestations, diseases and disorders that can affect the crop. Traditional varieties tend to fall short in at least one of these categories.

In fact, many traditional tomato varieties in southeastern Spain are endangered because of their high susceptibility to the three main viruses affecting tomatoes in the area: tomato mosaic virus (ToMV), tomato yellow leaf curl virus (TYLCV), and tomato spotted wilt virus (TSWV) [2]. Because of these viruses, in addition to other factors, traditional tomatoes are less attractive to farmers than modern hybrids, which are resistant to these viruses and are also far more productive. Yet, giving up on the traditional varieties could result in an irreversible loss of vital genetic variability [3]. For economic reasons, improving upon varieties with a limited market share is not a priority for seed companies and therefore should be undertaken by public institutions [4].

2. Genetic Variability Studies in Tomato

Morphological traits (such as the shape, size, and color of the different parts of the plant), and agronomical traits (such as yield, the ability to set in adverse conditions, and resistance to different biotic and abiotic stresses) are often very useful to distinguish traditional and commercial tomato varieties, even by visual inspection. However, there are other traits that can only be detected by using analytical methods. This is the case for the levels of different key components (such as sugars, acidity, or microelements), and the volatile compounds responsible for aroma. To study these traits, which are subtler than the more obvious differences, it is often necessary to use a specific methodology. Despite the fact that these subtle differences (such as fruit composition) are more difficult to see, they are, nevertheless, equally as important as the more obvious differences.

We already know that cultivated tomatoes have a narrow genetic base [5]. This narrow genetic base is probably due to the bottleneck effect that occurred during tomato domestication [5]. Based on estimates made with DNA markers, the relative species of wild tomatoes are much more variable than their cultivated counterpart at the whole genome level. It is estimated that the genomes of tomato cultivars contain <5% of the genetic variation of their wild counterparts [5]. In other words, cultivated tomatoes vary tremendously in their fruit size, shape, color, firmness, soluble solids, volatile aroma content, and acidity, but have little genetic variation elsewhere in their genome [6,7]. In one study, notable differences in parameters, such as flavor, texture and micronutrient content, between the traditional Muchamiel and De la pera varieties, and a modern F1 hybrid, were found [8]. Four years later, different aromatic profiles in Muchamiel accessions, De la pera accessions, and a F1 hybrid, were also found [9].

3. Molecular Markers in Variability Studies

Despite these previous findings and observations, it had not been possible to distinguish the phenotypic variation among the characterized accessions at the genetic level. For this reason, in parallel to the breeding program, our group began to study variability using different molecular markers, which would make it possible to distinguish between the different traditional varieties and to coherently organize them according to genotype. Our group believed this would be an informative and exceptionally powerful approach in the breeding program. Tomato was one of the first crops in which molecular markers were used to study genetic variability. In the genetic variability studies of tomato, a number of DNA markers, such as SSR [10,11] and AFLP [12], were successfully used. SSR [13,14] and (GATA)$_4$ probes [15,16] were also successfully used for variety identification.

Our group has performed a range of variability studies, mostly using accessions from closely related traditional European tomato varieties. In these studies, we have been able to confirm the efficacy of certain molecular markers in differentiating between varieties, and in grouping similar varieties together.

The results of the different marker analysis studies that we have carried out are detailed below, and are summarized in Table 1.

Table 1. Markers used in the genetic variability studies.

Marker	Number	Number of Bands	% of Polymorphic Bands	Usability
SRAP	26	384	60	Distinguish between cultivar types, wild relatives and 14 of 16 traditional cultivars studied
SSR	10 + 9	77	98	Distinguish between cultivar types, wild relatives and 23 of 34 traditional cultivars studied
AFLP	7	470	40	Distinguish between cultivar types, wild relatives and 24 of 31 traditional cultivars studied
(GATA)$_4$	N.A.	30	100	Distinguish between cultivar types, Spanish traditional cultivars and 4 of 10 Italian traditional cultivars studied
SNP	41	N.A.	76	Distinguish traditional cultivars from modern cultivars, hybrids and wild relatives

N.A.: Not applicable.

3.1. Sequence Related Amplified Polymorphism (SRAP)

The SRAP technique is based on an amplified polymerase chain reaction (PCR), and it is designed to detect polymorphisms in the DNA sequence [17].

In 2005, we carried out a variability study using SRAP and single sequence repeat (SSR) markers in traditional Spanish varieties, including the Muchamiel, the De la pera, the Moruno, commercial hybrids, and wild tomato species [18]. A total of 26 not specific for tomato primers were used. These primers made it possible to identify polymorphisms between the main varieties, and to distinguish between cultivar types and the wild relatives under study (De la pera, Muchamiel, Moruno, hybrids, *S. pimpinellifolium*, *S. chilense* and *S. peruvianum*). It was even possible to differentiate between the different accessions within each group. Although the SRAP markers are dominant and not very polymorphic, they have significant coverage throughout the genome and produce a more in-depth variability study than the SSRs used in the same analysis. SRAP, for example, made it possible to distinguish between the Mexican variety of Zapotec tomato and the Spanish varieties, which SSRs were unable to do.

3.2. Simple Sequence Repeat (SSR)

SSR markers, or microsatellites, are short tandem repeats in DNA sequences, mostly between two and four base pairs (bp) [19]. These markers are highly polymorphic because the repeated sequences vary significantly between the different genotypes.

In the study mentioned above [18], our group used 10 previously selected microsatellites for tomato. With only three SSRs, we were able to differentiate between the three main types of cultivated tomatoes under study (Muchamiel, De la pera and Moruno). Nevertheless, despite using 10 SSRs, we were unable to differentiate between the different cultivars within each varietal type, which could be done using SRAP. SSR markers are highly polymorphic, but they are concentrated in very specific loci within the genome, and they have very little genomic coverage. As a result, we were unable to find polymorphisms between highly similar accessions.

In a second study, we used up to 19 microsatellites selected for tomato, but we were unable to distinguish between all of the cultivars evaluated, even though there were clear phenotypical differences. Nevertheless, only four SSRs were necessary to differentiate between the three main varietal groups (Muchamiel, De la pera, and Moruno) [20].

The results of both studies demonstrate the minimal genetic variability among cultivated tomato varieties.

3.3. Amplified Fragment Length Polymorphisms (AFLPs)

AFLPs are molecular markers used to detect polymorphisms in the DNA sequence. After the restriction enzyme digestion of DNA, certain fragments are selected for amplification [21].

In a variability study conducted by our group [20], the obtained results were compared using 19 microsatellites and seven combinations of AFLP markers. The AFLPs made it possible to distinguish between the main varietal groups, although out of the 43 accessions studied, there were seven that could not be identified with certainty, and they were all traditional Spanish varieties. Using the SSR analysis, 11 accessions could not be differentiated, and, once again, they were all Spanish varieties. Curiously, the seven accessions that could not be identified using AFPLs were different from the 11 in the SSR analysis. This means that all of the cultivars used in the study could be identified using a combination of both markers.

Once again, the results showed the limited genetic variability among cultivated tomato varieties.

3.4. (GATA)$_4$ Probes

Labelled (GATA)$_4$ probes are molecular markers based on the hybridization of a short sequence (16 nucleotides, with the GATA motif repeated four times), which hybridizes when it finds a complementary sequence [22]. The (GATA)$_4$ technique was developed in 1995, and it has been used to successfully distinguish between varietal types, and also between different accessions [15,16,23].

In 2013, researchers at UMH and the *Università degli Studi di Napoli Federico II* (Naples, Italy) performed a variability study of traditional Spanish and Italian varieties, including accessions from the Muchamiel, and the De la pera type tomatoes, and the Italian varieties San Marzano, and Sorrento [24]. The (GATA)$_4$ probes could clearly detect differences between all of the Spanish accessions studied, but not the Italian accessions. Furthermore, the researchers found polymorphisms between different plants within a single accession in 14 of the 26 accessions studied (in a previous study, SSRs were only able to detect polymorphisms in two out of the 16 accessions studied). This level of polymorphism is not unusual, given that small-scale farmers have not traditionally considered crop uniformity as an essential factor in their selection process.

The results of this study confirmed that (GATA)$_4$ probes are a powerful tool for studying variability between closely related tomato accessions. The markers are able to distinguish between plants at a much deeper level than SSRs, SRAPs, and AFLPs.

3.5. Single Nucleotide Polymorphism (SNP)

A SNP is a variation in a single nucleotide in the DNA sequence [25]. SNPs are the most abundant polymorphisms in the genome, making them highly useful in both genetic variability and phylogenetic studies.

Our group studied 41 SNPs, obtained from expressed sequence tags (ESTs), in several different tomato varieties, including traditional varieties (mostly Spanish), commercial hybrids, and wild tomato varieties [26]. These markers were selected because they were readily detected on standard agarose or polyacrylamide gels, which was the equipment available in our laboratory at that time. This study found that it was not possible to differentiate between all of the traditional cultivars, although the researchers were able to distinguish between wild accessions and hybrid varieties.

The level of polymorphism found in this SNP study was therefore lower than the levels found in previous studies on the Muchamiel, De la pera, and Moruno varieties using SSRs, AFLPs, and SRAPs. This was perhaps due to the fact that the SNPs were obtained from ESTs, which are coding regions of DNA that show fewer mutations than non-coding regions and where it is typical to find more variability. Perhaps it would have been more interesting to study SNPs that are outside functional genes, which are highly polymorphic regions [27].

4. Traditional Tomato Variety Breeding Program

In 1998, the plant breeding team at the School of Engineering of Orihuela (EPSO), part of the Miguel Hernández University (UMH) in Elche, commenced a breeding program using marker-assisted backcrossing in order to simultaneously introduce genes that confer resistance to the three most relevant viruses found in traditional local tomato varieties, specifically in the Muchamiel and the De la pera types. We have used the genes *Tm-2a*, and *Sw-5*, which come from the wild tomato *Solanum peruvianum* L. [28,29] to confer resistance to ToMV, and TSWW, respectively, and the gene *Ty-1*, which confers resistance to TYLCV and originated in another wild tomato series, the *Solanum chilense* (Dunal) Reiche accession LA1969 [30].

We have performed the following steps in our program: agronomic characterization of the traditional varieties and of the sources of resistance; performance of crossings; performance of backcrossings; fixation of the resistance genes; selection of the best lines; and application for registration in the Spanish Register of Protected Plant Varieties.

As mentioned above, the first step in the breeding program was to characterize and select the most suitable plant material. Since the program aims to help preserve the genetic variability of certain traditional tomato varieties in southeastern Spain, it was essential to fully understand the plant material before proceeding.

5. Molecular Markers in Resistance Gene Introgression

The first step in our breeding program was to characterize different accessions of interest in order to select those that met the minimum quality standards. Based on these characterizations, we selected M18 Muchamiel accession, and P21 De la pera accession to be used as traditional parentals in the program. The next step was to manually cross these accessions with the F1 hybrid Anastasia (Seminis Vegetable Seeds), which is a source of resistance that contains the genes *Tm-2a*, *Sw-5*, and *Ty-1*.

The descendants obtained from the traditional varieties were repeatedly backcrossed with the initial parents in order to recover as much of the genome as possible. After each backcross, cleaved amplified polymorphic sequence (CAPS) molecular markers were used to select the plants with the resistance genes of interest. To do this, it was necessary to use specific markers that would efficiently recognize the presence or absence of these genes.

CAPS markers are short DNA fragments amplified by PCR and later digested by restriction enzymes, which produces a pattern of bands that makes it possible to distinguish between homozygous and heterozygous individuals. The CAPS markers used thus far in the breeding program are detailed below.

High selection pressure for desirable traditional cultivar characteristics (such as shape, and organoleptic quality) and good agronomic behavior (proper fruit set, sufficient uniformity among fruits, and yields) was applied during the backcrossing process. Only the best plants (between one and three per progeny) were selected for further backcrossing.

5.1. CAPS Linked to the Sw-5 Gene

The CAPS markers we used to select individuals resistant to TSWV have been successful thus far in the breeding program [31].

5.2. CAPS Linked to the Ty-1 Gene

The ApsF-2 marker was developed specifically for this breeding program from the isoenzyme marker Aps [32]. In earlier studies and in this breeding program, ApsF-2 has proven to be a suitable marker for confirming TYLCV tolerance.

5.3. CAPS Linked to the Tm-2ᵃ Gene

Initially, we tried to use two previously described CAPS markers [33,34]. After determining that these markers did not work correctly with our material, we designed new CAPS markers, using the allele sequences Tm-2ᵃ (AF536201) and tm-2 (AF536199), which were entered into the National Center for Biotechnology Information (NCBI) database [35].

5.4. Other Examples of Tomato Breeding Programs with MAS (Marker-Assisted Selection)

Tomato was among the first crop species for which genetic markers were suggested as indirect selection criteria for breeding purposes and for which molecular markers and maps were developed [36,37]. Private companies do not disclose the markers that they use. However, there seems to be a considerable use of markers for various purposes, including testing hybrid purity, screening breeding populations for disease resistance, and marker assisted backcross breeding [38]. The use of MAS in tomato breeding in public research institutions is well documented. The breeding programs that use MAS and send their results to the Register, are summarized in Table 2.

Table 2. Public research institutions with tomato breeding programs, coordinators and objectives.

Breeding Program	Coordinators	Traits
University of Florida (USA)	J.W. Scott and S.F. Hutton	*Fusarium oxysporum f.* sp. *lycopersici, Verticillium dahliae, Stemphyllium solani,* TSWV, TYLCV
University of North Carolina (USA)	R.G. Gardner	*Fusarium oxysporum f.* sp. *lycopersici, Verticillium dahliae, Alternaria solani,* TSWV
United States Department of Agriculture (ARS)	J.R. Stommel	Beta carotene content
Ohio State University (USA)	M. Francis	*Xanthomonas euvesicatoria*

6. Lines Obtained in the EPSO-UMH Breeding Program

When we consider that the characteristics of the traditional variety have been recovered (after five to eight backcross generations, depending on the line) and the pure-breeding lines have been obtained, selecting the plants that show homozygous resistance due to introgressed genes is necessary. These plants were selected using the three CAPS markers and the progenies were obtained by self-pollination during the last backcross generation.

In 2013, the EPSO-UMH breeding program obtained its first plant variety certificates for the lines UMH 1200 (a Muchamiel-type tomato), and UMH 1203 (De la pera), which both show homozygous resistance to the three viruses of interest (Table 3). The morphological characteristics and quality of these lines are similar to those found in the traditional varieties from which they are derived. Nevertheless, homozygous resistance to TYLCV considerably reduces production (up to 40%), particularly in TYLCV-free conditions, which are possible in greenhouse cultivation [39,40]. This decrease in production is due to the introgressed genes themselves and/or to the linkage drag associated with the resistance genes, particularly the *Ty-1* gene. This problem has been previously described in tomatoes for industrial use [41], in tobacco [42], and in tomatoes for fresh consumption [43].

Table 3. Breeding lines registered in the Register of Protected Plant Varieties, with their genotypes for the three virus resistance genes. RR: resistant homozygote, Rs: resistant heterozygote, ss: sensitive homozygote.

Varietal Type	Line	Resistance ToMV-TYLCV-TSWV	Sent to Registry	Title Obtained
Muchamiel	UMH 1200	RR-RR-RR	2011	2013
Muchamiel	UMH 1139	RR-ss-RR	2013	2017
Muchamiel	UMH 1101xIF	Rs-Rs-Rs	2014	2017
De la pera	UMH 1203	RR-RR-RR	2011	2013
De la pera	UMH 1422	RR-ss-ss	2013	2017
De la pera	UMH 1415	RR-ss-RR	2013	2017
De la pera	UMH 1353	RR-ss-RR	2013	2017
De la pera	UMH 1354	RR-ss-RR	2013	2017

In an attempt to avoid these problems, we have developed lines that are not resistant to TYLCV. Within the Muchamiel type, we have developed the lines UMH 1093, UMH 1127, and UMH 1139, which all have homozygous resistance to ToMV, and TSWV. UMH 1139 obtained a plant variety certificate in 2017. Within the De la pera type, we have developed the line UMH 1422 (with homozygous resistance to ToMV), and the lines UMH 1415, UMH 1353, and UMH 1354 (which all have homozygous resistance to ToMV and TSWV). All of these De la pera lines obtained plant variety certificates in 2017. The morphological characteristics and quality of these lines are similar to those found in the traditional varieties they are derived from, and there is no drop in production like that which occurs in UMH 1200, and in UMH 1203. These lines are therefore of great interest for cultivation purposes in TYLCV-free conditions, which can be achieved in a greenhouse with sufficient control measures, particularly in the spring season, when TYLCV occurs less frequently [44–46].

All of these improved lines possess homozygous resistance genes, and therefore, they can be cultivated by farmers year after year using a rigorous selection process.

In order to grow tomatoes outdoors in an environment with a high incidence rate of TYLCV, it was decided to develop hybrids with a heterozygous resistance to the three main viruses. Previous studies have shown that introducing resistance to TYLCV heterozygously has a reduced negative effect on yield and other traits [47]. The hybrid UMH 1101xIF, a Muchamiel-type tomato, was the first hybrid obtained in our breeding program, and more hybrids (from both Muchamiel and De la pera types) are currently being developed and will be ready to be presented in two years.

7. Massive Genotyping

In the last few years, the genetics team at UMH has performed numerous massive sequencing analyses using the Illumina platform. These analyses have made it possible to sequence a large number of traditional tomato varieties and accessions. In the most notable of these analyses, we use the 8K SolCap Illumina Infinium SNP chip, described as a part of the Solanaceae Coordinated Agricultural Project (SolCAP: http://solcap.msu.edu/) [48]. This chip is made up of 8784 SNPs distributed across 12 chromosomes, and it has been used for different purposes, both in the study of genetic variability (genetic association) and in the breeding program (reduction of linkage drag).

7.1. Genetic Association

As part of a collaboration with the University of Naples Federico II (Prof Rosa Rao), massive sequencing was performed on 42 traditional Spanish and Italian tomato varieties, and on three accessions of *Solanum lycopersicum* var. *cerasiforme*, and *Solanum pimpinellifolium*. The analyzed group of traditional varieties includes Muchamiel, De la pera, Valenciano, Morunos, Raf, Flor de Baladre, San Marzano, Sorrento, Costoluto, and Vesuvio. The genotypes obtained with the 8K SolCap chip will be used for future genetic association studies, taking advantage of the phenotyping performed in the group in order to study the variability among the traditional cultivars in greater depth.

7.2. Obtaining Plants with the Ty-1 Gene with Less Linkage Drag

Despite the breeding program's success, homozygous alleles for resistance, especially for the *Ty-1* gene, have been found to negatively affect both production levels and other parameters of agronomic quality [43]. This work does not clarify whether the negative effects on production and quality are directly due to the *Ty-1* gene, or whether they are caused by genes associated with the resistance gene. We do know that in the introgression of genes from wild species, large segments of chromosomes surrounding the resistant gene remain in the resulting line, even after many backcrosses [49]. In fact, it has been reported that two chromosomal inversions in *S. chilense* LA1969 (the *Ty-1* gene donor) give rise to a recombination suppression in the region of chromosome six, where the *Ty-1* gene is found in cultivated tomatoes [50].

Studying plants with a distinct number of backcrosses confirms these results. Despite backcrossing up to 10 times, a large fragment of chromosome six, in which the gene *Ty-1* is found, does not recombine. By selecting plants with TYLCV resistance, this fragment barely changes as it passes from generation to generation. Furthermore, unpublished results recently obtained by the UMH team, in collaboration with groups from the Institute of Plant Molecular and Cellular Biology (IBMCP) in Valencia, also show the negative effect of introgression on volatile compounds and other metabolites. All together, these results seem to indicate that it is the presence of DNA fragments associated with the *Ty-1* gene that is responsible for changes to the aromatic characteristics of the UMH tomato varieties. It is not clear whether this effect can be attributed to the *Ty-1* gene itself, or to other genes around it. Plants where the resistance gene is present but linkage drag was reduced by recombination may help in understanding this aspect.

8. Future Work

In the future, the main goal of the UMH breeding program is to introduce a resistance without altering the quality and productivity of the new lines of traditional tomato varieties. The *Ty-5* gene, for example, from the commercial hybrid Tyking, has shown good behavior against various mono- and bipartite begomoviruses, including TYLCV [51,52]. Using massive genotyping, next-generation sequencing (NGS) technologies, and bioinformatics, our group is currently collaborating with other groups from the Institute of Plant Molecular and Cellular Biology (IBMCP) in the search for descendants of UMH breeding program plants that self-fertilize and have been able to recombine in the zones adjacent to the *Ty-1* gene. This approach will allow us to obtain plants with different configurations in chromosome six and later study the phenotypic characteristics of these plants in the field. This process will be extremely useful in identifying the genes responsible for the most important productivity and quality parameters in tomato.

Acknowledgments: This work was partially supported by the Spanish MICINN through projects AGL2002-03329, AGL2005-03946, AGL2008-03822, AGL2011-26957 and the European Union (ACIF/2016/212). We thank Ansley Evans for the language review. We want to thank the anonymous reviewers for their suggestions.

Author Contributions: All the authors contributed to the work of the manuscript. P.C, S.G.-M. and J.J.R. conceived and wrote the paper.

Conflicts of Interest: The authors declare no conflict of interest.

References

1. Ruíz, J.J.; García-Martínez, S. Tomato varieties 'Muchamiel' and 'De la Pera' form the southeast of Spain: Genetic improvement to promote on-farm conservation. In *European Landrace: On-Farm Conservation, Management and Use. Biodiversity Technical Bulletin No 15*; Vetelainen, M., Negri, V., Maxted, N., Eds.; Bioversity International: Rome, Italy, 2009; pp. 171–176.
2. Picó, B.; Herraiz, J.; Ruiz, J.J.; Nuez, F. Widening the genetic basis of virus resistance in tomato. *Sci. Hortic.* **2002**, *94*, 73–89. [CrossRef]

3.   Nuez, F.; Roselló, S.; Picó, B. La conservación y recuperación de nuestro patrimonio hortícola. Mejorar para conservar. *Agrícola Vergel* **1998**, *194*, 74–80.

4.   Nuez, F.; Ruiz, J.J. *La Biodiversidad Agrícola Valenciana: Estrategias Para su Conservación y Utilización*; Universitat Politècnica de València: Valencia, Spain, 1999.

5.   Miller, J.C.; Tanksley, S.D. RFLP analysis of phylogenetic relationships and genetic variation in the genus Lycopersicon. *Theor. Appl. Genet.* **1990**, *80*, 437–448. [CrossRef] [PubMed]

6.   Tanksley, S.D. The Genetic, Developmental, and Molecular Bases of Fruit Size and Shape Variation in Tomato. *Plant Cell* **2004**, *16*, S181–S189. [CrossRef] [PubMed]

7.   Ruiz, J.J.; Martínez, N.; Valero, M.; García-Martínez, S.; Moral, R.; Serrano, M. Micronutrient composition and quality characteristics of traditional tomato cultivars in the south-east of Spain. *Commun. Soil Sci. Plant Anal.* **2005**, *36*, 649–660. [CrossRef]

8.   Ruiz, J.J.; Alonso, A.; García-Martínez, S.; Valero, M.; Blasco, P.; Ruiz-Bevia, F. Quantitative analisys of flavour volatiles detects differences among closely related traditional cultivars of tomato. *J. Sci. Food Agric.* **2005**, *85*, 54–60. [CrossRef]

9.   Alonso, A.; García-Aliaga, R.; García-Martínez, S.; Ruiz, J.J.; Carbonell-Barrachina, A.A. Characterization of Spanish tomatoes using aroma composition anddiscriminant analisys. *Food Sci. Technol. Int.* **2009**, *15*, 47–55. [CrossRef]

10.   Smulders, M.J.M.; Bredemeijer, G.; Rus-Kortekaas, W.; Arens, P.; Vosman, B. Use of short microsatellites from database sequences to generate polymorphisms among Lycopersiconesculentum cultivars and accessions of other Lycopersicon species. *Theor. Appl. Genet.* **1997**, *97*, 264–272. [CrossRef]

11.   Alvarez, A.E.; Van de Wiwl, C.C.M.; Smulders, M.J.M.; Vosman, B. Use of microsatellites to evaluate genetic diversity and species relationships in the genus Lycopersicon. *Theor. Appl. Genet.* **2001**, *103*, 1283–1292. [CrossRef]

12.   Park, Y.H.; West, M.A.L.; St. Clair, D.A. Evaluation of AFLPs for germplasm fingerprinting and assessment of genetic diversity in cultivars of tomato (*Lycopersicon esculentum* L.). *Genome* **2004**, *47*, 510–518. [CrossRef] [PubMed]

13.   Bredemeijer, G.M.M.; Cooke, R.J.; Ganal, M.W.; Peeters, R.; Isaac, P.; Noordijk, Y.; Rendell, S.; Jackson, J.; Röder, M.S.; Wndehake, K.; et al. Construction and testing of a microsatellite database containing more than 500 tomato varieties. *Theor. Appl. Genet.* **2002**, *105*, 1019–1026. [PubMed]

14.   He, C.; Poysa, V.; Yu, K. Development and characterization of simple sequence repeat (SSR) markers and their use in determining relationships among *Lycopersicon esculentum* cultivars. *Theor. Appl. Genet.* **2003**, *106*, 363–373. [CrossRef] [PubMed]

15.   Andreakis, N.; Giordano, I.; Pentangelo, A.; Fogliano, V.; Graziani, G.; Monti, L.M.; Rao, R. DNA fingerprinting and quality traits of Corbarino Cherry-like tomato landraces. *J. Agric. Food Chem.* **2004**, *52*, 3366–3371. [CrossRef] [PubMed]

16.   Rao, R.; Corrado, G.; Bianchi, M.; Di Mauro, A. (GATA)4 DNA fingerprinting identifies morphologically characterized San Marzano tomato plants. *Plant Breed.* **2006**, *125*, 173–176. [CrossRef]

17.   Li, G.; Quiros, C.F. Sequence-related amplified polymorphism (SRAP), a new marker system based on a simple PCR reaction: Its application to mapping and gene tagging in Brassica. *Theor. Appl. Genet.* **2001**, *103*, 455–461. [CrossRef]

18.   Ruiz, J.J.; García-Martínez, S.; Picó, B.; Gao, M.; Quiros, C.F. Genetic variability and relationship of closely related Spanish traditional cultivars of tomato as detected by SRAP and SSR markers. *J. Am. Soc. Hortic. Sci.* **2005**, *130*, 88–94.

19.   Tautz, D. Hypervariability of simple sequences as a general source for polymorphic DNA markers. *Nucleic Acids Res.* **1989**, *17*, 6463–6471. [CrossRef] [PubMed]

20.   García-Martínez, S.; Andreani, L.; García-Gusano, M.; Geuna, F.; Ruiz, J.J. Evaluation of amplified fragment length polymorphism and simple sequence repeats for tomato germplasm fingerprinting: Utility for grouping closely related traditional cultivars. *Genome* **2006**, *49*, 648–656. [CrossRef] [PubMed]

21.   Vos, P.; Hogers, R.; Bleecker, M.; Reijans, M.; van de Lee, T.; Hornes, M. AFLP: A new technique for DNA fingerprinting. *Nucleic Acid Res.* **1995**, *23*, 4407–4414. [CrossRef] [PubMed]

22.   Kaemmer, D.; Weising, K.; Bayermann, B.; Borner, T.; Epplen, J.T.; Kahl, G. Oligonucleotide fingerprinting of tomatoDNA. *Plant Breed.* **1995**, *114*, 12–17. [CrossRef]

23. Caramante, M.; Rao, R.; Monti, L.M.; Corrado, G. Discrimination of San Marzano accessions: A comparison of minisatellite, CAPS and SSR markers in relation to morphological traits. *Sci. Hortic.* **2009**, *120*, 560–564. [CrossRef]

24. García-Martínez, S.; Corrado, G.; Ruiz, J.J.; Rao, R. Diversity and structure of a sample of traditional Italian and Spanish tomato accesions. *Genet. Resour. Crop Evol.* **2013**, *60*, 789–798. [CrossRef]

25. Wang, D.G.; Fan, J.B.; Siao, C.J.; Berno, A.; Young, P.; Sapolsky, R.; Ghandour, G.; Perkins, N.; Winchester, E.; Spencer, J.; et al. Large-scale identification, mapping, and genotyping of single-nucleotide polymorphisms in the human genome. *Science* **1998**, *280*, 1077–1082. [CrossRef] [PubMed]

26. García-Gusano, M. Empleo de Marcadores Moleculares Para el Estudio de la Variabilidad Intraespecífica en Tomate (*Solanum lycopersicum* L.). Ph.D. Thesis, Miguel Hernandez University, Elche, Spain, 2007.

27. Le Corre, V.; Kremer, A. Genetic variability al neutral markers, quantitative trait loci and trait in a subdivided population under selection. *Genetics* **2003**, *164*, 1205–1219. [PubMed]

28. Alexander, L.J. Transfer of a dominant type of resistanceto the four known Ohio pathogenic strains of tobacco mosaicvirus (TMV), from *Lycopersicon peruvianum* to *L. esculentum*. *Phytopathology* **1963**, *53*, 869.

29. Stevens, M.R.; Scott, S.J.; Gergerich, R.C. Inheritance of a gene for resistance totomato spotted wilt virus (TSWV) from *Lycopersicon peruvianum* Mill. *Euphytica* **1992**, *59*, 9–17.

30. Zamir, D.; Ekstein-Michelson, I.; Zakay, Y.; Navot, N.; Zeidan, M.; Safatti, M.; Eshed, Y.; Harel, E.; Pleban, T.; van-Oss, H.; et al. Mapping and introgression of a Tomato yellow leaf curl virus tolerance gene, TY-1. *Theor. Appl. Genet.* **1994**, *88*, 141–146. [CrossRef] [PubMed]

31. Folkertsma, R.T.; Spassova, M.I.; Prins, M.; Stevens, M.R.; Hille, J.; Goldbach, R.W. Construction of a bacterial artificial chromosome (BAC) library of *Lycopersicon esculentum* cv. Stevens and its application to physically map the *Sw-5* locus. *Mol. Breed.* **1999**, *5*, 197–207. [CrossRef]

32. Rick, C.M.; Fobes, J.A. Association of an allozyme with nematodes resistance. *Tomato Genet. Coop. Rep.* **1974**, *24*, 25.

33. Dax, E.; Livneh, O.; Aliskevicius, E.; Edelbaum, O.; Kedar, N.; Gavish, N.; Milo, J.; Geffen, F.; Blumenthal, A.; Rabinowich, H.D.; et al. A SCAR marker linked to the ToMV resistance gene, *Tm-22*, in tomato. *Euphytica* **1998**, *101*, 73–77. [CrossRef]

34. Omori, T.; Murata, M.; Motoyoshi, F. Molecular characterization of the SCAR markers tightly linked to the Tm-2 locus of the genus Lycopersicon. *Theor. Appl. Genet.* **2000**, *101*, 64–69.

35. Lanfermeijer, F.C.; Dijkhuis, J.; Sturre, M.J.G.; de Haan, P.; Hille, J. Cloning and characterization of the durable tomato mosaic virus resistance gene Tm-22 from *Lycopersicon esculentum*. *Plant Mol. Biol.* **2003**, *52*, 1037–1049. [CrossRef] [PubMed]

36. Tanksley, S.D. Molecular markers in plant breeding. *Plant Mol. Biol. Rep.* **1983**, *1*, 3–8. [CrossRef]

37. Foolad, M. Genome mapping and molecular breeding of tomato. *Int. J. Plant Genom.* **2007**, *52*, 64358. [CrossRef] [PubMed]

38. Foolad, M.R.; Panthee, D.R. Marker-Assisted Selection in Tomato Breeding. *Crit. Rev. Plant Sci.* **2012**, *31*, 93–123. [CrossRef]

39. García-Martínez, S.; Grau, A.; Alonso, A.; Rubio, F.; Valero, M.; Ruiz, J.J. UMH 1200, a breeding line within the Muchamiel tomato type resistant to three viruses. *HortScience* **2011**, *46*, 1054–1055.

40. García-Martínez, S.; Grau, A.; Alonso, A.; Rubio, F.; Valero, M.; Ruiz, J.J. UMH 1203, a multiple virus-resistant fresh-market toamto breeding line for open-field conditions. *HortScience* **2012**, *47*, 124–125.

41. Tanksley, S.D.; Bernachi, D.; BeckBunn, T.; Emmatty, D.; Eshed, Y.; Inai, S. Yield and quality evaluations on a pair of processing tomato lines nearly isogenic for the Tm2a gene for resistance to the Tobacco mosaic virus. *Euphytica* **1998**, *99*, 77–83. [CrossRef]

42. Lewis, R.S.; Linger, L.R.; Wolff, M.F.; Wernsman, E.A. The negative influence of N-mediated TMV resistance on yield in tobacco: Linkage drag versus pleitropy. *Theor. Appl. Genet.* **2007**, *115*, 169–178. [CrossRef] [PubMed]

43. Rubio, F.; Alonso, A.; García-Martínez, S.; Ruiz, J.J. Introgression of virus-resistance genes into traditional Spanish tomato cultivars (*Solanum lycopersicum* L.): Effects on yield and quality. *Sci. Hortic.* **2016**, *198*, 183–190. [CrossRef]

44. García-Martínez, S.; Grau, A.; Alonso, A.; Rubio, F.; Valero, M.; Ruiz, J.J. UMH 1422 and UMH 1415: Two fresh-market tomato breeding lines resistant to Tomato Mosaic Virus and Tomato Spotted Wilt Virus. *HortScience* **2014**, *49*, 1465–1466.

45. García-Martínez, S.; Grau, A.; Alonso, A.; Rubio, F.; Carbonell, P.; Ruiz, J.J. UMH 916, UMH 972, UMH 1093, UMH 1127 and UMH 1139: Four fresh-market breeding lines resistant to viruses within the Muchamiel tomato type. *HortScience* **2015**, *50*, 927–929.

46. García-Martínez, S.; Grau, A.; Alonso, A.; Rubio, F.; Carbonell, P.; Ruiz, J.J. New breeding lines resistant to Tomato Mosaic Virus and Tomato Spotted Wilt Virus within the 'De la Pera' tomato type: UMH 1353 and UMH 1354. *HortScience* **2016**, *51*, 456–458.

47. Alonso, A.; García-Martínez, S.; Arroyo, A.; García-Gusano, M.; Grau, A.; Giménez-Ros, M.; Romano, M.E.; Valero, M.; Ruiz, J.J. Efecto de la introducción de resistencia genética a TYLCV (gen Ty-1) en caracteresproductivos y de calidad en tomate. *Actas Hortic.* **2008**, *51*, 175–176.

48. Sim, S.C.; Durstewitz, G.; Plieske, J.; Wieseke, R.; Ganal, M.W.; Van Deynze, A.; Hamilton, J.P.; Buell, C.R.; Causse, M.; Wijeratne, S.; et al. Development of a large SNP genotyping array and generation of high-density genetic maps in tomato. *PLoS ONE* **2012**, *7*, e40563. [CrossRef] [PubMed]

49. Young, N.D.; Tanksley, S.D. RFLP analisys of the size of chromosomal segments retained around the TM-2 locus of tomato during backcross breeding. *Theor. Appl. Genet.* **1989**, *77*, 353–359. [CrossRef] [PubMed]

50. Verlaan, M.G.; Szinay, D.; Hutton, S.F.; de Jong, H.; Kormelink, R.; Visser, G.F. Chromosomal rearrangements between tomato and Solanum chilense hamper mapping and breeding of the TYLCV resistance gene Ty-1. *Plant J.* **2011**, *68*, 1093–1103. [CrossRef] [PubMed]

51. Hutton, S.F.; Scott, J.W.; Schuster, D.J. Recessive Resistance to Tomato yellow leaf curl virus from the Tomato Cultivar Tyking is located in the Same Region as Ty-5 on Chromosome 4. *HortScience* **2012**, *47*, 324–327.

52. Pereira-Carvalho, R.C.; Díaz-Pendón, J.A.; Fonseca, M.E.N.; Boiteux, L.S.; Fernández-Muñoz, R.; Moriones, E.; Resende, R.O. Recessive Resistance Derived from Tomato cv. Tyking-Limits Drastically the Spread of Tomato Yellow Leaf Curl Virus. *Viruses* **2015**, *7*, 2518–2533. [CrossRef] [PubMed]

diversity

MDPI

Review

Harnessing Genetic Diversity of Wild Gene Pools to Enhance Wheat Crop Production and Sustainability: Challenges and Opportunities

Carla Ceoloni *, Ljiljana Kuzmanović, Roberto Ruggeri, Francesco Rossini, Paola Forte, Alessia Cuccurullo and Alessandra Bitti

Department of Agricultural and Forest Sciences (DAFNE), University of Tuscia, 01100 Viterbo, Italy; ljiljanakuzmanovic@gmail.com (L.K.); r.ruggeri@unitus.it (R.R.); rossini@unitus.it (F.R.); pforte64@gmail.com (P.F.); alessia2105@gmail.com (A.C.); bittia@unitus.it (A.B.)
* Correspondence: ceoloni@unitus.it; Tel.: +39-761-357-202

Received: 24 October 2017; Accepted: 26 November 2017; Published: 1 December 2017

Abstract: Wild species are extremely rich resources of useful genes not available in the cultivated gene pool. For species providing staple food to mankind, such as the cultivated *Triticum* species, including hexaploid bread wheat (*Triticum aestivum*, 6x) and tetraploid durum wheat (*T. durum*, 4x), widening the genetic base is a priority and primary target to cope with the many challenges that the crop has to face. These include recent climate changes, as well as actual and projected demographic growth, contrasting with reduction of arable land and water reserves. All of these environmental and societal modifications pose major constraints to the required production increase in the wheat crop. A sustainable approach to address this task implies resorting to non-conventional breeding strategies, such as "chromosome engineering". This is based on cytogenetic methodologies, which ultimately allow for the incorporation into wheat chromosomes of targeted, and ideally small, chromosomal segments from the genome of wild relatives, containing the gene(s) of interest. Chromosome engineering has been successfully applied to introduce into wheat genes/QTL for resistance to biotic and abiotic stresses, quality attributes, and even yield-related traits. In recent years, a substantial upsurge in effective alien gene exploitation for wheat improvement has come from modern technologies, including use of molecular markers, molecular cytogenetic techniques, and sequencing, which have greatly expanded our knowledge and ability to finely manipulate wheat and alien genomes. Examples will be provided of various types of stable introgressions, including pyramiding of different alien genes/QTL, into the background of bread and durum wheat genotypes, representing valuable materials for both species to respond to the needed novelty in current and future breeding programs. Challenging contexts, such as that inherent to the 4x nature of durum wheat when compared to 6x bread wheat, or created by presence of alien genes affecting segregation of wheat-alien recombinant chromosomes, will also be illustrated.

Keywords: *Triticum*; crop wild relatives; alien gene transfer; alien gene pyramiding; chromosome engineering; segregation distortion; new plant breeding techniques

---

1. Global Crop Demand: the Need for New Strategies

In recent years, scientists and experts from different fields are increasingly focusing on the challenging and threatening projection that in the next 50 years we will need to produce as much food as has been consumed over our entire human history. The growing population will cause exponentially rising demand, and a warming climate, water scarcity, and arable land shrinkage will make the challenge more difficult.

With a world acreage of over 220 million hectares in 2016–2017 [1], wheat covers more of the Earth than any other crop, and, among the major staple cereal crops, it is the only one that is adapted to low temperatures, hence can be grown during the cool season. This gives it a unique position in many different crop rotations across the globe with rice, cotton, soybean, and corn. Because of its adaptability, ease of grain storage, and ease of converting grain into flour, wheat is a major global diet component, providing more than 20% of calories for human consumption, a figure that jumps to 50% in places like Africa and parts of Asia.

The "Green revolutions" of the 20th century [2,3] had dramatic results on increasing food production: as to wheat, India alone doubled its harvest from 1965 to 1972, and the worldwide wheat production systems have been, in general, highly successful in increasing yield with respect to that existing at the beginning of the century [4]. The high rate of yield gains, which was particularly prominent in the second half of the century, was consistent for both the bread wheat [4] and the durum wheat [5] crop. However, on one hand, projections by 2050 forecast a considerable gap between demand and supply, particularly if climate-change-induced stresses are considered [6,7]; on the other hand, in a sustainable perspective, pathways that mimic past trends for increasing yield to meet global food demand are no longer practicable, as they would imply severe effects on the environment [8]. Instead, trajectories of global agricultural development that are directed to greater achievement of the technology improvement and technology transfer to less-developed and lower-yielding countries, are expected to enable the preservation of global biodiversity and to minimize major environmental impacts of agriculture, such as those due to greenhouse gas emissions and nitrogen use [8].

In fact, for wheat and all of the major cereal crops, yield increases have been plateauing in recent years, as if yield potential had approached its ceiling [7,8]. However, since depressed rates of progress do not seem to be majorly due to biological limits in the system [7], space for effectual interventions to counter yield stagnation does exist. Recent comprehensive analyses suggest that, as in the past 20–30 years, future yield progress will depend more on breeding than on new developments in crop agronomy [7,9]. A large genetic basis is pre-requisite for successful breeding; yet, current breeding materials contain only a fraction of the useful genetic variation available within crop related gene pools, which is perhaps comparable to the visible portion of an iceberg. As for other crops, wheat relatives, particularly wild Triticeae species, provide a vast reservoir for most, if not all, agronomically important traits [10,11], for which a unique array of cytogenetic materials have in many cases allowed for the identification of the carrier chromosome or chromosome arm. From such pre-breeding stocks, including addition and substitution lines of single alien chromosomes into the wheat genome, the most successful avenue to make the desired genes/QTL exploitable in breeding, has consistently proved to be the transfer of only chromosome segments from the alien source, applying well established cytogenetic methodologies of "chromosome engineering" [12]. Through this approach, new allelic combinations for the target traits can be generated via meiotic pairing and recombination between the alien chromosome, or just a region of it, and its wheat counterpart, promoting the process by *Ph* (*Pairing homoeologous*) wheat mutants (available in both bread and durum wheat, see [13]), when the pairing partners share only partial homology (i.e., homoeology). This is often the case when the donor chromosome belongs to more distant wheat relatives, which, in turn, possess a lot of still untapped genetic variation. A homoeologous recombination-based strategy has the double advantage of providing well-compensated wheat-alien products, and of minimizing the size of the alien segment and consequent linkage drag, which is often responsible for the reduced fitness of sizable introgressions from wild donor species (see, e.g., [14–20]). Meeting these two requisites has in fact almost invariably resulted in successful wheat-alien transfers, even in the case of durum wheat (*Triticum durum* Desf., 2n = 4x = 28, genome AABB), whose tetraploid nature reduces tolerance toward extensive manipulations when compared to the hexaploid bread wheat (*T. aestivum* L. 2n = 6x = 42, genome AABBDD) case [15,21]. In the last years, tremendous scientific and technological advances are greatly expanding our knowledge of the highly complex genomes of cultivated wheats and of related Triticeae species, and hence our ability to manipulate them in a directed manner. In the following, the behaviour and impact on the recipient wheat genome of a variety of alien segment introgressions will be illustrated. They represent examples of "smart" chromosome engineering efforts, through which multiple breeding targets have been pursued and corresponding alien genes/QTL pyramided into wheat genotypes, also in challenging contexts, e.g., those that are determined by the tetraploid condition of durum wheat (see above), or by the presence of alien genes that affect the segregation of wheat-alien recombinant chromosomes.

2. Stacking Different Alien Segments

In chromosome engineering experiments, an array of wheat-alien chromosomes, with varying recombinant breakpoint positions, and hence amounts of alien chromatin, can be obtained. By genetic (molecular markers) and physical (e.g., FISH or GISH) mapping, introgressions of minimal size still containing the target gene(s) can be selected, and these in the largest majority of cases exert the best overall performance in a breeding perspective. By exploiting high-resolution tools, this goal could be achieved in several instances, both in bread and also durum wheat (see several chapters in [11] and references therein; see also [15–20]). As to the latter species, successful examples not only include the transfer of genes for resistance to diseases [14,22,23] and abiotic stresses [24], but also grain quality [14,25–28] and yield-related traits [29,30].

For three such transfers, each involving an alien chromosome segment that is spanning around 20% of the recipient wheat arm, an attempt was made to stack them in a single genotype. The individual transfer lines contained on wheat chromosome arms, namely 7AL, 3BS, and 1DS, distal portions of the homoeologous *Th. ponticum* $7el_1$L arm, with the *Lr19* (leaf rust resistance) + *Sr25* (stem rust resistance) + *Yp* (yellow pigment content = YPC) genes [22], of the *Ae. longissima* $3S^lS$ arm, with the powdery mildew resistance gene *Pm13* [14], and of the bread wheat 1DS arm, harbouring the *Gli-D1/Glu-D3* loci, with positive effects on gluten quality [25,31]. Already at early stages of the work [32], transmission of two out the three recombinant chromosomes appeared to be normal through both germlines in F_2 cross progenies. At advanced breeding stages, the result could be confirmed also for segregation of all three wheat-alien chromosomes, for which no significant difference with respect to the expected genotypic classes was detected by a X^2 test of F_2 progenies of the triple recombinant (RRR) after crosses and backcrosses (BC) with adapted and good-yielding varieties (Figure 1). Identification of the different genotypes was enabled by the use of polymorphic PCR-based markers associated to the alien and wheat homoeologous segments (Figure 2).

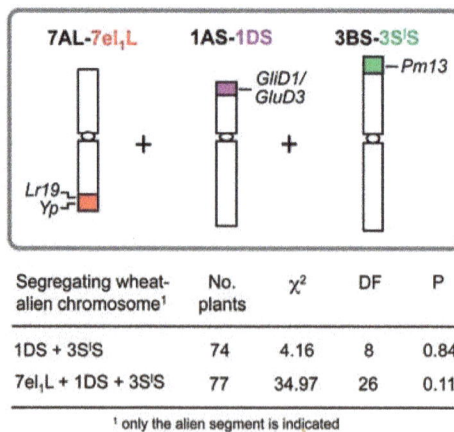

Segregating wheat-alien chromosome[1]	No. plants	χ^2	DF	P
1DS + $3S^lS$	74	4.16	8	0.84
$7el_1$L + 1DS + $3S^lS$	77	34.97	26	0.11

[1] only the alien segment is indicated

Figure 1. Wheat-alien recombinant chromosomes transferred into durum wheat and the simultaneous transmission ability of two and all three of them in F_2 progenies of selfed heterozygotes.

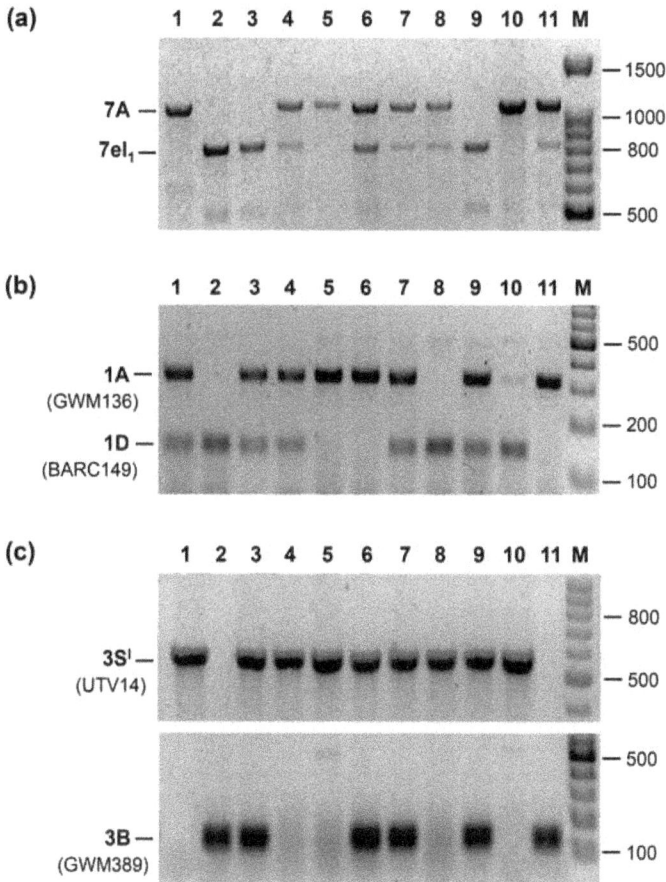

Figure 2. Profiles of PCR-based markers used to select multiple recombinant genotypes in durum wheat segregating progenies. (a) Multiplex PCR with STS (sequence tagged site) primer combination [33] generates a codominant assay for the presence of *Lr19* and associated genes into the same 7el$_1$L *Th. ponticum* segment, including *Yp* and *Sr25* (see text): lanes 2, 3, 9 (7el$_1$L homozygous carriers, HOM+); lanes 1, 5, 10 (7el$_1$L homozygous non-carriers,, HOM−); 4, 6, 7, 8, 11 (7AL-7el$_1$L heterozygotes, HET); (b) Multiplex PCR with primers for SSR (simple sequence repeat) markers associated to the 1AS and 1DS segments, the latter including *GliD1/GluD3* genes [25,26]: lanes 2, 8, 10 (1DS HOM+); lanes 5, 6, 11 (1DS HOM−); lanes 1, 3, 4, 7, 9 (1DS-1AS HET); (c) A *Pm13* (3S^{l}S)-associated STS marker (UTV14, [34]), combined with a 3BS-linked SSR marker (GWM389): lanes 1, 4, 5, 8, 10 (3S^{l}S HOM+); lanes 2, 11 (3S^{l}S HOM−); lanes 3, 6, 7, 9 (3S^{l}S-3BS HET). M = 100 bp ladder (relevant base pair figures on the right side of gels).

2.1. Agronomic Evaluation of Multiple Recombinant Lines

Although the transmission ability of wheat-alien recombinant chromosomes represents a first, important aspect to assess their impact on the recipient genotype (see also ahead), the effect on yield remains the most critical parameter in a breeding perspective. In the majority of cases, yield reduction due to linkage drag is a major problem when introgressing alien chromatin into durum wheat [14,29,30,35]. The three transfer chromosomes that are described above (Figure 1) had shown no major effects on yield, also when combined in a single genotype. However, this evidence was mostly based on small-scale field tests, or it was somewhat biased by leaf-rust epidemics, which could have favoured *Lr19*-bearing genotypes [25]. In the 2015 season, a large-scale comparative field trial, including several currently grown national varieties, selected recombinant genotypes (R117) bearing one, two, or three of the alien segments, and the cultivated varieties mostly involved in the pedigree of the R117 lines (i.e., the Italian cv. Simeto and the French cv. Karur), was carried out in Viterbo (Central Italy).

The analysis of co-variance (ANCOVA), focused on the R117 lines and the latter two varieties, taken as controls, showed significant differences between the genotypes for several of the traits that were measured (Table 1).

For most of the phenotypic attributes (see HD, PH, GNS, and TGW), recombinant (R117) genotypes generally displayed a closer resemblance with cv. Karur, as expected from its major contribution to the R117 pedigree (3 BCs preceding four self generations, i.e., BC_3F_4 progenies).

As for the main productivity traits (SNM2, BM2, GYM2, GNM2, and HI), the values of R117 lines resulted to be essentially at the level of control cultivars, or superior to them (see GYM2, GNM2, HI). This important result shows that the presence of one, two, or even three alien chromosome segments of relatively small size (Figure 1), originating from genetic pools more or less distant from durum wheat, did not cause any yield penalty on the recipient genotype. Instead, in the case of genotypes R117-11-20 (triple recombinant) and R117-11-8 (recombinant for the *Th. ponticum* segment only), yield per unit area (GYM2) was increased of about 10% vs. Karur and 24–30% vs. Simeto. It can be noted that this increase was principally due to increase in grain number (GNM2; +3–55%), biomass (BM2; +2–16%), and consequently, harvest index (HI; +8–10%), the latter trait being increased in all of the recombinants except R117-9-59 (Table 1). Furthermore, introgression of different alien segments significantly increased some of the spike fertility traits: GNS in recombinants R117-11-20 and R117-2-21 (+5–33%), GYS in R117-11-20 (+7–36%), and GNSP in R117-11-20 and R117-11-8 (+4–20%). This suggests that the increase of yield per spike was a key determinant of the final yield increase, especially of the two most productive recombinants (R117-11-20 and R117-11-8).

Enhancement of yield per unit area and per spike of the triple recombinant R117-11-20 validates, in fact, preliminary results from the earlier field study that was carried out on spaced plants of the triple recombinant genotype, before its BCs with cv. Karur [25]. For example, spike fertility traits, such as GNS and GNSP, which in the case of R117-11-20 reached 10% and 20% increases, respectively, were increased by 3–10% in the original triple recombinant [25]. Similarly, the observed increase of up to 30% of grain yield per unit area of R117-11-20 genotype is in line with the 60% yield increase that was formerly exhibited in one location of the Central-West Italy [25]. However, with respect to the previous trials, results from the 2015 season were obtained in a leaf rust-free environment, which was subjected to an appropriate disease management: this emphasizes the positive effect of the introgressed segments on yield *per se*, irrespective of the advantage conferred by the *Lr19* gene, present in all recombinant genotypes (Table 1).

Table 1. Mean values of yield-related traits[1] recorded on recombinant and control durum wheat genotypes in 2015 growing season and p-values from the ANCOVA analysis. Letters in each row correspond to the ranking of Tukey HSD test at p < 0.05.

Genotype/Trait[1]	R117-11-20		R117-9-11		R117-9-71		R117-2-21		R117-9-59		R117-11-8		cv. Karur		cv. Simeto		ANCOVA p-Value[2]
Lr19 + Yp	+		+		+		+		+		+		—		—		
Glu-D3	+		+		+		+		+		+		—		—		
Pm13	+		+		+		—		—		—		—		—		
HD	115.0	b	118.3	a	117.7	ab	116.3	ab	117.3	ab	116.3	ab	115.7	ab	106.7	c	0.000 ***
PH	73.3	ab	66.7	c	66.7	c	72.3	abc	67.0	c	70.0	bc	71.3	bc	78.3	a	0.000 ***
SNM2	195.4		182.4		158.8		170.6		147.1		195.4		200.7		181.7		0.332
BM2	759.0		710.8		628.0		628.8		559.6		763.9		741.4		659.5		0.458
GYM2	319.8		281.1		249.6		266.4		214.2		323.8		291.1		248.9		0.254
GNM2	6328.5		6031.5		5192.9		5432.1		4334.1		6293.8		6133.0		4070.0		0.096
HI	0.42		0.40		0.40		0.42		0.38		0.42		0.39		0.38		0.164
GNS	58.6	a	52.0	ab	54.6	a	55.9	a	52.4	ab	52.9	ab	53.4	ab	45.2	b	0.001 **
GYS	3.0	a	2.4	b	2.6	ab	2.7	ab	2.6	ab	2.7	ab	2.5	ab	2.8	ab	0.035 *
GNSP	3.0	a	2.6	b	2.8	ab	2.8	ab	2.6	b	2.9	a	2.5	b	2.8	b	0.000 ***
TGW	50.8	bc	46.6	c	47.3	bc	49.0	bc	49.5	bc	51.4	b	47.1	c	61.1	a	0.000 ***

[1] HD, heading date; PH, plant height; SNM2, spike number m^{-2}; BM2, biomass m^{-2}; GYM2, grain yield m^{-2}; GNM2, grain number m^{-2}; HI, harvest index; GNS, grain number spike^{-1}; GYS, grain yield spike^{-1}; GNSP, grain number spikelet^{-1}; TGW, thousand grain weight; [2] *, ** and *** indicate significant F values at $p < 0.05$, $p < 0.01$ and $p < 0.001$, respectively.

2.2. Breeding Potential of the Triple Recombinant

From the breeding point of view, the R117-11-20 line, homozygous for the 7A-7el$_1$, 1A-1D and 3B-3S^l recombinant chromosomes (Figure 1), resulted in the most promising and most productive out of the six recombinant genotypes when compared to control varieties. The sister line tested, namely R117-9-11, had a good, though not equally outstanding, performance, indicating a particularly favourable allelic combination in the background genotype of R117-11-20. The latter, as the original triple recombinant used in BCs to cv. Karur, contains in its pedigree also cv. Simeto. A comparison among the three genotypes highlights the differences for the most relevant yield-related traits associated with the triple alien introgression (Table 2).

Table 2. Percentage differences between R117-11-20 triple recombinant line and the two control cultivars Karur and Simeto for the relevant yield traits (see Table 1 for symbols).

Trait	Difference vs. Karur	Difference vs. Simeto
SNM2	−3%	8%
BM2	2%	15%
GYM2	10%	28%
GNM2	3%	55%
HI	8%	11%
TGW	8%	−17%
GNS	10%	30%
GYS	20%	7%

Almost all of traits considered, directly or indirectly involved in final yield formation, showed increased values in the novel recombinant with respect to the two varieties, with positive effects being more evident vs. Simeto, the older cultivar of the two (Table 2). Final yield (GYM2) of R117-11-20 was remarkably higher than that of both the controls, yet the traits that contributed the most to this result seem to be different for each cultivar. When compared to Karur, spike yield parameters, i.e., GNS, GYS, and TGW, were those displaying higher values in the R117-11-20 recombinant (+10%, 20%, 8%, respectively). On the other hand, in comparison with Simeto, not only grain number spike^{-1} (GNS, +30%), but also spikes number m^{-2} (SNM2, +8%), seem to have caused the considerable increase of grain number (GNM2) and grain yield (GYM2) per unit area. R117-11-20 is a valid example of unconventional breeding, where intraspecific and interspecific variation have been usefully combined to give enhanced yield potential and novel quality attributes to the durum wheat crop.

3. Assembling Genes from Different Alien Sources into the Same Segment

An alternative strategy to enrich a given wheat genotype with multiple alien traits consists of bringing together into a single alien chromosome segment genetic material originating from diverse though closely related sources. The re-engineered chromosomes with the desired novel assembly are the result of recombination between alien segments (either homologous or homoeologous) previously incorporated into different parental translocation lines. Various beneficial genes can thus be accumulated, and unwanted genes possibly eliminated. Particularly noteworthy examples of this strategy concern gene combinations from perennial Triticeae species of the *Thinopyrum* genus, one of the richest sources of valuable genes/QTL for wheat improvement [36,37]. By means of *ph1b*-mediated recombination between a bread wheat-*Th. ponticum* T4 translocation (70% of the alien 7el$_1$L arm into wheat 7DL) and a homoeologous *Th. intermedium* segment, also on 7DL, the highly effective leaf rust resistance gene *Lr19* from the former and the barley yellow dwarf virus resistance gene *Bdv2* from the latter were recently combined [17,38]. Similarly, and in this case exploiting the high degree of homology relating two group 7 chromosomes (named 7el$_1$ and 7el$_2$) of different *Th. ponticum* accessions, the *Lr19* gene (located on the 7el$_1$L arm) was pyramided, both in bread wheat [39–41] and in durum wheat [41], with *Fhb-7el2*, a major QTL for resistance to Fusarium head blight (FHB) mapped on 7el$_2$L (recently renamed *Fhb7* [42]).

A remarkable successful attempt [43] involved the bread wheat T4 translocation line, containing *Lr19* and associated genes/QTL on the *Th. ponticum* 7el$_1$L portion, and the bread wheat cv. Chinese Spring (CS) 7E(7D) substitution line, which was used as donor parent of an exceptionally effective FHB resistance locus associated with the *Th. elongatum* 7EL arm (*Fhb-7EL*). In this case, due to the close homoeology relating the 7el$_1$L and 7EL arms, as confirmed by cytological (meiotic chromosome pairing) and molecular (genetic map synteny and collinearity) evidence [43], even in the absence of any genetic pairing promotion (e.g., *ph1* mutation), target genes from the two *Thinopyrum* species could be brought together in a single alien segment on wheat 7DL with relative ease. The array of segregating progeny from the cross between the two donor lines has been recently widened. The 114 total plants tested with appropriate molecular markers (Figure 3), have revealed a total frequency of recombinant types of over 14%, which is higher than the 10.6% that was previously estimated on a smaller sample [43]. Of these, equal number of T4 and 7E types were isolated in the cross progeny of (CS7E(7D) × T4) F$_1$s with normal bread wheat cultivars, used as pollen parents (Figure 3).

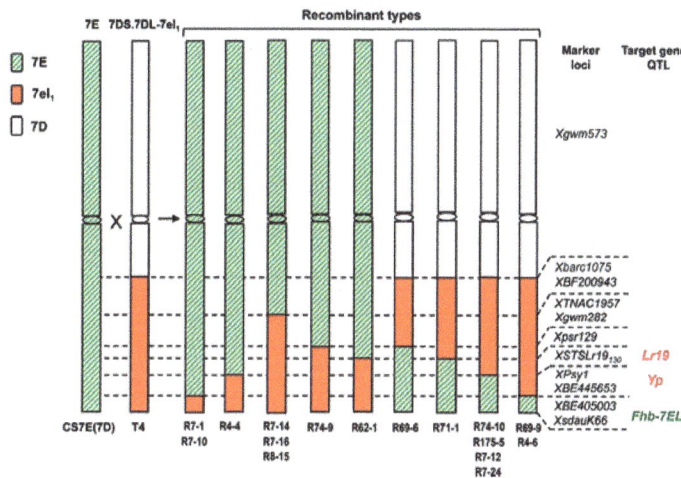

Figure 3. Cytogenetic maps of group-7 chromosomes of parental and recombinant genotypes identified in cross progenies of (CS7E(7D) × T4) F$_1$s with normal bread wheat cultivars, based on molecular marker and GISH analyses. 7EL-7el$_1$L and 7DL-7el$_1$L breakpoints are indicated by dashed lines. Only a subset of tested markers is reported; for additional marker data see [43].

In a breeding perspective, the T4 recombinant types are the most suitable. Among them, particularly promising are those that carry all of the target genes, such as R74-10 (and so R175-5, R7-12 and R7-24), bearing *Lr19* from *Th. ponticum* 7el$_1$L, combined with the *Fhb-7EL* major QTL from *Th. elongatum*, conferring an outstanding resistance to Fusarium most threatening diseases worldwide, i.e., FHB as well as crown rot [43]. This group of recombinants carries a 7EL allele for the YPC phenotype (*Psy1* gene), which is associated with reduced flour yellowness, and is hence more desirable for bread wheat breeding [43]. On the other hand, recombinant types R69-9 and R4-6, with less 7EL distal chromatin (Figure 3), possess a 7el$_1$L allele for *Psy1* gene, determining a higher YPC [43]; thus, they are the most suited donor lines for the ongoing transfer of the *Lr19* + *Fhb-7EL* combination into durum wheat.

Identification and characterization of the various recombinant types was enabled by use of GISH (genomic in situ hybridization) and by several molecular markers (see Figure 3 and [43]), which highlighted the different 7EL-7el$_1$L breakpoints. The most polymorphic and user-friendly of such markers (e.g., Figure 4) represent a valuable tool for assisted selection (MAS) of recombinant chromosomes with the novel gene combinations in breeding programs.

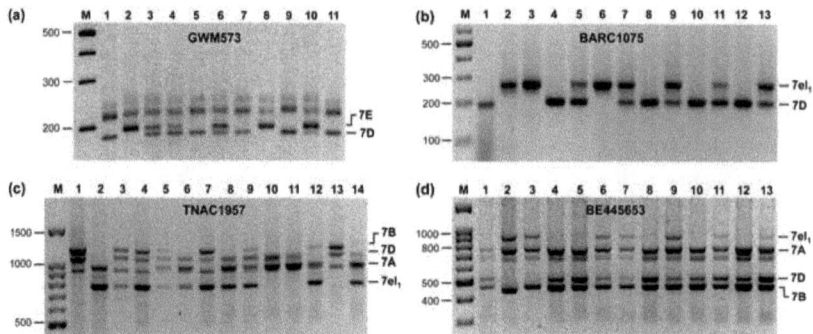

Figure 4. Profiles of some PCR-based markers used to select recombinant chromosomes with gene combinations from *Th. ponticum* 7el$_1$L and *Th. elongatum* 7EL arms in cross progenies of (CS7E(7D) × T4) F$_1$s with normal bread wheat cultivars (CS and Blasco). Group-7 wheat and *Thinopyrum* spp. alleles are indicated. (**a**) GWM573, a short arm-linked marker (7DS and 7ES alleles), allows discrimination between T4 and 7E chromosome types (both parental and recombinant): lane 1, Blasco (7DS); lane 2, CS7E(7D), 7ES; lanes 3–11, segregating progeny, homozygous (HOM+) for the 7ES allele (8), for the 7DS allele (i.e., 7ES HOM−: 5, 7, 9, 11) or heterozygous (7ES-7DS HET: 3, 4, 6, 10); (**b**) BARC1075 highlights a 7el$_1$L and a 7DL allele: lane 1, Blasco (7DL); lane 2, T4 (7el$_1$L); lanes 3–13, 7el$_1$L HOM+ (3, 6); 7el$_1$L HOM− (4, 8, 10, 12) and 7el$_1$L-7DL HET (5, 7, 9, 11, 13) segregates; (**c**) TNAC1957: lane 1, Blasco (7DL allele); lane 2, T4 (7el$_1$L allele); lanes 3–14: 7el$_1$L HOM+ (12); 7el$_1$L HOM− (10, 11, 13) and 7el$_1$L-7DL HET (3–9, 14) segregates for the two critical alleles; (**d**) BE445653: lane 1, Blasco (7DL allele); lane 2, T4 (7el$_1$L allele); lanes 3–13, 7el$_1$L HOM+ (3), 7el$_1$L HOM− (4, 5, 8, 10, 12) and 7el$_1$L-7DL HET (6, 7, 9, 11, 13) segregates for the two critical alleles. M = 100 bp ladder (relevant base pair figures on the left side of gels).

4. An Intriguing Issue in Wheat-Alien Gene Transfer: Segregation Distortion

A fairly frequently observed phenomenon upon the introduction of an alien chromosome or a rearranged wheat-alien chromosome into the wheat genome is its abnormal segregation in progenies from heterozygotes. Segregation distortion (SD) with respect to expected Mendelian ratios is of widespread occurrence in plant and animal species, for which it is an important evolutionary force [44–47]. SD is likely to represent one result of genetic incompatibilities between parental genomes that have been separated by reproductive barriers [48]. In this view, it is not surprising to find the level of SD to increase, both in the number of SD regions (SDRs) within a chromosomal set, and in the number of markers within each SDR, in intraspecific cross progeny involving genetically and geographically more distant parents [49]. In contrast to the latter work, which was carried out on wheat, a recent study on F$_2$ populations generated from several *Arabidopsis thaliana* accessions showed little correlation between the degree of genetic differentiation between the parental accessions and the probability of observing allelic distortion in their progeny [50]. Nonetheless, many cases of SD were discovered in interspecific contexts, although this evidence, as suggested by Seymur et al. [50], may not be due to a higher occurrence of the phenomenon in such contexts, but because the severity of distortion is more extreme (hence more easily detectable), and more rapidly reaching fixation in the absence of species-specific modifiers.

Since SD can affect accuracy of linkage map construction and also have breeding implications, it has been addressed in several studies involving crop species, including wheat (e.g., [47,49,51–53]). In progeny of wheat-alien combinations, one of the most likely causes for the skewed transmission of complete or segmental alien chromosomes of various wild species (most studied those from

Aegilops and *Thinopyrum* genera) has been suggested to be gametophytic competition during zygote formation, with male gametogenesis being affected in the majority, though not all of the cases [36,54,55]. The phenomenon, however, proved to be highly variable even for a given SDR containing a segregation distortion (*Sd*) gene, going from selective or even exclusive retention in the wheat background through generations (preferential transmission), to a more or less dramatic self-elimination. Such changes in the direction and magnitude of the effects of *Sd* genes have been associated to allelic variation at several wheat "responder" loci, with the background genetic/chromosomal environment thus strongly conditioning the SD phenotype [56–64].

For *Sd* genes that prevailingly or exclusively determine the preferential transmission of the carrier chromosome, the term *Gc* (gametocidal) is usually adopted, which is suggestive of the underlying mechanism of action. When in heterozygous or hemizygous condition, such genes in fact "kill" gametes lacking them, by causing chromosome breakages and various consequent aberrations, mainly in the course of post-meiotic divisions of micro- and macro-sporogenesis [54,55,65]. Both male and female gametes without the *Gc* gene (e.g., normal wheat types) become abortive, and as a result, only gametes with the *Gc* gene (and thus, with the alien or wheat-alien chromosome) are transmitted to the next generation. As such, *Gc* genes are typical selfish genetic elements [48]. Besides those that were identified on various chromosomes of many *Aegilops* species [54,55,65], *Sd* genes with more or less marked characteristics of *Gc* elements were detected on group 7 chromosomes of polyploid *Thinopyrum* species, including *Th. intermedium* [66] and *Th. ponticum* [36,58,59,61,64]. The *Th. ponticum* largely homologous $7el_1$L and $7el_2$L chromosome arms, both contain one or more *Sd* genes, which have been studied rather intensively, primarily because of the linked beneficial genes of breeding relevance (see above in this section, and Section 2). The SD phenotype determined by *Sd* gene(s)/SDR(s) that is linked to the $7el_1$L segment of translocation line T4 and of other bread and durum wheat lines possessing the same or shorter $7el_1$L portions, exemplifies the highly variable, probably background-dependent, outcomes, as described earlier (reviewed in [36]). On the other hand, the effects of the $7el_2$L-linked *Sd* gene(s) appear relatively more consistent through studies and backgrounds. What seems to be the result of a gametocidal type of action, confers to the entire $7el_2$ chromosome or portions of it a marked preferential transmission from heterozygous wheat carrier lines. However, not always the same germline appears to be affected: female gametes were reported to be exclusively involved in the first observations [58,59], while male gametes were later indicated as being implicated in the almost uniparental segregation [64]. In the latter study, the phenomenon was analysed in recombinant inbred lines (RILs) from the cross between wheat substitution lines for the complete $7el_2$ and $7el_1$ chromosomes in cv. Thatcher background. A survey of the RILs progeny with molecular markers revealed a clear-cut disproportion of $7el_2$-linked marker alleles preferentially transmitted through the male germline. The effect of $7el_1$-associated *Sd* genes was virtually undetectable when co-present with $7el_2$, evidently stronger/dominant over $7el_1$ for its gametocidal action, similarly to what observed for the interaction of *Gc* genes from different *Aegilops* species (reviewed in [54]).

The prevailing or almost exclusive SD/Gc effect of $7el_2$ was confirmed in a subsequent study, in which *Sd* genes/SDRs that are associated to $7el_1$L and $7el_2$L arms were brought together in the cross progeny of [KS24-1 (7DL·$7el_2$L centric translocation line) × T4 (7DS·7DL-$7el_1$L translocation line)] × cv. Blasco [41]. In this context, a further different picture emerged, again likely influenced by the background genotype. While in F_1s after pollination with Blasco a preferential transmission of $7el_2$L marker alleles through female gametes was observed, transmission of recombinant $7el_1$L-$7el_2$L chromosomes in the resulting F_2 progenies apparently followed a "bimodal" behaviour, being normal for some of them, while significantly deviating from the 1:2:1 ratio for others (Table 3). The relative abundance of the segregating genotypes in the latter progenies suggests that preferential transmission probably occurred through both germlines, as homozygous recombinant segregates (HOM+) were very frequent, more than the heterozygotes (HET) in some progenies, while the homozygous non-recombinant segregates (HOM−) were almost absent in all of them (Table 3).

When comparing the allele type of the several markers tested for progeny genotyping, it appeared that, irrespective of other $7el_1L$ vs. $7el_2L$ regions, the one comprising PSP3123, BARC172, and BF200943 (highlighted in Figure 5), was critical in determining the SD outcome; in fact, all of the recombinant chromosomes possessing $7el_2L$ alleles at these marker loci were preferentially transmitted (Table 3 and Figure 5).

As shown in Figure 5, the work of Forte et al. [41] enabled a higher map resolution than previously achieved [64], separating this SDR from a more proximal one, containing the PSR129 marker, associated, in turn, to $7el_1L$-linked SD effects in various studies and materials (both bread and durum wheat recombinant lines, see [36] for a review). Within the $7el_2L$ SDR, it seems of interest to note the presence of a wheat EST (Expressed Sequence Tag), namely BF200943, for whose nucleotide sequence a BLAST search (https://blast.ncbi.nlm.nih.gov/Blast.cgi) indicated a high homology with a protein product belonging to the mitogen-activated protein kinase (MAPK) cascades of several Triticeae species. MAPKs are ubiquitous signaling modules in eukaryotes, which are involved in the regulation of different developmental processes. They play a crucial role in proliferation, differentiation, and cycle regulation in germ cells of different mammalian species [67]. MAPKs are also involved in almost every aspect of plant growth and development. As to plant reproduction, MAPKs play vital roles in various stages, including anther, pollen, ovule, and seed development [68].

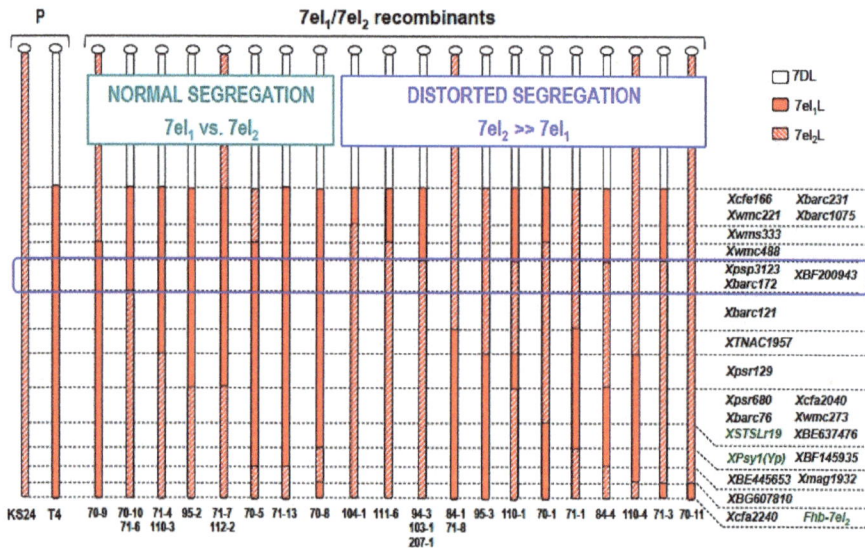

Figure 5. Cytogenetic maps of the $7DL$-$7el_1L$/$7el_2L$ arms of bread wheat–*Th. ponticum* recombinant chromosomes, isolated from KS24-1/T4/cv. Blasco crosses, based on molecular marker analysis and genomic in situ hybridisation (GISH). Dotted lines indicate the $7DL$/$7el_1L$/$7el_2L$ breakpoints. The location of the $7el_2L$ region that is likely to have the main effect on segregation distortion observed in this context (see text), is bounded by the violet box. P = parental lines. Modified from Forte et al., 2014 [41].

Table 3. Segregation in F$_2$ progenies from the KS24-1/T4/cv. Blasco cross and allele type of some of the markers used for progeny genotyping (see also Figure 5).

Rec. No.	Chrom. Type	Marker Alleles				F$_2$ Segregation				
		BF200943 PSP3123 BARC172	BARC 121	TNAC 1957	PSR 129	HOM+	HET	HOM−	X^2 1:2:1	p Value [1]
70-5	T4	7el$_1$	7el$_1$	7el$_1$	7el$_1$	9	19	7	0.49	0.78
70-8	T4	"	"	"	"	6	24	13	2.86	0.24
70-9	KS24	"	"	"	"	8	10	5	1.18	0.55
71-7	KS24	"	"	"	"	5	11	10	2.54	0.28
71-13	T4	"	"	"	"	4	7	5	0.38	0.83
95-2	T4	"	"	"	"	4	7	1	1.83	0.40
110-3	T4	"	"	"	7el$_2$	5	11	5	0.04	0.98
70-10	T4	"	7el$_2$	7el$_2$	"	2	8	2	1.33	0.51
71-6	T4	"	"	"	"	3	8	1	2.00	0.37
70-1	T4	"	"	"	"	5	14	0	6.89	0.03 *
70-11	KS24	7el$_2$	"	"	"	9	13	0	8.09	0.02 *
71-3	T4	"	"	"	"	8	16	0	8.00	0.02 *
84-4	T4	"	"	"	"	21	13	0	27.82	0.000 ***
94-3	T4	"	"	"	"	9	18	0	9.02	0.01 *
111-6	T4	"	"	"	7el$_1$	16	10	3	14.45	0.001 **
110-1	T4	"	"	"	7el$_1$	10	8	0	7.33	0.03 *
110-4	KS24	"	"	"	"	10	13	1	6.92	0.03 *
71-1	T4	"	"	7el$_1$	"	8	11	0	7.21	0.03 *
71-8	KS24	"	"	"	"	8	8	0	8.00	0.02 *
84-1	KS24	"	"	"	"	9	9	0	6.79	0.03 *

[1] *, ** and *** indicate significant differences at p < 0.05, p < 0.01 and p < 0.001, respectively.

At this stage of knowledge, a possible involvement in the observed SD phenotype of a 7el$_2$ putative variant allele of the BF200493 gene can only be a matter of speculation. Indeed, despite the apparent ubiquity of segregation distortion, the relative contribution of different biological processes, such as epistatic incompatibilities, gametic selection, or meiotic drive, is often unclear, especially in plants [50]. Even when the mechanism is relatively well described, as in the case of *Gc* factors [54,55], still, little information is available on the molecular basis of the phenomenon. Studies of a range of different Gc chromosomes/factors suggest that there are at least two types of elements that are associated with the mechanism responsible for their preferential transmission: the breaker element, causing extensive chromosomal breakage, and hence the lethality of gametes not carrying the Gc chromosome/factor, and the inhibitor element, preventing these effects in gametes retaining the Gc chromosome/factor (reviewed in [54]).

The two phenotypes that are controlled by such elements were hypothesized to be at the basis of the preferential transmission of the *Ae. sharonensis* 4S^{sh} chromosome, with the breaker element (*Gc2* or *GcB*) being initially mapped to the distal end of the long arm [51], and later more precisely assigned to the region immediately proximal to a block of sub-telomeric heterochromatin on 4S^{sh}L arm [69]. Recently, comparative genomic studies, coupled with deletion mapping and targeted-capture sequencing, have enabled the identification of unique single nucleotide polymorphisms (SNPs) that are specific to the short *Ae. sharonensis* 4S^{sh}L segment introgressed into wheat and associated to the Gc phenotype, of which 18 represent candidate genes for the 4S^{sh}L breaker element [70]. One of these SNPs showed sequence conservation with a protein that is involved in transposition to unrelated chromosomal sites. This finding strengthens the transposon theory that was suggested for *Gc* genes of wheat and related species by Tsujimoto and Tsunewaki [71], which was based on hybrid dysgenesis in *Drosophila melanogaster*, caused by mobile P elements. In fact, in both of the systems, the observed symptoms include sterility, lethality, mutations, and chromosome breakage; hence, also for the 4S^{sh}L Gc case, the two-loci theory, with the involvement of mobile element(s), appears as a likely hypothesis: the breaker element might be a transposon similar to the *Drosophila* P elements, and the inhibitor would be located close to the breaker in the sub-telomeric region [69,70].

The one just described remains so far the best studied case of genes underlying SD phenotypes in wheat and related species. However, when considering that such genes are widely spread, if not omnipresent, in plants, in which they must have played a major evolutionary role in terms of karyotype diversification and speciation by sexual isolation, it seems reasonable to expect a variety of elements and mechanisms responsible for these phenotypes between and within species [36,55,65]. In this frame, epigenetic phenomena should also be contemplated. In fact, since Barbara McClintock's intuition that "species crosses are another potent source of genome modification that could yield TE activation" [72], a large body of evidence has demonstrated that the merging of parental species in hybrids and allopolyploids can lead to a "genomic shock", which triggers the dysregulation of normal cell functions and various forms of pertubations, almost invariably resulting in activation of previously silent transposable elements (TEs) [73–75]. Besides that in complete interspecific hybrids, epigenetic changes, particularly the alteration of DNA methylation patterns, have been observed in introgression lines of rice [76,77] and wheat [78,79] involving even sub-chromosomal amounts of alien chromatin. In some such contexts, sequences that were corresponding to TEs were shown to be the main target of altered DNA methylation [79].

Alteration in the pattern of allele/chromosome segregation is among the phenomena correlated to chromatin modifications, as clearly demonstrated in the *Segregation Distorter* (*SD*) system of *Drosophila* [44]. In line with this, changes in the pattern of segregation distortion of parental alleles, besides that in the position of meiotic recombination, were found to be induced in an inter-subspecific cross of *Oryza sativa* L. by modification of inactive chromatin states [52]. As a whole, the available evidence suggests that the introduction of alien material into a recipient background can affect and/or induce a variety of SD phenotypes, not only causing new cis-trans genetic interactions (see the background-dependent effects above recalled), but also through epigenetic alterations.

In this view, the disruption of the genetic and epigenetic makeup of an alien and/or recipient chromosome may imply the outbreak of SD phenomena not exhibited by the original structure(s). This may be the case for the pattern of segregation displayed by the bread wheat-*Thinopyrum* recombinant types described earlier (see § 3), carrying 7E and $7el_1$ portions onto or in place of wheat chromosome 7D (Figure 3).

In contrast to the good fit (>0.5) to the normal 1:2:1 segregation ratio that was detected in F_2 progeny from 7D-7E double monosomics with different backgrounds (Table 4), which confirms early observations on the absence of marked SD associated to the complete chromosome 7E [80], recombinant 7E chromosomes, partly substituted on their long (L) arm (7EL) by distal portions of $7el_1L$, displayed significantly deviated ratios. These were in favour of segregates that lacked the alien segment (HOM−), indicating a suicidal behaviour of gametes (not ascertained whether male or female or both) carrying an altered 7E, with the deviation being of considerable extent even for the R7-1 chromosome, in which the length of the "alien" $7el_1L$ segment is minimal (Figure 3). On the other hand, all of the recombinant chromosomes of the T4 type, i.e., carrying a 70% long $7el_1L$ segment with 7EL insertions of variable length (Figure 3), including some particularly amenable for breeding exploitation (see § 3), showed, irrespective of the background genotype, normal segregation, not differently from the original T4 segment (Table 4).

Table 4. Segregation ratios in F_2 progenies from the cross of $7E/7el_1$ recombinants and parental lines (T4 and the 7E(7D) substitution line), taken as control, with normal bread wheat cultivars (CS or Blasco). Maps of recombinant chromosomes are reported in Figure 3.

Recomb. Genotype	Chromosome Type	F_2 Segregation				
		HOM+	HET	HOM−	$X^2_{1:2:1}$	*p* Value [1]
R69-9/CS	T4/distal 7EL	3	4	3	0.36	0.83
R69-9/2*Blasco		8	8	8	2.67	0.26
R74-10/2*Blasco	"	21	32	15	1.29	0.52
R74-10/4*Blasco		10	21	5	3.18	0.20
R175-5/CS	"	14	16	6	4.00	0.14
R175-5/2*Blasco		6	8	10	4.00	0.14
R71-1/CS	"	8	20	8	0.44	0.80
R71-1/2*Blasco		4	4	5	2.07	0.36
R69-6/Blasco	"	9	12	3	3.0	0.22
R7-1/Blasco	7E/distal $7el_1L$	7	27	26	12.63	0.002 **
R4-4/Blasco	"	8	21	26	14.85	0.000 ***
R62-1/CS	"	8	15	17	6.55	0.04 *
R74-9/Blasco	"	4	14	46	75.38	0.000 ***
R7-14/Blasco	"	6	20	21	10.61	0.005 **
Parental lines						
T4 (Thatcher)/CS		7	17	5	1.14	0.57
T4 (Thatcher)/CS//Blasco		6	10	8	1.00	0.61
7E(7D) substit. (CS)/CS		9	14	10	0.82	0.66
7E(7D) substit. (CS)/Blasco		11	14	10	1.33	0.51

[1] *, ** and *** indicate significant differences at $p < 0.05$, $p < 0.01$ and $p < 0.001$, respectively.

For both the 7D-$7el_1$/7E (Table 4) and the 7D-$7el_1$/$7el_2$ (Table 3) recombinant types showing abnormal segregation, no appreciable alteration of plant morphology, seed set, and seed development was observed among different segregates (carriers or non-carriers of the recombinant chromosomes), and in comparison with control lines. Thus, no limitation to their possible utilization in breeding is caused by the *Sd/Gc* factors they harbour. By contrast, *Gc* genes of some *Aegilops* species were found to also induce chromosome breakage in zygotic cells, often resulting in a varying degree of sterility and/or seed shrivelling (reviewed in [54]). However, these effects seem to be dependent on several factors, including the degree of "strength" and penetrance of the particular *Gc* gene, the type of gamete lacking the *Gc* gene(s), as well as the background genotype, as above mentioned for the overall SD phenotype. In cases where the presence of *Gc* genes, either linked to the target alien genes [81], or present in the initial hybrid genotype [82], did show undesirable effects on plant fertility, strategies have been adopted to overcome these effects, in one instance breaking the linkage with the desirable genes through *ph1b*-induced homoeologous recombination [81], in the other making use in the transfer scheme of an "anti-gametocidal" wheat mutant, whose mutated, hence ineffective, *Gc* locus [83], replaced the active one [82].

5. The Way Forward

To thwart the impoverishment of crop genetic base, resulting from domestication and recurrent selection in the course of the breeding process, the exploitation of genetic variation from even distant relatives of crop species has since long attracted considerable interest. In recent years, there has been a growing consensus about an important role that crop wild relatives (CWRs) can play in crop improvement. CWRs are more and more looked at as the nexus between food security, climate change adaptation, and biodiversity conservation issues [84]. This consciousness, coupled with the availability of a wide array of appropriate characterization and selection tools from fast developing fields of molecular genetics, cytogenetics, and genomics, has given a boost to effective strategies for the exploitation of the enormous, still unlocked potential of CWRs. One such strategy, recently called 'introgressiomics' [85], consists of a whole-genome introgression approach, that is transfer of chromosome segments from the entire genome of a given donor (mostly wild) species into the crop species background, irrespective of any traits that the wild relative might carry. The "step change" in this otherwise long-standing procedure of creation of interspecific F$_1$ hybrids and derived amphidiploids or backcross progenies, is application of high-throughput screening technology to detect and characterize the obtained introgressions, as well as of efficient phenotyping to maximize exploitation of their potential. This would enable linking the recombinant genotypes to a variety of agronomically important target traits, of current, but also foreseen and unforeseen value, in line with a pre-emptive breeding perspective [85]. A recent successful example of this initially un-focused generation of genetic diversity accompanied by high-throughput genotyping involves wheat and one of its wild relatives, namely *Amblyopyrum muticum* (syn. *Aegilops mutica*) [86]. A high number of genome-wide, segmental *Am. muticum* introgressions into wheat could be detected by using a subset of a previously developed array of single nucleotide polymorphism (SNP) markers (SNPs), showing polymorphism between ten wild wheat relatives and wheat genotypes [87].

On the other hand, if specific target traits are identified, and so is the chromosomal or even sub-chromosomal location of the genes underlying them, a focused strategy can be successfully adopted to incorporate the smallest possible alien segments containing them into the crop genotype. In this, multiple chromosomal segments deriving from different alien sources and bearing different genes/QTL can then be pyramided, either on different recipient chromosomes, or stacked in a single, complex portion, as in the wheat-alien recombinant lines described above (see § 2 and 3). Also, in the latter cases, the characterization and selection phases were helped by the use of appropriate analytical tools.

In all of the cases, the common objective of any interspecific, breeding-addressed transfer strategy is to exploit at best the alien donor trait(s), while minimizing possible linkage drag, which in principle represents a significant deterrent to using wild species in many crop improvement programs. To achieve this, a deeper knowledge of the alien genetic material, as well as the availability of an array of different transfer products for a given project objective, are essential pre-requisites. For instance, in the recombinant chromosomes in which alleles for various traits originating from different *Thinopyrum* species have been combined into bread wheat (Figure 3), knowledge of the differential effect of grain pigment content ascribable to the *Th. elongatum* and *Th. ponticum Psy1* alleles [43] has been critical for the choice of the best candidate genotype for bread wheat or durum wheat improvement (see § 3). On the other hand, unexpected outcomes may result from the introgression of alien genetic material into a cultivated background. Effects that are not detected in the donor species, but may become manifested when its genome, or a portion of it, is inserted in the new, nuclear or chromosomal, "environment", include the SD/Gc phenotypes above described (see § 4), and also positive effects on yield, as those conferred by *Th. ponticum* genes/QTL present in $7el_1L$ segments incorporated into the bread wheat translocation line T4 or various durum wheat recombinant lines ([29,30] and references therein). Indeed, even *Gc* genes may have positive effects in a breeding perspective. On one hand, the preferential transmission of chromosomes with a gametocidal locus (*Gc* chromosomes), at least for those *Gc* genes with "mild" or no detrimental effects on the plant phenotype (see § 4), ensures maintenance through subsequent generations of traits naturally, or even on purpose (e.g., by a GM approach), linked to the *Gc* locus, avoiding the need for trait selection. On the other hand, in several instances the ability of *Gc* genes to cause chromosomal breakages has been exploited to induce translocations into wheat of chromosomal segments belonging to relatively distant relatives, which are recalcitrant to homoeologous pairing promoting systems ([51,55,70] and references therein).

It seems thus clear that a comprehensive evaluation of the materials under study, including efficient genotyping and phenotyping, offers good promise for the practical use of alien genetic material, and of CWRs in particular. As suggested for the "introgressiomics" approach [85], any technology-assisted chromosome engineering of a crop genome with alien introgressions may well fit the definition that is given to 'new plant breeding techniques', such as cisgenesis and genome editing [88,89]. Both of the approaches allow to overcome interspecific barriers and eliminate CWRs-associated linkage drag; however, in contrast to the randomness of the insertion using a cisgenic gene delivery, genome editing techniques allow to edit, delete, replace, or insert, in targeted sites, specific genomic sequences of interest, even addressing multiple genes simultaneously [90,91]. Thus, genome editing, particularly through the highly versatile CRISP/Cas approach or additional future technological refinements, is predicted to revolutionize plant breeding. However, both cisgenesis and genome editing require knowledge of the genetic basis of the trait(s) of interest, still lacking or incomplete for several species of agricultural importance, particularly the CWRs. Moreover, legal uncertainty and the still widespread consumer concern about these technologies restrict their current usefulness. A wise reasoning would probably be that, while awaiting for wider genomic information on at least the most relevant species and accessions of CWRs to come, and legal/social bottlenecks for the implementation of high-precision biotechnological approaches, particularly genome editing, into the breeding pipeline to be overcome, the available "smart" strategies of chromosome engineering, whether being focused or initially unfocused, should be taken advantage of and valorized. As the examples that are reported in the present review hopefully contribute to demonstrate, they prove to be "ripe" for an effective "rewilding" [92] of our main crops, with wheat being one of the most significant, hence targeted, species. To this aim, investments in pre-breeding programs and human capacity development are highly recommended [84]. The latter issue is particularly addressed to counter the largely reported decline in available expertise in certain fields, including cytogenetics, botany, and taxonomy, which could effectively complement skill in recent technologies in harnessing at our best the treasure contained in CWRs. No doubt, such a holistic approach offers the highest potential for developing new generations of crops, with enhanced capacity and plasticity to perform well in changing and challenging environmental, agricultural, and societal scenarios.

Acknowledgments: Financial support from Lazio region—FILAS project "MIGLIORA" is gratefully acknowledged.

Author Contributions: Ljiljana Kuzmanović, Paola Forte, Alessia Cuccurullo and Alessandra Bitti performed the molecular marker-assisted selection and GISH analyses of the various wheat-alien recombinant lines; Roberto Ruggeri and Francesco Rossini were responsible for field tests; Ljiljana Kuzmanović, Roberto Ruggeri and Francesco Rossini performed the phenotyping of yield-related traits and analyzed the data; Ljiljana Kuzmanović, Roberto Ruggeri and Francesco Rossini contributed to the layout and editing of the manuscript; Carla Ceoloni coordinated the research and wrote the manuscript.

Conflicts of Interest: The authors declare no conflict of interest.

References

1. Statista—The Statistical Portal: Global Grain Acreage by Type—2011/12 to 2016/17. Available online: https://www.statista.com/statistics/272536/acreage-of-grain-worldwide-by-type/ (accessed on 30 July 2017).
2. Giorgi, B.; Porfiri, O. (Eds.) The Varieties of Strampelli: A Milestone in Wheat Breeding in Italy and in the World. In Proceedings of the Special Session of the 41st Annual Congress of the Italian Society of Agricultural Genetics (SIGA), Abbadia di Fiastra, Tolentino, Italy, 26 September 1997; Tipografia San Giuseppe: Pollenza, Italy, 1998; pp. 1–63, (In Italian with English Abstracts).
3. Borlaug, N.E. Sixty-two years of fighting hunger: Personal recollections. *Euphytica* **2007**, *157*, 287–297. [CrossRef]
4. Calderini, D.F.; Slafer, G.A. Changes in yield and yield stability in wheat during the 20th century. *Field Crops Res.* **1998**, *57*, 335–347. [CrossRef]
5. De Vita, P.; Nicosia, O.L.D.; Nigro, F.; Platani, C.; Riefolo, C.; Di Fonzo, N.; Cattivelli, L. Breeding progress in morpho-physiological, agronomical and qualitative traits of durum wheat cultivars released in Italy during the 20th century. *Eur. J. Agron.* **2007**, *26*, 39–53. [CrossRef]
6. Fischer, T.; Byerlee, D.; Edmeades, G. *Crop Yields and Global Food Security: Will Yield Increase Continue to Feed the World?* ACIAR Monograph No. 158; Australian Centre for International Agricultural Research: Canberra, Australia, 2014; p. xxii + 634.
7. Lobell, D.B.; Gourdji, S.M. The influence of climate change on global crop productivity. *Plant Physiol.* **2012**, *160*, 1686–1697. [CrossRef] [PubMed]
8. Tilman, D.; Balzer, C.; Hill, J.; Befort, B.L. Global food demand and the sustainable intensification of agriculture. *Proc. Natl. Acad. Sci. USA* **2011**, *108*, 20260–20264. [CrossRef] [PubMed]
9. Hawkesford, M.; Araus, J.; Park, R.; Calderini, D.; Miralles, D.; Shen, T.; Zhang, J.; Parry, M.A.J. Prospects of doubling global wheat yields. *Food Energy Secur.* **2013**, *2*, 34–48. [CrossRef]
10. Mujeeb-Kazi, A.; Kazi, A.G.; Dundas, I.; Rasheed, A.; Ogbonnaya, F.; Kishii, M.; Bonnett, D.; Wang, R.R.-C.; Xu, S.; Chen, P.; et al. Genetic diversity for wheat improvement as a conduit to food security. In *Advances in Agronomy*; Sparks, D.L., Ed.; Academic Press: Burlington, NI, USA, 2013; Volume 122, pp. 179–257.
11. Molnár-Láng, M.; Ceoloni, C.; Doležel, J. (Eds.) *Alien Introgression in Wheat—Cytogenetics, Molecular Biology, and Genomics*; Springer: Cham, Switzerland, 2015; pp. 1–385. [CrossRef]
12. Sears, E.R. Transfer of alien genetic material to wheat. In *Wheat Science-Today and Tomorrow*; Evans, L.T., Peacock, W.J., Eds.; Cambridge University Press: Cambridge, UK, 1981; pp. 75–89.
13. Naranjo, T.; Benavente, E. The mode and regulation of chromosome pairing in wheat-alien hybrids (*Ph* genes, an updated view). In *Alien Introgression in Wheat—Cytogenetics, Molecular Biology, and Genomics*; Molnár-Láng, M., Ceoloni, C., Doležel, J., Eds.; Springer: Cham, Switzerland, 2015; pp. 133–162.
14. Ceoloni, C.; Biagetti, M.; Ciaffi, M.; Forte, P.; Pasquini, M. Wheat chromosome engineering at the 4x level: The potential of different alien gene transfers into durum wheat. *Euphytica* **1996**, *89*, 87–97. [CrossRef]
15. Ceoloni, C.; Jauhar, P.P. Chromosome engineering of the durum wheat genome: Strategies and applications of potential breeding value. In *Genetic Resources, Chromosome Engineering, and Crop Improvement: Cereals*; Singh, R.J., Jauhar, P.P., Eds.; CRC Press: Boca Raton, FL, USA, 2006; pp. 27–59.
16. Qi, L.; Friebe, B.; Zhang, P.; Gill, B.S. Homoeologous recombination, chromosome engineering and crop improvement. *Chromosome Res.* **2007**, *15*, 3–19. [CrossRef] [PubMed]
17. Ayala-Navarrete, L.; Mechanicos, A.A.; Gibson, J.M.; Singh, D.; Bariana, H.S.; Fletcher, J.; Shorter, S.; Larkin, P.J. The *Pontin* series of recombinant alien translocations in bread wheat: Single translocations integrating combinations of *Bdv2*, *Lr19* and *Sr25* disease-resistance genes from *Thinopyrum intermedium* and *Th. ponticum*. *Theor. Appl. Genet.* **2013**, *126*, 2467–2475. [CrossRef] [PubMed]
18. Niu, Z.; Klindworth, D.L.; Friesen, T.L.; Chao, S.; Ohm, J.B.; Xu, S.S. Development and characterization of wheat lines carrying stem rust resistance gene *Sr43* derived from *Thinopyrum ponticum*. *Theor. Appl. Genet.* **2014**, *127*, 969–980. [CrossRef] [PubMed]
19. Danilova, T.V.; Zhang, G.; Liu, W.; Friebe, B.; Gill, B.S. Homoeologous recombination-based transfer and molecular cytogenetic mapping of a wheat streak mosaic virus and *Triticum* mosaic virus resistance gene *Wsm3* from *Thinopyrum intermedium*. *Theor. Appl. Genet.* **2017**, *130*, 549–555. [CrossRef] [PubMed]

20. Liu, W.; Koo, D.-H.; Xia, Q.; Li, C.; Bai, F.; Song, Y.; Friebe, B.; Gill, B.S. Homoeologous recombination-based transfer and molecular cytogenetic mapping of powdery mildew-resistant gene *Pm57* from *Aegilops searsii* into wheat. *Theor. Appl. Genet.* **2017**, *130*, 841–848. [CrossRef] [PubMed]

21. Ceoloni, C.; Pasquini, M.; Simeone, R. The cytogenetic contribution to the analysis and manipulation of the durum wheat genome. In *Durum Wheat Breeding: Current Approaches and Future Strategies*; Royo, C., Nachit, M.N., Di Fonzo, N., Araus, J.L., Pfeiffer, W.H., Slafer, G.A., Eds.; Haworth Press: New York, NY, USA, 2005; pp. 165–196.

22. Ceoloni, C.; Forte, P.; Gennaro, A.; Micali, S.; Carozza, R.; Bitti, A. Recent developments in durum wheat chromosome engineering. *Cytogenet. Genome Res.* **2005**, *109*, 328–344. [CrossRef] [PubMed]

23. Klindworth, D.L.; Niu, Z.; Chao, S.; Friesen, T.L.; Jin, Y.; Faris, J.D.; Cai, X.; Xu, S.S. Introgression and characterization of a goatgrass gene for a high level of resistance to Ug99 stem rust in tetraploid wheat. *G3 Genes Genomes Genet.* **2012**, *2*, 665–673. [CrossRef] [PubMed]

24. Han, C.; Zhang, P.; Ryan, P.R.; Rathjen, T.M.; Yan; Z.H.; Delhaize, E. Introgression of genes from bread wheat enhances the aluminium tolerance of durum wheat. *Theor. Appl. Genet.* **2016**, *129*, 729–739. [CrossRef] [PubMed]

25. Gennaro, A.; Forte, P.; Carozza, R.; Savo Sardaro, M.L.; Ferri, D.; Bitti, A.; Borrelli, G.M.; D'Egidio, M.G.; Ceoloni, C. Pyramiding different alien chromosome segments in durum wheat: Feasibility and breeding potential. *Isr. J. Plant Sci.* **2007**, *55*, 267–276. [CrossRef]

26. Gennaro, A.; Forte, P.; Panichi, D.; Lafiandra, D.; Pagnotta, M.A.; D'Egidio, M.G.; Ceoloni, C. Stacking small segments of the 1D chromosome of bread wheat containing major gluten quality genes into durum wheat: Transfer strategy and breeding prospects. *Mol. Breed.* **2012**, *30*, 149–167. [CrossRef]

27. Morris, C.F.; Simeone, M.N.; King, G.E.; Lafiandra, D. Transfer of soft kernel texture from *Triticum aestivum* to durum wheat, *Triticum turgidum* ssp. *durum. Crop Sci.* **2011**, *51*, 114–122. [CrossRef]

28. Gazza, L.; Sgrulletta, D.; Cammerata, A.; Gazzelloni, G.; Galassi, E.; Pogna, N. Breeding and quality of soft-textured durum wheat. In Proceedings of the International Symposium on Genetics and Breeding of Durum Wheat, Rome, Italy, 27–30 May 2013; Options Méditerranéennes, Séries A; No. 110. Porceddu, E., Damania, A.B., Qualset, C.O., Eds.; 2014; pp. 511–516.

29. Kuzmanović, L.; Gennaro, A.; Benedettelli, S.; Dodd, I.C.; Quarrie, S.A.; Ceoloni, C. Structural-functional dissection and characterization of yield-contributing traits originating from a group 7 chromosome of the wheatgrass species *Thinopyrum ponticum* after transfer into durum wheat. *J. Exp. Bot.* **2014**, *65*, 509–525. [CrossRef] [PubMed]

30. Kuzmanović, L.; Ruggeri, R.; Virili, M.E.; Rossini, F.; Ceoloni, C. Effects of *Thinopyrum ponticum* chromosome segments transferred into durum wheat on yield components and related morpho-physiological traits in Mediterranean rain-fed conditions. *Field Crops Res.* **2016**, *186*, 86–98. [CrossRef]

31. Ceoloni, C.; Margiotta, B.; Colaprico, G.; D'Egidio, M.G.; Carozza, R.; Lafiandra, D. Introgression of D-genome associated gluten protein genes into durum wheat. In Proceedings of the 10th International Wheat Genetics Symposium, Paestum, Italy, 1–6 September 2003; S.I.M.I: Rome, Italy, 2003; pp. 1320–1322.

32. Micali, S.; Forte, P.; Bitti, A.; D'Ovidio, R.; Ceoloni, C. Chromosome engineering as a tool for effectively introgressing multiple useful genes from alien Triticeae into durum wheat. In Proceedings of the 10th International Wheat Genetics Symposium, Paestum, Italy, 1–6 September 2003; pp. 896–898.

33. Gennaro, A.; Koebner, R.M.D.; Ceoloni, C. A candidate for *Lr19*, an exotic gene conditioning leaf rust resistance in wheat. *Funct. Integr. Genom.* **2009**, *9*, 325–334. [CrossRef] [PubMed]

34. Cenci, A.; D'Ovidio, R.; Tanzarella, O.A.; Ceoloni, C.; Porceddu, E. Identification of molecular markers linked to *Pm13*, an *Aegilops longissima* gene conferring resistance to powdery mildew of wheat. *Theor. Appl. Genet.* **1999**, *98*, 448–454. [CrossRef]

35. Klindworth, D.L.; Hareland, G.A.; Elias, E.M.; Xu, S.S. Attempted compensation for linkage drag affecting agronomic characteristics of durum wheat 1AS/1DL translocation lines. *Crop Sci.* **2013**, *53*, 422–429. [CrossRef]

36. Ceoloni, C.; Kuzmanović, L.; Gennaro, A.; Forte, P.; Giorgi, D.; Grossi, M.R.; Bitti, A. Genomes, chromosomes and genes of perennial Triticeae of the genus *Thinopyrum*: The value of their transfer into wheat for gains in cytogenomic knowledge and 'precision' breeding. In *Advances in Genomics of Plant Genetic Resources*; Tuberosa, R., Graner, A., Frison, E., Eds.; Springer: Dordrecht, The Netherlands, 2014; pp. 333–358.

37. Ceoloni, C.; Kuzmanović, L.; Forte, P.; Virili, M.E.; Bitti, A. Wheat-perennial Triticeae introgressions: Major achievements and prospects. In *Alien Introgression in Wheat—Cytogenetics, Molecular Biology, and Genomics*; Molnár-Láng, M., Ceoloni, C., Doležel, J., Eds.; Springer: Cham, Switzerland, 2015; pp. 273–313.

38. Ayala-Navarrete, L.; Bariana, H.S.; Singh, R.P.; Gibson, J.M.; Mechanicos, A.A.; Larkin, P.J. Trigenomic chromosomes by recombination of *Thinopyrum intermedium* and *Th. ponticum* translocations in wheat. *Theor. Appl. Genet.* **2007**, *116*, 63–75. [CrossRef] [PubMed]

39. Shen, X.; Ohm, H. Molecular mapping of *Thinopyrum*-derived Fusarium head blight resistance in common wheat. *Mol. Breed.* **2007**, *20*, 131–140. [CrossRef]

40. Zhang, X.L.; Shen, X.R.; Hao, Y.F.; Cai, J.J.; Ohm, H.W.; Kong, L. A genetic map of *Lophopyrum ponticum* chromosome 7E, harboring resistance genes to Fusarium head blight and leaf rust. *Theor. Appl. Genet.* **2011**, *122*, 263–270. [CrossRef] [PubMed]

41. Forte, P.; Virili, M.E.; Kuzmanović, L.; Moscetti, I.; Gennaro, A.; D'Ovidio, R.; Ceoloni, C. A novel assembly of *Thinopyrum ponticum* genes into the durum wheat genome: Pyramiding Fusarium head blight resistance onto recombinant lines previously engineered for other beneficial traits from the same alien species. *Mol. Breed.* **2014**, *34*, 1701–1716. [CrossRef]

42. Guo, J.; Zhang, X.; Hou, Y.; Cai, J.; Shen, X.; Zhou, T.; Xu, H.; Ohm, H.W.; Wang, H.; Li, A.; et al. High-density mapping of the major FHB resistance gene *Fhb7* derived from *Thinopyrum ponticum* and its pyramiding with *Fhb1* by marker-assisted selection. *Theor. Appl. Genet.* **2015**, *128*, 2301–2316. [CrossRef] [PubMed]

43. Ceoloni, C.; Forte, P.; Kuzmanović, L.; Tundo, S.; Moscetti, I.; De Vita, P.; Virili, M.E.; D'Ovidio, R. Cytogenetic mapping of a major locus for resistance to Fusarium head blight and crown rot of wheat on *Thinopyrum elongatum* 7EL and its pyramiding with valuable genes from a *Th. ponticum* homoeologous arm onto bread wheat 7DL. *Theor. Appl. Genet.* **2017**, *130*, 2005–2024. [CrossRef] [PubMed]

44. Lyttle, T.W. Segregation distorters. *Ann. Rev. Genet.* **1991**, *25*, 511–557. [CrossRef] [PubMed]

45. Hurst, G.D.D.; Werren, J.H. The role of selfish genetic elements in eukaryotic evolution. *Nat. Rev. Genet.* **2001**, *2*, 597–606. [CrossRef] [PubMed]

46. Taylor, D.R.; Ingvarsson, P.K. Common features of segregation distortion in plants and animals. *Genetica* **2003**, *117*, 27–35. [CrossRef] [PubMed]

47. Liu, X.; Guo, L.; You, J.; Liu, X.; He, Y.; Yuan, J.; Liu, G.; Feng, Z. Progress of segregation distortion in genetic mapping of plants. *Res. J. Agron.* **2010**, *4*, 78–83. [CrossRef]

48. Burt, A.; Trivers, R. Selfish Genetic Elements. In *Genes in Conflict—The Biology of Selfish Genetic Elements*; Belknap Press: Cambridge, UK, 2006; pp. 1–18.

49. Li, C.; Bai, G.; Chao, S.; Wang, Z. A High-density SNP and SSR consensus map reveals segregation distortion regions in wheat. *BioMed Res. Int.* **2015**, *2015*, 830618. [CrossRef] [PubMed]

50. Seymour, D.K.; Chae, E.; Ariöz, B.I.; Koenig, D.; Weigel, D. The genetic architecture of recurrent segregation distortion in *Arabidopsis thaliana*. *bioRxiv* **2017**. [CrossRef]

51. Endo, T.R. The gametocidal chromosome as a tool for chromosome manipulation in wheat. *Chromosome Res.* **2007**, *15*, 67–75. [CrossRef] [PubMed]

52. Habu, Y.; Ando, T.; Ito, S.; Nagaki, K.; Kishimoto, N.; Taguchi-Shiobara, F.; Numa, H.; Yamaguchi, K.; Shigenobu, S.; Murata, M.; et al. Epigenomic modification in rice controls meiotic recombination and segregation distortion. *Mol. Breed.* **2015**, *35*, 103. [CrossRef]

53. Kwiatek, M.T.; Wisniewska, H.; Slusarkiewicz-Jarzina, A.; Majka, J.; Majka, M.; Belter, J.; Pudelska, H. Gametocidal factor transferred from *Aegilops geniculata* Roth can be adapted for large-scale chromosome manipulations in cereals. *Front. Plant Sci.* **2017**, *8*, 409. [CrossRef] [PubMed]

54. Tsujimoto, H. Gametocidal genes in wheat as the inducer of chromosome breakage. In *Frontiers of Wheat Bioscience*; Memorial Issue, Wheat Information Service No. 100; Tsunewaki, K., Ed.; Kihara Memorial Yokohama Foundation: Yokohama, Japan, 2005; pp. 33–48.

55. Endo, T.R. Gametocidal genes. In *Alien Introgression in Wheat—Cytogenetics, Molecular Biology, and Genomics*; Molnár-Láng, M., Ceoloni, C., Doležel, J., Eds.; Springer: Cham, Switzerland, 2015; pp. 121–131.

56. Knott, D.R. The transfer of genes for disease resistance from alien species to wheat by induced translocations. In *Mutation Breeding for Disease Resistance*; IAEA: Vienna, Austria, 1971; pp. 67–77.

57. McIntosh, R.A.; Dyck, P.L.; Green, G.J. Inheritance of leaf rust and stem rust resistance in wheat cultivars Agent and Agatha. *Aust. J. Agric. Res.* **1976**, *28*, 37–45. [CrossRef]

58. Kibirige-Sebunya, I.; Knott, D.R. Transfer of stem rust resistance to wheat from an *Agropyron* chromosome having a gametocidal effect. *Can. J. Genet. Cytol.* **1983**, *25*, 215–221. [CrossRef]

59. Scoles, G.J.; Kibirige-Sebunya, I.N. Preferential abortion of gametes in wheat induced by an *Agropyron* chromosome. *Can. J. Genet. Cytol.* **1983**, *25*, 1–6. [CrossRef]

60. Marais, G.F. Genetic control of a response to the segregation distortion allele, *Sd-1d*, in the common wheat line 'Indis'. *Euphytica* **1992**, *60*, 89–95.

61. Prins, R.; Marais, G.F. A genetic study of the gametocidal effect of the *Lr19* translocation of common wheat. *S. Afr. J. Plant Soil* **1999**, *16*, 10–14. [CrossRef]

62. Marais, G.F.; Marais, A.S.; Groenwald, J.Z. Evaluation and reduction of *Lr19-149*, a recombined form of the *Lr19* translocation of wheat. *Euphytica* **2001**, *121*, 289–295. [CrossRef]

63. Groenwald, J.Z.; Fourie, M.; Marais, A.S.; Marais, G.F. Extention and use of a physical map of the *Thinopyrum*-derived *Lr19* translocation. *Theor. Appl. Genet.* **2005**, *112*, 131–138. [CrossRef] [PubMed]

64. Cai, J.; Zhang, X.; Wang, B.; Yan, M.; Qi, Y.; Kong, L. A genetic analysis of segregation distortion revealed by molecular markers in *Lophopyrum ponticum* chromosome 7E. *J. Genet.* **2011**, *90*, 373–376. [CrossRef] [PubMed]

65. Endo, T.R. Gametocidal chromosomes and their induction of chromosome mutations in wheat. *Jpn. J. Genet.* **1990**, *65*, 135–152. [CrossRef]

66. Kong, L.; Anderson, J.M.; Ohm, H.W. Segregation distortion in common wheat of a segment of *Thinopyrum intermedium* chromosome 7E carrying *Bdv3* and development of a *Bdv3* marker. *Plant Breed.* **2008**, *128*, 591–597. [CrossRef]

67. Sun, Q.Y.; Breitbart, H.; Schatten, H. Role of the MAPK cascade in mammalian germ cells. *Reprod. Fertil. Dev.* **2000**, *11*, 443–450. [CrossRef]

68. Xu, J.; Zhang, S. Mitogen-activated protein kinase cascades in signaling plant growth and development. *Trends Plant Sci.* **2015**, *20*, 56–64. [CrossRef] [PubMed]

69. Knight, E.; Binnie, A.; Draeger, T.; Moscou, M.; Rey, M.-D.; Sucher, J.; Mehra, S.; King, I.; Moore, G. Mapping the 'breaker' element of the gametocidal locus proximal to a block of sub-telomeric heterochromatin on the long arm of chromosome 4S^{sh} of *Aegilops sharonensis*. *Theor. Appl. Genet.* **2015**, *128*, 1049–1059. [CrossRef] [PubMed]

70. Grewal, S.; Gardiner, L-J.; Ndreca, B.; Knight, E.; Moore, G.; King, I.P.; King, J. Comparative mapping and targeted-capture sequencing of the gametocidal loci in *Aegilops sharonensis*. *Plant Genome* **2017**, *10*. [CrossRef] [PubMed]

71. Tsujimoto, H.; Tsunewaki, K. Hybrid dysgenesis in common wheat caused by gametocidal genes. *Jpn. J. Genet.* **1985**, *60*, 565–578. [CrossRef]

72. McClintock, B. The significance of responses of the genome to challenge. *Science* **1984**, *226*, 792–801. [CrossRef] [PubMed]

73. Shaked, H.; Kashkush, K.; Ozkan, H.; Feldman, M.; Levy, A.A. Sequence elimination and cytosine methylation are rapid and reproducible responses of the genome to wide hybridization and allopolyploidy in wheat. *Plant Cell* **2001**, *13*, 1749–1759. [CrossRef] [PubMed]

74. Feldman, M.; Levy, A.A. Genome evolution in allopolyploid wheat—A revolutionary reprogramming followed by gradual changes. *J. Genet. Genom.* **2009**, *36*, 511–518. [CrossRef]

75. Levy, A.A. Transposons in plant speciation. In *Plant Transposons and Genome Dynamics in Evolution*, 1st ed.; Fedoroff, N.V., Ed.; John Wiley & Sons, Inc.: Hoboken, NJ, USA, 2013; pp. 165–179.

76. Liu, Z.; Wang, Y.; Shen, Y.; Guo, W.; Hao, S.; Liu, B. Extensive alterations in DNA methylation and transcription in rice caused by introgression from *Zizania latifolia*. *Plant Mol. Biol.* **2004**, *54*, 571–582. [CrossRef] [PubMed]

77. Dong, Z.Y.; Wang, Y.M.; Zhang, Z.J.; Shen, Y.; Lin, X.Y.; Ou, X.F.; Han, F.P.; Liu, B. Extent and pattern of DNA methylation alteration in rice lines derived from introgressive hybridization of rice and *Zizania latifolia* Griseb. *Theor. Appl. Genet.* **2006**, *113*, 196–205. [CrossRef] [PubMed]

78. Zhang, Y.; Liu, Z.H.; Liu, C.; Yang, Z.J.; Deng, K.J.; Peng, J.H.; Zhou, J.P.; Li, G.R.; Tang, Z.X.; Ren, Z.L. Analysis of DNA methylation variation in wheat genetic background after alien chromatin introduction based on methylation-sensitive amplification polymorphism. *Sci. Bull.* **2008**, *53*, 58–69. [CrossRef]

79. Fu, S.; Sun, C.; Yang, M.; Fei, Y.; Tan, F.; Yan, B.; Ren, Z.; Tang, Z. Genetic and epigenetic variation induced by wheat-rye 2R and 5R monosomic addition lines. *PLoS ONE* **2013**, *8*, e54057. [CrossRef] [PubMed]

80. Dvorak, J. Homoeology between *Agropyron elongatum* chromosomes and *Triticum aestivum* chromosomes. *Can. J. Genet. Cytol.* **1980**, *22*, 237–259. [CrossRef]

81. Marais, G.F.; Bekker, T.A.; Eksteen, A.; McCallum, B.; Fetch, T.; Marais, A.S. Attempts to remove gametocidal genes co-transferred to common wheat with rust resistance from *Aegilops speltoides*. *Euphytica* **2010**, *171*, 71–85. [CrossRef]

82. Millet, E.; Manisterski, J.; Ben-Yehuda, P.; Distelfeld, A.; Deek, J.; Wan, A.; Chen, X.; Steffenson, B.J. Introgression of leaf rust and stripe rust resistance from Sharon goatgrass (*Aegilops sharonensis* Eig) into bread wheat (*Triticum aestivum* L.). *Genome* **2014**, *57*, 309–316. [CrossRef] [PubMed]

83. Friebe, B.; Zhang, P.; Nasuda, S.; Gill, B.S. Characterization of a knockout mutation at the *Gc2* locus in wheat. *Chromosoma* **2003**, *111*, 509–517. [CrossRef] [PubMed]

84. Dempewolf, H.; Baute, G.; Anderson, J.; Kilian, B.; Smith, C.; Guarino, L. Past and future use of wild relatives in crop breeding. *Crop Sci.* **2017**, *57*, 1070–1082. [CrossRef]

85. Prohens, J.; Gramazio, P.; Plazas, M.; Dempewolf, H.; Kilian, B.; Díez, M.J.; Fita, A.; Herraiz, F.J.; Rodríguez-Burruezo, A.; Soler, S.; et al. Introgressiomics: A new approach for using crop wild relatives in breeding for adaptation to climate change. *Euphytica* **2017**, *213*, 158. [CrossRef]

86. King, J.; Grewal, S.; Yang, C.; Hubbart, S.; Scholefield, D.; Ashling, S.; Edwards, K.J.; Allen, A.M.; Burridge, A.; Bloor, C.; et al. A step change in the transfer of interspecific variation into wheat from *Amblyopyrum muticum*. *Plant Biotechnol. J.* **2017**, *15*, 217–226. [CrossRef] [PubMed]

87. Winfield, M.O.; Allen, A.M.; Burridge, A.J.; Barker, G.L.A.; Benbow, H.R.; Wilkinson, P.A.; Coghill, J.; Waterfall, C.; Davassi, A.; Scopes, G.; et al. High-density SNP genotyping array for hexaploid wheat and its secondary and tertiary gene pool. *Plant Biotechnol. J.* **2015**, *13*, 733–742. [CrossRef] [PubMed]

88. Lusser, M.; Parisi, C.; Plan, D.; Rodríguez-Cerezo, E. *New Plant Breeding Techniques: State-of-the-Art and Prospects for Commercial Development*; Reference Report by the Joint Research Centre of the European Commission; Publications Office of the European Union: Luxembourg, 2011. [CrossRef]

89. Cardi, T. Cisgenesis and genome editing: Combining concepts and efforts for a smarter use of genetic resources in crop breeding. *Plant Breed.* **2016**, *135*, 139–147. [CrossRef]

90. Lowder, L.G.; Zhang, D.; Baltes, N.J.; Paul, J.W.; Tang, X.; Zheng, X.; Voytas, D.F.; Hsieh, T.F.; Zhang, Y.; Qi, Y. A CRISPR/Cas9 toolbox for multiplexed plant genome editing and transcriptional regulation. *Plant Physiol.* **2015**, *169*, 971–985. [CrossRef] [PubMed]

91. Ma, X.; Zhang, Q.; Zhu, Q.; Liu, W.; Chen, Y.; Qiu, R.; Wang, B.; Yang, Z.; Li, H.; Lin, Y.; et al. A robust CRISPR/Cas9 system for convenient, high-efficiency multiplex genome editing in monocot and dicot plants. *Mol. Plant* **2015**, *8*, 1274–1284. [CrossRef] [PubMed]

92. Palmgren, M.G.; Edenbrandt, A.K.; Vedel, S.E.; Andersen, M.M.; Landes, X.; Osterberg, J.T.; Falhof, J.; Olsen, L.I.; Christensen, S.B.; Sandoe, P.; et al. Are we ready for back-to-nature crop breeding? *Trends Plant Sci.* **2015**, *20*, 155–164. [CrossRef] [PubMed]

diversity

MDPI

Review

Barley Developmental Mutants: The High Road to Understand the Cereal Spike Morphology

Valeria Terzi [1,*], **Giorgio Tumino** [1], **Donata Pagani** [1], **Fulvia Rizza** [1], **Roberta Ghizzoni** [1], **Caterina Morcia** [1] and **Antonio Michele Stanca** [2]

[1] CREA—GB, Research Centre for Genomics and Bioinformatics, Fiorenzuola d'Arda 29017, Italy; giorgiotumino@hotmail.it (G.T.); donata.pagani@crea.gov.it (D.P.); fulvia.rizza@crea.gov.it (F.R.); roberta.ghizzoni@crea.gov.it (R.G.); caterina.morcia@crea.gov.it (C.M.)
[2] Department of Agricultural and Food Science, University of Modena and Reggio Emilia, Reggio Emilia 42122, Italy; michele@stanca.it
* Correspondence: valeria.terzi@crea.gov.it; Tel.: +39-0523-983758

Academic Editors: Rosa Rao and Giandomenico Corrado
Received: 23 February 2017; Accepted: 6 May 2017; Published: 11 May 2017

Abstract: A better understanding of the developmental plan of a cereal spike is of relevance when designing the plant for the future, in which innovative traits can be implemented through pre-breeding strategies. Barley developmental mutants can be a Mendelian solution for identifying genes controlling key steps in the establishment of the spike morphology. Among cereals, barley (*Hordeum vulgare* L.) is one of the best investigated crop plants and is a model species for the *Triticeae* tribe, thanks to several characteristics, including, among others, its adaptability to a wide range of environments, its diploid genome, and its self-pollinating mating system, as well as the availability of its genome sequence and a wide array of genomic resources. Among them, large collections of natural and induced mutants have been developed since the 1920s, with the aim of understanding developmental and physiological processes and exploiting mutation breeding in crop improvement. The collections are not only comprehensive in terms of single Mendelian spike mutants, but with regards to double and triple mutants derived from crosses between simple mutants, as well as near isogenic lines (NILs) that are useful for genetic studies. In recent years the integration of the most advanced omic technologies with historical mutation-genetics research has helped in the isolation and validation of some of the genes involved in spike development. New interrogatives have raised the question about how the behavior of a single developmental gene in different genetic backgrounds can help in understanding phenomena like expressivity, penetrance, phenotypic plasticity, and instability. In this paper, some genetic and epigenetic studies on this topic are reviewed.

Keywords: homeotic mutants; *Hordeum vulgare*; spike architecture; genomics

1. The Cereal Spike

A cereal spike is an important plant organ, being the single biggest source of food for humankind. This food and feed source must be further improved; the FAO indicates that a 50% increase of cereal production (from 2.1 to 3 billion tonnes) is needed to meet the demand of the increasing population [1]. This means that the grain number of cereal spikes must be improved in the near future, together with a biomass increase [2]. As discussed by Sreenivasulu and Schnurbusch [3], grain number enhancement can be theoretically obtained through modifications of the spike fertility and morphology. Due to the implications in the grain production and yield, the genetic dissection of the developmental plan of this storage sink is therefore of relevance when designing the cereal for the future. In this frame, collections of morphological barley mutants can help improve the understanding of the cereal spike development process. Some examples of the use of barley genetic resources to elucidate spike development will be reviewed.

2. Why Barley?

Among cereals, barley is a model organism from both a genetic and genomic point of view: barley is characterized by a high degree of natural variation and by its adaptability to several different cultivation environments. Several morphological and physiological forms have evolved, including winter, spring, two-rowed, six-rowed, awned, awnless, hooded, naked, and covered grain, malting, feed (grain and forage), and food types. Its diploid genome, whose sequence is available [4], and the self-pollinating mating system, together with the availability of genetic and genomic resources, make this plant a reference model. Barley is characterized by strong genetic variability for developmental traits and this characteristic is important in ensuring its broad adaptability.

3. Genetic Dissection of the Key Developmental Trait of Barley Spike

Historically, barley genetic studies have their foundations in Mendelian mutants, characterized by an altered physiology and/or morphology. Starting in the 1920s, thousands of different barley mutants—both natural or obtained via mutagenesis—have been collected world-wide and designed as Barley Genetic Stocks (BGS). The BGS lines are conserved in different gene banks and information about them can be found in the "Barley Genetic Stocks Database" [5] which is illustrated with images. These genetic resources provide one of the most efficient tools to study the developmental process of the spike, to identify the single genes involved, and to understand their regulation and interactions. The collections are not only comprehensive in terms of single Mendelian spike mutants, but with regards to double and triple mutants (Figure 1), derived from crosses between single mutants, as well as near isogenic lines (NILs), which are useful tools for genetic studies.

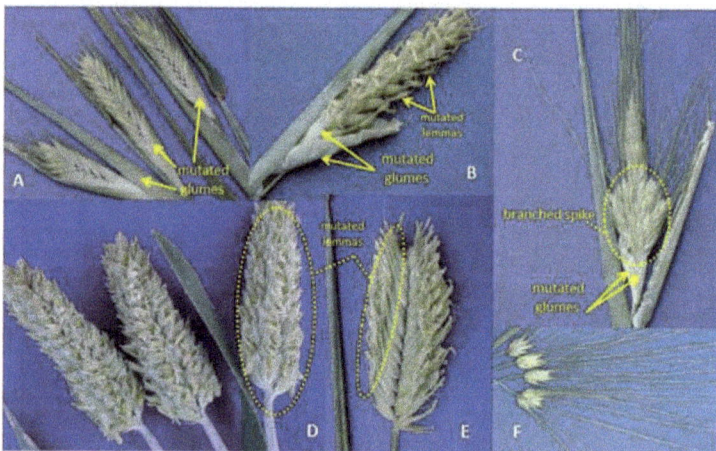

Figure 1. Phenotypes of some double mutants, obtained by crossing single ones: (**A**) = *Awnless/third outer glume:* the double mutant is characterized by the presence of a very large bract (third outer glume) subtending the lowest spikelet, followed by some large glumes; (**B**) = *Hooded/third outer glume:* the lemma develops a trifurcate structure, similar to a "hood", that includes a central deformed floret with two lateral wings. Moreover, large glumes are present at the spike basis; (**C**) = *Branched/third outer glume:* the spike is ramified and large glumes are present at its basis; (**D**) = *Hooded/low number of tillers;* (**E**) = *Hooded/wide outer glume:* the lemmas bear the hoods and the spikes are compact; and (**F**) = *Many glumes on the lateral/wide outer glume:* in the plant population derived from this cross, a mutant characterized by large glumes and "reduced number of internode rachis" (rnir) has been found.

In recent years, the integration of the most advanced omics technologies with historical mutation-genetics research has helped in the isolation and validation of some of the genes involved in spike development. Interestingly, some genes that have been known for a long time to be responsible for the mutant phenotype have recently been cloned and their functions have been elucidated [6]. Several of these genes are involved in mutations that, selected by early farmers, transformed wild plants into domesticated ones, representing an important contribution to the development of the ancient agrarian societies [7,8].

4. The Phytomer Model: Key to Understand the Ontogeny of Barley Inflorescence

The mature inflorescence of cultivated barley (spike or ear or head) consists of the floral stem (rachis) and floral units (spikelets). This indeterminate inflorescence is a raceme in which one central and two lateral spikelets are positioned in groups of three at each rachis internode. Each spikelet is enclosed and consists of a floret subtended by two bracts (outer glumes) that are the upper inner palea and the lower lemma. The top region of the lemma can bear the awn. The floret consists of a carpel with one single ovary, two styles with plumose stigmata, three stamens, and two lodicules [9].

Phytomeric models based on anatomical, histological, and genetic analyses have been proposed by Bossinger et al. [10]. More recently, Forster et al. [11] revised the Bossinger's model, starting from the visual analysis of developmental mutants—already described or induced in "Optic" barley. The disruptive effect of mutations on the metameric structure of this plant organ has been demonstrated to be a powerful tool for understanding a complex structure like the spike. The observations were made for mutants at different growth stages and a simpler explanation has been adopted for the interpretation of each mutant structure [11]. According to both models, central to the grass architecture is the presence and repetition of phytomer units, linked together by nodal structures. Each node is composed of two half nodes. Classically, two types of phytomers are reported: the "vegetative" type 1 phytomer and the "generative" type 2, a special structure present at branching. Additionally, the inflorescence has a phytomeric structure, even if this is particularly difficult to explain. According to Bossinger et al. [10], the restriction of the internode elongation of the rachis and rachilla is a characteristic of barley spike. The first two organs that are positioned on the rachilla axis are the subtending glumes, which Bossinger considered to be a unique organ, corresponding to a type 2 phytomer. The lemma is considered to be a leaf-like structure (type 1 phytomer), whereas palea belongs to a type 2 phytomer. Lodicules and stamens can be reconducted to variants of a type 1 phytomer, whereas the carpel is a special terminal type 1 phytomer. This model has been simplified by Forster et al. [11], who suggested that both vegetative and generative structures can be explained by a single repeating phytomer unit. The organs of the barley plant-and of the spike- can therefore be divided into two different types: single or paired. Specific organs can be derived from the fusion of paired structures. Central in the Forster's model, is therefore the concept that a plant organ can be the result of growth activation/suppression in specific regions of the phytomer, and even the result of the association of linked phytomers. Therefore, the plant development could be led by the positioning of phytomeric units and by switching their growth on and off.

5. The Brittle Rachis

The most important trait selected by humans during the cereal domestication process and related to the evolution of barley spike is the transformation of a brittle spike into a non-brittle spike. In wild spike, at maturity, the formation of "constriction grooves" that result in the disarticulation of each rachis node and the free dispersion of seeds can be seen (Figure 2).

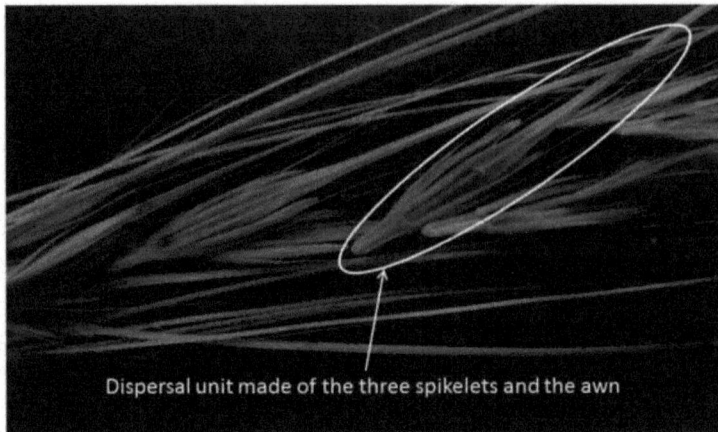

Dispersal unit made of the three spikelets and the awn

Figure 2. Close-up of a *Hordeum vulgare* spp. *spontaneum* spike disarticulated at maturity. In wild *Hordeum* species, the three spikelets and their slender awns form a light dispersal unit that permits both anemochory and zoochory.

The loss of this natural grain dispersal allows the quantity and quality of harvestable grain to increase. Using classical genetic approaches, two linked loci, mapped on barley chromosome 3H, *brittle rachis 1 (Btr1)* and *brittle rachis 2 (Btr2)*, are involved in the mutation from the brittle rachis into a non-brittle phenotype. The conversion is possible in the presence of the dominant alleles of both genes. In major details, non-brittle, domesticated barleys have a 1bp deletion in *Btr1* or an 11bp deletion in *Btr2*. Recently, molecular studies have better elucidated the evolution of this key domestication trait. The paper from Pourkheirandish et al. [12] is a milestone in this direction: the authors hypothesize that the anthropogenic selection operated in favor of the mutated forms of a signal transducing receptor and its protein ligand. The two gene products, BTR1 and BTR2, act together to control the cell wall thickening in the disarticulation zone of the rachis node, through molecular mechanisms that are not fully understood. Moreover, the authors, on the basis of both DNA sequences and archaeo-botanical data, demonstrated the independent origins of barley domestication. By tracing the evolutionary history of allelic variation in both genes, it can be concluded that the *Btr1*- and *Btr2*-type barleys emerged independently in different environments and at different times. According to genetic studies, two "transition zones" were found, characterized by a high level of changes between the *Btr1*- and *Btr2*-types: the area between Iran and Afghanistan, and the Levant and the Southern part of the Mediterranean Sea. According to the archaeological record, the cultivation of wild barley, before its domestication, was present in the Southern Levant area [8]. More recently, Civan and Brown [13] discovered a third type of non-brittle genotype, carrying the change of a leucine into a proline in the BTR1 aminoacid sequence.

Using a strategy based on the development of a high-resolution population, a new locus, *thresh-1*, present in *Hordeum spontaneum* and involved in the threshability phenotype, has been identified and mapped on chromosome 1H [14]. The candidate genes identified control the plant cell wall composition.

6. The Row Number

Triticeae inflorescence bears one-three spikelets with a single flower at each rachis internode. A barley spike is characterized by the development of one central and two lateral spikelets at each rachis internode. The row number of the barley spike depends on the fertility of the lateral spikelets. Six-rowed spikes have fertile lateral spikelets, whereas two-rowed spikes have sterile lateral spikelets.

Six-rowed genotypes produce more grains per spike, compared with two-rowed ones. The two-rowed state is ancestral, being found in the wild progenitor of cultivated barley (*Hordeum vulgare* ssp. *spontaneum*), where the sterile spikelets form part of the seed dispersal mechanism.

Up to now, five independent loci—i.e., *six-rowed spike1* (*vrs1*), *vrs2*, *vrs3*, *vrs4*, and *Intermedium spike-c* (*Int-c*)—have been identified as being involved in the six-rowed phenotype [15]. *Vrs1* encodes a homeodomain-leucine zipper class I transcription factor that is a negative regulator of lateral spikelet fertility [16], whereas *int-c* is an ortholog of the *TEOSINTE BRANCHED1* maize gene. It has been shown that the allelic combinations of both *vrs1* and *int-c* can modify lateral spikelet development. Six-rowed barleys generally bear loss-of-function *vrs1.a*, together with *Int-c.a*. On the contrary, two-rowed phenotypes have a functional *Vrs1.b* accompanied by *int-c.b*. In brief, more than ten different and independent *INT* genes have been identified that can influence the effect of the *Vrs1* locus, determining modulation in the size and fertility of lateral spikelets. Ramsay et al. [17] identified the primary function of *INT* loci as being related to the growth of the axillary organs of the plant. The other three *vrs* loci determine varying levels of lateral spikelet fertility. In particular, *vrs4* mutants show complete lateral spikelet fertility, as well as the possibility to produce additional spikelets and florets. Koppolu et al. [15] characterized *vrs4*, finding orthology with the maize transcription factor RAMOSA2, involved in inflorescence development in grasses. Expression studies indicated that HvRA2 is a central player in inflorescence development, being involved not only in the regulation of triple spikelet meristems determinacy, but also in the control of *Hordeum* specific row type determination. Moreover, *vrs4* regulates the expression of *vrs1*, and therefore, the *Hordeum* specific row type determination, which is either two- or six-rowed.

A third class of row-type is known as *Labile*-barley (*Hordeum vulgare* L. convar. *Labile* (Schiem.) Mansf.) and it was initially considered to be an irregular row-type of Abyssinian accession. The *Labile*-barley has a variable number of fertile spikelets at each rachis internode (zero to three fertile spikelets/rachis internode) and therefore, its phenotype is intermediate between a two- and a six-rowed type. A deep phenotypic description, using scanning electron microscopy, of spikelet fertility in *Labile*-barleys has been presented by Youssef et al. [18]. These authors observed, in *Labile*-barleys, some arrested central floral primordia during the stamen development. The re-sequencing of *vrs1* and *int-c* loci in 219 *Labile* accessions showed that these genotypes have a six-rowed genetic background, but reduced lateral spikelet fertility due to the recessive *labile* (*lab*) locus presence [18]. Recently, Helmy et al. [19] studied how *vrs2*, which encodes a SHORT INTERNODES (SHI) transcriptional regulator, contributes to barley inflorescence and shoot development. *Vrs2* in the floral organ positively regulates auxin (IAA) biosynthesis and indirectly (via IAA) negatively regulates both cytokinins and the conversion of bioactive gibberellic acid forms. There is a gradient of *vrs2* expression along the length of the inflorescence: the highest expression of this gene is detected at the base of the spike and decreases toward the apex. The same gradient is observed for the hormonal level. Therefore, this gene maintains auxin homeostasis and gradients during normal spike development. Moreover, *vrs2* transcripts are abundant at the tips of spikelet and floret primordia, suggesting a further role in the formation of axillary structures from the main shoot. In the *Vrs2.e* mutant, hormone patterns are disrupted, promoting the formation of six-rowed and supernumerary spikelets [20].

7. Branched Spike

A canonical barley spike has a branchless shape. However, mutants characterized by branched spikes (Figure 3) have been described as naturally occurring since ancient times.

A poly-row-and-branched spike (*prbs*) mutation has been described as being involved in the inflorescence differentiation from a panicle into a spike. This mutation can alter the inflorescence morphology in two ways: (a) determining the conversion of the rudimentary lateral spikelets specific to two-rowed genotypes into fertile spikelets; and (b) determining the development of additional spikelets in the middle of the spike, resulting in a branched spike. Shang et al. [21], starting from morphological observations of developing immature spikes of the mutant and descendants with branched spikes, showed that the *prbs* gene has a key role in the spikelet development at the triple-mound stage. In mutant *Prbs*, new meristems develop at the flanks of lateral spikelets and the meristems located in the middle spikelet were changed into branch meristems, from which branched spikes are formed. The same authors mapped the *prbs* gene to chromosome 3H and demonstrated that this gene is not allelic to *vrs4* [19]. *Vrs4* has also been found to be involved in another mutant phenotype derived from a particular development of the node. Poursarebani et al. [22] demonstrated that *vrs4* is involved in the regulation of *compositum2* expression, a gene found to be orthologous to the *branched head*^{t} (*bh*^{t}) locus regulating spike-branching in tetraploid 'Miracle-Wheat'.

Figure 3. Branched mutant spike.

The branching of the spike can have an epigenetic basis. On this subject, Brown and Bregitzer [23] observed that *Ds-miR172* mutants have an abnormal spikelets development plan, with the conversion of glumes to partially developed florets in apical regions of spikes. The spike basal region has an abnormal branching phenotype, as a result of the irregular development of spikelet meristems. Each branch is formed by multiple, abnormal spikelets and other floral organs, instead of a single spikelet. A similar phenotype was found in maize (*Ts4* mutant), and even in this genotype, the mutation affects an orthologous *miR172*.

8. Spike Density

The reduction or increase of the rachis internode length results in different spike densities (Figure 4). Classical genetic studies have identified several loci as being involved in the modulation of spike density, such as *dense spike*, *zeocriton*, *lax spike*, and *laxatum*. This is a recessive mutation (*lax-a*) with a pleiotropic effect: long rachis internode, large base of lemma awns, and the transformation of lodicules into two additional stamens. Consequently, the *Laxatum* mutant shows five anthers instead of the regular three [13,24]. A panel of major genes is involved in the control of spike density, as demonstrated by the genetic analysis of several morphological mutants. A class of such mutants controls, through *dense spike* (*dsp*) genes, the rachis internode length, resulting in dense or compact spikes. One of the *dsp* genes (*dsp.ar*) was mapped by high resolution bi-parental mapping to a 0.37 cM interval between markers SC57808 (*Hv_SPL14*)–CAPSK06413 residing on chromosome 7H. This region putatively hosts more than 800 genes, as deduced by a comparison with barley, rice, sorghum, and *Brachypodium* collinear regions [25]. However, the classical map-based cloning of the gene *dsp.ar* is complicated because of the unfavorable relationship between the genetic and physical distances at the target locus [23]. In the same position of *dsp.ar*, Taketa et al. [26] have mapped *dsp.1*. The same authors have positioned *lks.2* (short awns) on chromosome 7. Houston et al. [27]) exploited an allelic series of *Zeo* mutants and demonstrated that dense spikes are largely caused by polymorphisms in the microRNA172 (miR172)-binding site of the *HvAP2* gene. This is an ortholog of an *APETALA2* transcription factor. When examined in detail, the authors demonstrated that HvAP2 turnover driven by microRNA 172 regulates the length of a critical developmental window required for the elongation of the inflorescence internodes. In other words, the increase or decrease of this developmental temporal window can dramatically impact the spike density [27].

Figure 4. Different spike density: lax spike (**left**) and dense spike (**right**).

9. Cleistogamy

Cultivated barley is an autogamous plant. In the barley flower, the stigmas become receptive before anther extrusion and, when the anthers become ready for pollination, the stigmas are able to capture sufficient self-pollen without fertilization by windborne non-self pollen. On the contrary, in some wild accessions, the extrusion of the anthers is so pronounced that the rate of outcrossing is strongly increased in comparison with cultivated barley. There are natural variants of cultivated barley in which it closed flowering can be observed, due to the fact that palea and lemma remain tightly closed throughout the period of pollen release. Such a phenomenon is known as cleistogamy. Cleistogamous flowers typically have smaller lodicules in comparison with the non-cleistogamous types. A canonical barley floret is therefore strictly cleistogamous; however, mutants characterized by a variation in the cleistogamy level have been described and have been the starting point for the identification of the *cly1* gene. This is the same gene as the previously cited *HvAP2*, involved in the spike density phenotype. The cleistogamous state in barley is recessive and is under the control of a single gene at the *cleistogamy 1* (*cly1*) locus, which maps to the long arm of chromosome 2H. Nair et al. [28] have isolated *cly1* by positional cloning. This gene is a transcription factor containing two AP2 domains and a putative *miR172* targeting site, and codes for an AP2-protein that inhibits the development of the lodicule, avoiding the opening of the flower and the subsequent pollen dispersal and cross-pollination. *Cly1* gene expression is therefore epigenetically regulated. In non-cleistogamous barleys, *cly1* mRNA is subjected to a cleavage directed by miR172, resulting in low levels of the AP2 protein and floret opening. On the contrary, in cleistogamous genotypes, *cly1* mRNA is not cleaved and therefore, high levels of the AP2 protein result in the failure of lodicule swelling. However, there are genotypes in which alternative mechanisms of regulation can impact on the cleistogamy level. Wang et al. [29] showed that the cv. SV235 is cleistogamous, and in this genotype, the downregulation of *cly1* is unrelated to miR172-directed mRNA degradation, and is caused by an epiallele that represses transcription.

10. Lemma and Awn

The barley floret is protected by two leafy organs, the lemma and the palea, both considered to be reduced vegetative leaves. The upper part of the lemma forms the awn. Several mutations can perturbate the canonical development of these structures [30]. Among these, there is the dominant *Hooded* (*K*) mutation. In such mutants, a flower develops on the lemma instead of the awn (Figure 5).

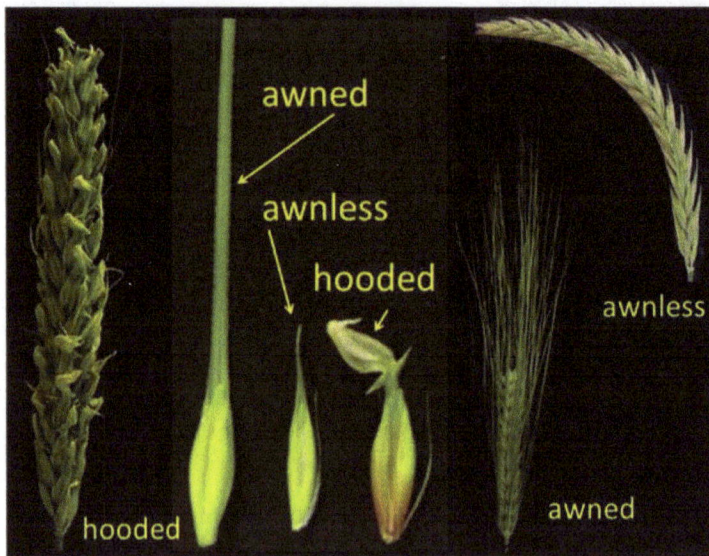

Figure 5. *Hooded* spike in comparison with awned and awnless ones.

A duplication of 305 bp in intron IV of the homeobox gene *Bkn3*—belonging to the knox family and well known to have a pivotal role in the development of leaf primordia- is responsible for the *Hooded* phenotype [31]. In a *Hooded* mutant, the *Bkn3* gene is overexpressed. Four proteins that bind the intron-located regulatory element (Kap intron-binding proteins) have been identified by Osnato et al. [32]. Two of these proteins, Barley Ethylene Response Factor1 (BERF1) and Barley Ethylene Insensitive Like1 (BEIL1), should mediate the fine-tuning of *Bkn3* expression by ethylene. In mutagenized *KK* seeds, five genetic loci (*suK*) have been identified as able to suppress *K* expression in transcription, leading to a phenotype characterized by the replacement of the ectopic *K* flower with awns shorter than the wild type [31].

The awn, an apical extension from the lemma of the spikelet, is a relevant photosynthetic organ that plays important roles in determining the grain size and yield. Wide natural variation for the awn length and shape has been observed. More than 700 short awn (*Lks*) and breviaristatum mutants have been characterized with the tools of classical genetics. Two main groups of awn-mutants can be identified: one characterized by phenotypic variation in the awn only, and the other in which the mutated phenotype is not only restricted to the awn, but also extends to several other plant organs. Among the several loci involved in awn development, You et al. [33] studied the short awn 2 (*lks2*) gene, which produces awns that are about 50% shorter than normal, and this is a natural variant restricted to Eastern Asia. Positional cloning revealed that *lks2* encodes an *SHI*-family transcription factor. Histological observations of longitudinal awn sections showed that the *Lks2* short-awn phenotype resulted from a reduced number of cells.

11. Naked Seed

The great majority of cultivated barleys have covered (hulled) caryopses, in which outer lemma and inner palea are strictly adherent to the pericarp epidermis at maturity. However, few genotypes have free-threshing called naked (hulless) caryopses. Hulled barley is mainly used in animal feed because of its higher yield, mainly due to the fact that the hull protects embryos from damage during mechanical harvesting. Even barley for brewing is covered, because the presence of the glumes provides a filtration medium in the separation of fermentable extract (wort) during malt processing. Obviously, covered grain is an adaptive trait in the wild, but during the domestication process, naked barley has been selected for direct human consumption. Barley domestication has been proposed to have originated more than 10,000 years before present [34], whereas naked barley appeared around 8000 years before present in Neolithic agricultural settlement sites in the Near East and western India, and quickly spread to Europe, Africa, and Asia. Although naked barley is today distributed worldwide, it is more frequent in East Asia, especially in the highlands of Nepal and Tibet. It has recently been used to clone the gene *nud*, which controls the covered/naked caryopsis. The gene has been mapped on chromosome arm 7HL and its greatest level of expression has been localized to the testa. The *nud* gene is homologous to the *Arabidopsis WIN1/SHN1* transcription factor gene, which is involved in the lipid biosynthesis pathway. In barley, the hulled caryopsis is therefore controlled by an *Ethylene Response Factor* (*ERF*) family transcription factor gene regulating the lipid biosynthesis pathway [35]. Briefly, the *nud* gene regulates the deposition of lipids on the epidermis of the pericarp. In covered barleys, this lipid layer is present and favours the adhesion of the hull to the caryopsis surface. On the contrary, in naked barleys, the lipid layer is missing and no adhesion is ensured between the hull and caryopsis, resulting in the free-threshing of the hull at maturity.

12. The Genetic Background Effect

Starting from the observation of barley mutants, new interrogatives have raised the question about how the behavior of a single developmental gene in different genetic backgrounds could help in understanding phenomena like expressivity, penetrance, phenotypic plasticity, and instability. An example of expressivity modulation has been found in genetic materials bearing a lemma mutation. The "*Leafy lemma*" phenotype was isolated in 1990 at the Istituto Sperimentale per la Cerealicoltura (Fiorenzuola d'Arda, Italy), in a plot in which the recessive mutant short awn (*lk2*) was grown. In the mutant, the lemma is transformed in a leaf-like structure. The transition zone of the lemma is particularly similar to the ligule-auricle region. Genetic analysis was carried out by crossing the mutant with several wild type genotypes and the segregation ratio was 15:1 [30]. The *Leafy lemma* mutant is characterized by the transformation of the lemma in a leaf-like structure, with a consequent increase in the seed size and photosynthetically active area (Figure 6).

Classical genetic information is available for the *Leafy lemma* mutant [30], indicating the involvement of two independent genes in the mutant phenotype. In the frame of the CREA-GB's mutant collection, 27 pairs of sister lines (wt/*Leafy lemma*) were generated in a backcross program in which the *lel* mutation has been introgressed in the 'Kaskade' background. These lines carry a *lel* mutation in different genetic backgrounds derived from different combinations of the parental genomes. In the lines carrying the mutation, a wide variation in the size of the mutated lemma can be observed. In other words, the mutated lemma ranges from a small foliar-like structure to a large one, depending on the genetic background characteristic of each line. Figure 7 reports the mean areas of the mutant lemmas in the 27 sister mutant lines.

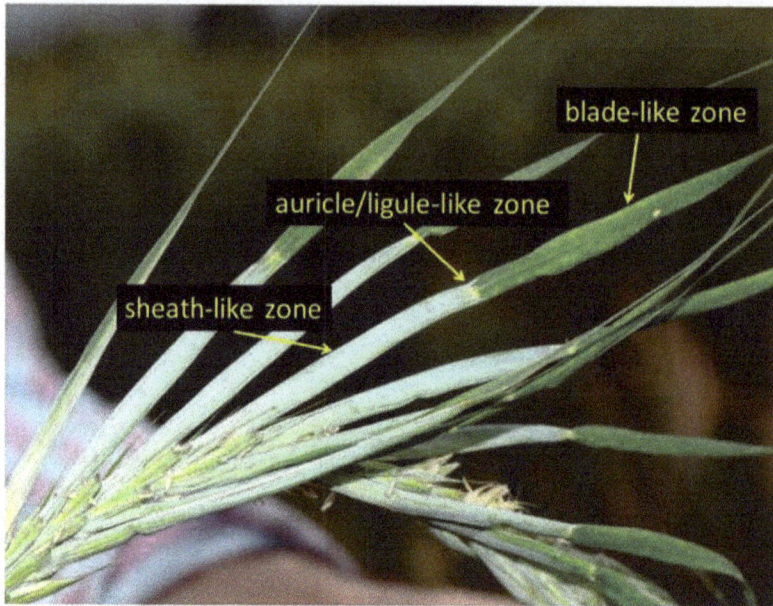

Figure 6. Spike of *Leafy lemma* mutant.

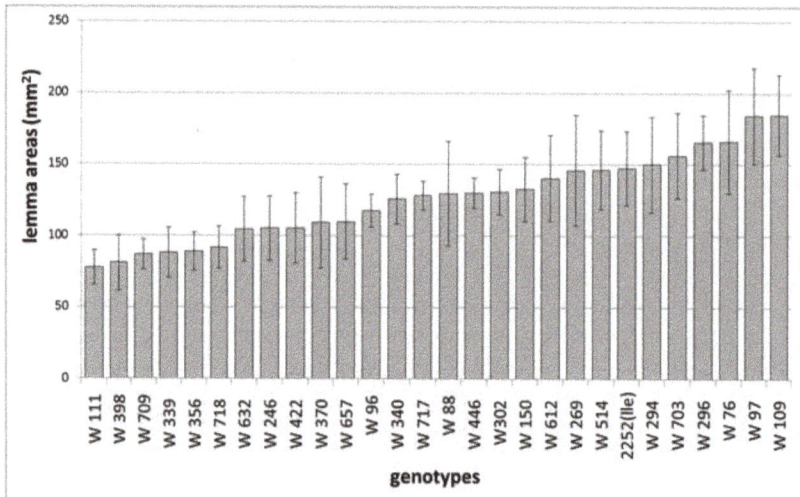

Figure 7. Different expressivity of the mutation—in terms of mutated lemma size—in sister lines derived from the cross between *Leafy lemma* mutant and Kaskade cultivar. The *x*-axis illustrates the labels of the different sister lines, whereas the *y*-axis represents the areas (expressed as mm^2) in which the mutated lemmas are reported.

In this genetic system, the expressivity of *lel* genes are clearly dependent on the genetic background, although the molecular basis of this epistatic effect is totally unclear. It is well known that genetic background effects contribute to the phenotypic consequence of a mutation, even if little is known about how these effects modify genetic systems. In the Drosophila model system, Dworkin et al. [36], working on the background effect on the expressivity of a *Scallopped* mutation, found that the phenotype is mediated through the misregulation of a series of developmental patterning genes, the epistatic interaction between mutated genes is background dependent, and the phenotypic variations correlate with qualitative and quantitative differences in downstream gene expression.

The genetic background—and the interaction among some specific genes—can be responsible not only for phenotypic variations, but also for other complex effects, like phenotypic instability. A very interesting example is presented in the work of Siuksta et al. [37]. Working on barley double mutants, obtained by crossing single ones, these authors focused their attention on *Hv-Hd/tw2* double mutants, characterized by inherited phenotypic instability. Some of the *Hv-Hd/tw2* mutants are phenotypically instable and this phenomenon is evident through several generations. Inflorescence variation in these double mutants includes the development of ectopic flowers, which range from negligible outgrowths to flowers with sterile organs, together with several new features of the spike, such as bract/leaf-like structures and naked gaps in the spike. This same phenomenon of phenotypic instability has been observed in our CREA-GB's collection [38], in which double and triple mutants have been developed by crossing the *Hooded* genotype with other homeotic single mutants. In this collection, rare "*Seeded in Hood*" have been found, in which small seeds in some hoods are developed. *Seeded in Hood* is a spontaneous phenotype that occurred in a plot of the double mutant *deficiens (Vrs1.t) x Hooded (K)"*. In the *deficiens* background, the size of the hood is bigger than in the other *Hooded* mutants and, probably due to this characteristic, some fertile florets can develop in the hood (Figure 8).

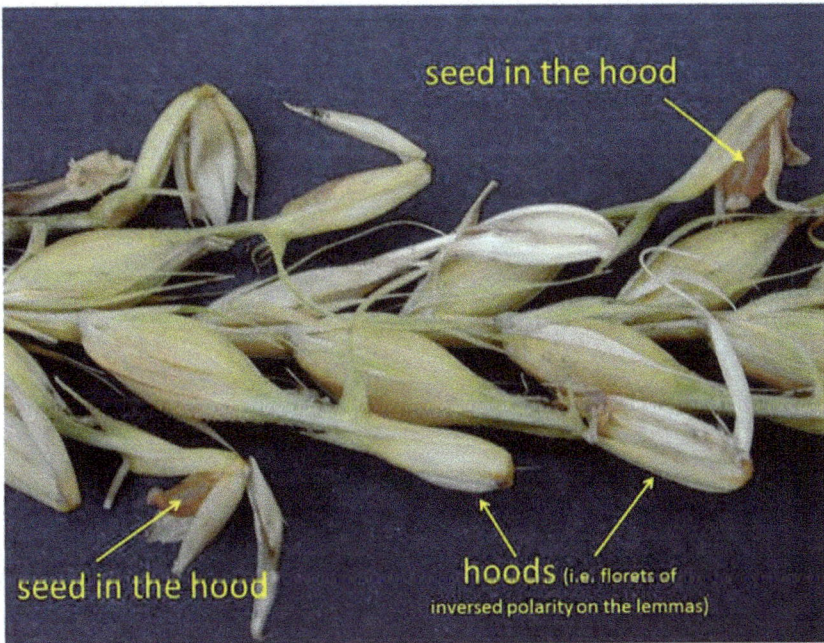

Figure 8. Two small seeds in the hoods of the "*Seeded in Hood*" double mutant, derived from the cross *deficiens/Hooded*.

This fact is evidenciated by the presence of small seeds in the hood. These small seeds have been grown in a glasshouse and the F1 plants grow normally as the *Hooded* phenotype. The F2 seeds derived from the F1 plants, also grown in a glasshouse, generated complex segregant populations, characterized by extreme phenotypic instability, with different *Hooded* spikes, including *Elevated hooded*, *Branched hooded*, and the first stage of hood development (*Trifurcatum*) (Figure 9).

Figure 9. Some phenotypes of the F2 progenies derived from the segregation of the "*Seeded in Hood*" caryopsis. Top left is a hooded branched spike, top right is a trifurcatum spike, and bottom shows hooded and elevated hooded spikes.

Siuksta et al. [37] suggested that unbalanced hormonal pathways can be a key factor to explain such a phenomenon. In particular, auxin interaction with the *Kn1* gene and other homeotic genes can determine an imbalance in auxin distribution, resulting in phenotypic instability. To demonstrate this hypothesis, Siutska et al. [37] treated flower/spike structures of double mutant lines with auxin inhibitors and with 2,4-D. They observed a normalization effect on the phenotype caused by auxin inhibitors—and an opposite effect of 2,4-D-, indicating that ectopic auxin hyper-accumulation probably plays a role. However, there marked variability among different double mutant lines in the response to 2,4-D and auxin inhibitors. This fact can be explained by the role played by the different genetic backgrounds resulting from segregation. In conclusion, double and triple mutants can be a very promising system to elucidate the molecular mechanisms at the basis of phenomena already described by classical genetics, like expressivity, which still remain largely unclear.

13. Conclusive Remarks

The study of barley morphological mutants is useful at two different levels:

- from an applied point of view, to identify loci that control traits of agronomic and qualitative relevance for pre-breeding and breeding programs;

\- from a speculative point of view, as unique tools to better understand the developmental plan of a crop and the major forces driving its evolution.

Regarding this last point, it is well known that genetic changes can occur at a nucleotide scale (single-nucleotide variations-SNV-, insertions/deletions—InDels), at a gene scale (e.g., copy number variations-CNVs), and at a chromosomal scale. At a nucleotide scale, for example, a single SNV created the cleistogamous flower type in barley. Structural variations (SV), at chromosomal a scale, are another relevant source of genetic diversity [39] and, among the mechanisms responsible for SV, transposable element (TE) dynamics are relevant. TE jumping can lead to genome rearrangements and changes in the genome size. The activity of TEs, in particular the DNA repair after TE excision, is a known source of mutations, which can accelerate genome evolution [40]. Gene enhancers and promoters can be altered by TE activities, as well as gene coding regions. TEs are highly represented in the barley genome, constituting more than 60% of the whole genome sequence, and are involved in the creation of novel genetic diversity. For example, the duplication of the *HvHox2* gene, due to TE activity, and its neo-funzionalization, created a row-type spike [16]. If TE dynamics can heavily modify the genome structure, it has been suggested that chromatin modifications and, in general, epigenetic regulation, can shape the TE-induced effects [41]. Verisimilarly, a combination of these mechanisms is at the basis of the continuum of morphological variants that can be observed in a barley spike. However, this is only speculation, and needs experimental demonstrations to be proved. To this aim, the recent development of "fast-forward genetics", based on NGS technologies, offers a new and powerful tool to geneticists and developmental biologists. The legacy of barley developmental mutants, obtained by meticulous research over half a century, together with derived mapping populations, can now be exploited in mapping-by-sequencing studies. Some examples of such an approach are already available. Mascher et al. [42], studying the multinoded phenotype, demonstrated that exome sequencing performed on phenotypic bulks of a mapping population is a feasible strategy to identify candidate genes involved in the mutated phenotype. Liu et al. [43] applied genotyping by sequencing (GBS) on a recombinant inbred line population (GPMx) derived from a cross between a two-rowed and a six-rowed barley. These authors identified three Quantitative Trait Loci (QTL) linked to plant height, the first in a region encompassing the spike architecture gene *Vrs1* on chromosome 2H, the second in an uncharacterized centromeric region on chromosome 3H, and the third in a region of chromosome 5H coinciding with the previously described dwarfing gene *Breviaristatum-e* (*Ari-e*).

Even genome-wide reverse genetics in barley open new perspectives to understand the gene-function relationships. New genetic materials can be developed with an alternative approach to the classical isolation of mutants, i.e., new mutations can be created through targeted mutagenesis and genome editing [44]. Even TILLING (Targeting Induced Local lesions IN Genomes) populations can be useful for forward and reverse approaches, and can provide multiple independent alleles of a single gene with varying levels of phenotypic expressivity. TILLING populations have been developed in Barke barley cultivar and have been used on the one hand to screen morphological mutants (forward genetics), and on the other to determine the molecular mutation frequency of candidate genes (reverse genetics) [45]. A detailed TILLING analysis for the vHox1 gene, controlling the row-type morphology, identified a set of nucleotide positions that, if mutated, alters the row spike morphology, indicating the existence of functionally relevant sites of the HvHOX1 protein.

In conclusion, spike morphology is associated with row type, grain density, spike length and grain number, and is a target of central importance in crop improvement. The barley developmental mutants have demonstrated their high value for the identification and functional characterization of key genes that can be important in prebreeding for ideal plant architecture (IPA), as a means to enhance the yield potential of existing elite varieties and to design the plant for the future.

Acknowledgments: Work funded by the Italian MiPAF projects "RGV-FAO". The technical assistance of Renzo Alberici in photographing plant materials is gratefully acknowledged.

Author Contributions: A.M.S. conceived the creation of the CREA-GB barley developmental mutant collection and D.P. contributed to its conservation. V.T. is now responsible for CREA-GB mutant collection. F.R. and G.T. characterized some mutated photosynthetic structure. R.G. and C.M. contributed to *Lel* sister lines characterization. All the authors contributed to the concept and layout of the manuscript and V.T. wrote the manuscript.

Conflicts of Interest: The authors declare no conflict of interest.

References

1. How to Feed the World in 2050. Available online: http://www.fao.org/fileadmin/templates/wsfs/docs/expert_paper/How_to_Feed_the_World_in_2050.pdf (accessed on 9 May 2017).
2. Fischer, R.A.; Edmeades, G.O. Breeding and cereal yield progress. *Crop Sci.* **2010**, *50*, 85–98. [CrossRef]
3. Sreenivasulu, N.; Schnurbusch, T. A genetic playground for enhancing grain number in cereals. *Trends Plant Sci.* **2012**, *17*, 91–101. [CrossRef] [PubMed]
4. Mascher, M.; Gundlach, H.; Himmelbach, A.; Beier, S.; Twardziok, S.O.; Wicker, T.; Radchuk, V.; Dockter, C.; Hedley, P.E.; Russell, J.; et al. A chromosome conformation capture ordered sequence of the barley genome. *Nature* **2017**, *544*, 427–433. [CrossRef] [PubMed]
5. International Database for Barley Genes and Barley Genetic Stocks. Available online: www.nordgen.org/bgs/ (accessed on 9 May 2017).
6. Druka, A.; Franckowiak, J.; Lundqvist, U.; Bonar, N.; Alexander, J.; Houston, K.; Radovic, S.; Shahinnia, F.; Vendramin, V.; Morgante, M.; et al. Genetic Dissection of Barley Morphology and Development. *Plant Physiol.* **2011**, *155*, 617–627. [CrossRef] [PubMed]
7. Sakuma, S.; Salomon, B.; Komatsuda, T. The Domestication Syndrome Genes Responsible for the Major Changes in Plant Form in the Triticeae Crops. *Plant Cell Physiol.* **2011**, *52*, 738–749. [CrossRef] [PubMed]
8. Mascher, M.; Schuenemann, V.J.; Davidovich, U.; Marom, N.; Himmelbach, A.; Hübner, S.; Korol, A.; David, M.; Reiter, E.; Riehl, S.; et al. Genomic analysis of 6000-year-old cultivated grain illuminates the domestication history of barley. *Nat. Genet.* **2016**, *48*, 1089–1093. [CrossRef] [PubMed]
9. Kirby, E.J.M.; Appleyard, M. *Cereal Development Guide*, 2nd ed.; Arable Unit, National Agricultural Centre: Warwickshire, UK, 1986; pp. 53–59.
10. Bossinger, G.; Rohde, W.; Lundqvist, U.; Salamini, F. Genetics of barley development: Mutant phenotypes and molecular aspects. In *Barley: Genetics, Biochemistry, Molecular Biology and Biotechnology*; Shewry, P.R., Ed.; CAB International: Wallingford, UK, 1992; pp. 231–264.
11. Forster, B.P.; Franckowiak, J.D.; Lundqvist, U.; Lyon, J.; Pitkethly, I.; Thomas, W.T.B. The Barley Phytomer. *Ann. Bot.* **2007**, *100*, 725–733. [CrossRef] [PubMed]
12. Pourkheirandish, M.; Hensel, G.; Kilian, B.; Kumlehn, J.; Sato, K.; Komatsuda, T. Evolution of the Grain Dispersal System in Barley. *Cell* **2015**, *162*, 527–539. [CrossRef] [PubMed]
13. Civáň, P.; Brown, T.A. A novel mutation conferring the non brittle phenotype of cultivated barley. *New Phytol.* **2017**, *214*, 468–472.
14. Schmalenbach, I.; March, T.J.; Bringezu, T.; Waugh, R.; Pillen, K. High-Resolution Genotyping of Wild Barley Introgression Lines and Fine-Mapping of the Threshability Locus *thresh-1* Using the Illumina GoldenGate Assay. *G3 Genes Genomes Genet.* **2011**, *1*, 187–196. [CrossRef] [PubMed]
15. Koppolu, R.; Anwar, N.; Sakuma, S.; Tagiri, A.; Lundqvist, U.; Pourkheirandish, M.; Rutten, T.; Seiler, C.; Himmelbach, A.; Ariyadasa, R.; et al. Six-rowed spike4 (*Vrs4*) controls spikelet determinacy and row-type in barley. *Proc. Natl. Acad. Sci. USA* **2013**, *110*, 13198–13203. [CrossRef] [PubMed]
16. Komatsuda, T.; Pourkheirandish, M.; He, C.; Azhaguvel, P.; Kanamori, H.; Perovic, D.; Stein, N.; Graner, A.; Wicker, T.; Tagiri, A.; et al. Six-rowed barley originated from a mutation in a homeodomain-leucine zipper I-class homeobox gene. *Proc. Natl. Acad. Sci. USA* **2007**, *104*, 1424–1429. [CrossRef] [PubMed]
17. Ramsay, L.; Comadran, J.; Druka, A.; Marshall, D.F.; Thomas, W.T.W.; Macaulay, M.; MacKenzie, K.; Simpson, C.; Fuller, J.; Bonar, N.; et al. *INTERMEDIUM-C*, a modifier of lateral spikelet fertility in barley, is an ortholog of the maize domestication gene *TEOSINTE BRANCHED 1*. *Nat. Genet.* **2011**, *43*, 169–173. [CrossRef] [PubMed]
18. Youssef, H.M.; Koppolu, R.; Rutten, T.; Korzun, V.; Schweizer, P.; Schnurbusch, T. Genetic mapping of the *labile* (*lab*) gene: A recessive locus causing irregular spikelet fertility in *labile*-barley (*Hordeum vulgare* convar. *labile*). *Theor. Appl. Genet.* **2014**, *127*, 1123–1131. [CrossRef] [PubMed]

19. Helmy, M.Y.; Eggert, K.; Koppolu, R.; Alqudah, H.M.; Poursarebani, N.; Fazeli, A.; Sakuma, S.; Tagiri, A.; Rutten, T.; Govind, G.; et al. VRS2 regulates hormone-mediated inflorescence patterning in barley. *Nat. Genet.* **2017**, *49*, 157–161.

20. Boden, S.A. How hormones regulate floral architecture in barley. *Nat. Gen.* **2017**, *49*, 8–9. [CrossRef] [PubMed]

21. Shang, Y.; Zhu, J.; Hua, W.; Wang, J.; Jia, Q.; Yang, J. Characterization and mapping of a *Prbs* gene controlling spike development in *Hordeum vulgare* L. *Genes Genome* **2014**, *36*, 275–282. [CrossRef]

22. Poursarebani, N.; Seidensticker, T.; Koppolu, R.; Trautewig, C.; Gawroński, P.; Bini, F.; Govind, G.; Rutten, R.; Sakuma, S.; Tagiri, A.; et al. The genetic basis of composite spike form in barley and 'Miracle-Wheat'. *Genetics* **2015**, *115*, 176628. [CrossRef] [PubMed]

23. Brown, R.H.; Bregitzer, P. A *Ds* Insertional Mutant of a Barley *miR172* Gene results in Indeterminate Spikelet Development. *Crop Sci.* **2011**, *51*, 1664–1672. [CrossRef]

24. Jost, M.; Taketa, S.; Mascher, M.; Himmelbach, A.; You, T.; Shahinnia, F.; Rutten, T.; Druka, A.; Schmutzer, T.; Steuernagel, B.; et al. A homolog of Blade-On-Petiole 1 and 2 (BOP1/2) controls internode length and homeotic changes of the barley inflorescence. *Plant Physiol.* **2016**, *171*, 1113–1127. [CrossRef] [PubMed]

25. Shahinnia, F.; Druka, A.; Franckowiak, J.; Morgante, M.; Waugh, R.; Stein, N. High resolution mapping of *Dense spike-ar* (*dsp.ar*) to the genetic centromere of barley chromosome 7H. *Theor. Appl. Genet.* **2012**, *124*, 373–384. [CrossRef] [PubMed]

26. Taketa, S.; Yuo, T.; Sakurai, Y.; Miyake, S.; Ichii, M. Molecular mapping of the short awn 2 (*lks2*) and dense spike 1 (*dsp1*) genes on barley chromosome 7H. *Breed. Sci.* **2011**, *61*, 80. [CrossRef]

27. Houston, K.; McKim, S.M.; Comadran, J.; Bonar, N.; Druka, I.; Uzrek, N.; Cirillo, E.; Guzy-Wrobelska, J.; Collins, N.C.; Halpin, C.; et al. Variation in the interaction between alleles of HvAPETALA2 and microRNA172 determines the density of grains on the barley inflorescence. *Proc. Natl. Acad. Sci. USA* **2013**, *110*, 16675–16680. [CrossRef] [PubMed]

28. Nair, S.K.; Wang, N.; Turuspekov, Y.; Pourkheirandish, M.; Sinsuwongwat, S.; Chen, G.; Sameri, M.; Tagiri, A.; Honda, I.; Watanabe, Y.; et al. Cleistogamous flowering in barley arises from the suppression of microRNA-guided *HvAP2* mRNA cleavage. *Proc. Natl. Acad. Sci. USA* **2010**, *107*, 490–495. [CrossRef] [PubMed]

29. Wang, N.; Ning, S.; Wu, J.; Tagiri, A.; Komatsuda, T. An Epiallele at *cly1* Affects the Expression of Floret Closing (Cleistogamy) in Barley. *Genetics* **2015**, *199*, 95–104. [CrossRef] [PubMed]

30. Pozzi, C.; Faccioli, P.; Terzi, V.; Stanca, A.M.; Cerioli, S.; Castiglioni, P.; Fink, R.; Capone, R.; Müller, K.J.; Bossinger, G.; et al. Genetics of mutations affecting the development of a barley floral bract. *Genetics* **2000**, *154*, 1335–1346. [PubMed]

31. Roig, C.; Pozzi, C.; Santi, L.; Müller, J.; Wang, Y.; Stile, M.R.; Rossini, L.; Stanca, A.M.; Salamini, F. Genetics of barley hooded suppression. *Genetics* **2004**, *167*, 439–448. [CrossRef] [PubMed]

32. Osnato, M.; Stile, M.R.; Wang, Y.; Meynard, D.; Curiale, S.; Guiderdoni, E.; Liu, Y.; Horner, D.S.; Ouwerkerk, B.F.; Pozzi, C.; et al. Cross Talk between the KNOX and Ethylene Pathways Is Mediated by Intron-Binding Transcription Factors in Barley. *Plant Physiol.* **2010**, *154*, 4, 1616–1632. [CrossRef] [PubMed]

33. Yuo, T.; Yamashita, Y.; Kanamori, H.; Matsumoto, T.; Lundqvist, U.; Sato, K.; Ichii, M.; Jobling, S.A.; Taketa, S. A SHORT INTERNODES (SHI) family transcription factor gene regulates awn elongation and pistil morphology in barley. *J. Exp. Bot.* **2012**, *63*, 5223–5232. [CrossRef] [PubMed]

34. Von Bothmer, R.; Komatsuda, T. Barley origin and related species. In *Barley: Production, Improvement and Uses*; Ullrich, S.E., Ed.; Wiley Blackwell: Oxford, UK, 2011; pp. 14–62.

35. Taketa, S.; Amano, S.; Tsujino, Y.; Sato, T.; Saisho, D.; Kakeda, K.; Nomura, M.; Suzuki, T.; Matsumoto, T.; Sato, K.; et al. Barley grain with adhering hulls is controlled by an ERF family transcription factor gene regulating a lipid biosynthesis pathway. *Proc. Natl. Acad. Sci. USA* **2008**, *105*, 4062–4067. [CrossRef] [PubMed]

36. Dworkin, I.; Kennerly, E.; Tack, D.; Hutchinson, J.; Brown, J.; Mahaffey, J.; Gibson, G. Genomic Consequences of Background Effects on *scalloped* Mutant Expressivity in the Wing of *Drosophila melanogaster*. *Genetics* **2009**, *181*, 1065–1076. [CrossRef] [PubMed]

37. Šiukšta, R.; Vaitkūnienė, V.; Kaselytė, G.; Okockytė, V.; Žukauskaitė, J.; Žvingila, D.; Rančelis, V. Inherited phenotype instability of inflorescence and floral organ development in homeotic barley double mutants and its specific modification by auxin inhibitors and 2,4-D. *Ann. Bot.* **2015**, *115*, 651–663. [CrossRef] [PubMed]

38. Stanca, A.M.; Tumino, G.; Pagani, D.; Rizza, F.; Alberici, R.; Lundqvist, U.; Morcia, C.; Tondelli, A.; Terzi, V. The "Italian" barley genetics mutant collection: Conservation, development of new mutants and use. In *Advance in Barley Sciences*; Zhang, G., Li, C., Liu, X., Eds.; Springer Science & Business Media: New York, NY, USA, 2013; pp. 30–35.

39. Kim, N.S. The genomes and transposable elements in plants: Are they friends or foes? *Genes Genom.* **2017**, *39*, 359. [CrossRef]

40. Wicker, T.; Yu, Y.; Haberer, G.; Mayer, K.F.; Marri, P.R.; Rounsley, S.; Roffler, S. DNA transposon activity is associated with increased mutation rates in genes of rice and other grasses. *Nat. Commun.* **2016**, *7*. [CrossRef] [PubMed]

41. Springer, N.M.; Lisch, D.; Li, Q. Creating order from chaos: Epigenome dynamics in plants with complex genomes. *Plant Cell* **2016**, *28*, 314–325. [CrossRef] [PubMed]

42. Mascher, M.; Jost, M.; Kuon, J.-E.; Himmelbach, A.; Aβfalg, A.; Beier, S.; Scholz, U.; Graner, A.; Stein, N. Mapping by sequencing accelerates forward genetics in barley. *Genome Biol.* **2014**, *15*, R78. [CrossRef] [PubMed]

43. Liu, H.; Bayer, M.; Druka, A.; Russell, J.R.; Hackett, C.A.; Poland, J.; Ramsay, L.; Hedley, P.E.; Waugh, R. An evaluation of genotyping by sequencing (GBS) to map the *Breviaristatum-e (ari-e)* locus in cultivated barley. *BMC Genom.* **2014**, *15*, 104–115. [CrossRef] [PubMed]

44. Wendt, T.; Holm, P.B.; Starker, C.G.; Christian, M.; Voytas, D.F.; Brinch-Pedersen, H.; Holme, I.B. TAL effector nucleases induce mutations at a pre-selected location in the genome of primary barley transformants. *Plant Mol. Boil.* **2013**, *83*, 279–285. [CrossRef] [PubMed]

45. Gottwald, S.; Bauer, P.; Komatsuda, T.; Lundquist, U.; Stein, N. TILLING in the two-rowed barley cultivar "Barke" reveals preferred sites of functional diversity in the gene *HvHoxI. BMC Res. Not.* **2009**, *2*, 258.

diversity

MDPI

Article

Olive Tree (*Olea europaea* L.) Diversity in Traditional Small Farms of Ficalho, Portugal

Maria Manuela Veloso [1,2,]*, Maria Cristina Simões-Costa [2,3], Luís C. Carneiro [1], Joana B. Guimarães [1], Célia Mateus [4], Pedro Fevereiro [5,6] and Cândido Pinto-Ricardo [5]

[1] Instituto Nacional de Investigação Agrária e Veterinária, Unidade de Investigação de Biotecnologia e Recursos Genéticos, Quinta do Marquês, 2784-505 Oeiras, Portugal; luis.c.carneiro@gmail.com (L.C.C.); joana.guimaraes@iniav.pt (J.B.G.)

[2] LEAF, Linking Landscape, Environment, Agriculture and Food, Instituto Superior de Agronomia, Universidade de Lisboa, Tapada da Ajuda, 1349-017 Lisboa, Portugal; simoescosta@isa.ulisboa.pt

[3] Departamento de Recursos Naturais, Ambiente e Território, Instituto Superior de Agronomia, Universidade de Lisboa, Tapada da Ajuda, 1349-017 Lisboa, Portugal

[4] Instituto Nacional de Investigação Agrária e Veterinária, Unidade de Investigação em Sistemas Agrários-Produção e Sustentabilidade, Quinta do Marquês, 2784-505 Oeiras, Portugal; celia.mateus@iniav.pt

[5] Instituto de Tecnologia Química e Biológica, Universidade Nova de Lisboa, Av. da República, Apt. 127, 2781-901 Oeiras, Portugal; psalema@itqb.unl.pt (P.F.); ricardo@itqb.unl.pt (C.P.-R.)

[6] Departamento Biologia Vegetal, Faculdade de Ciências, Universidade de Lisboa, Campo Grande, 1749-016 Lisboa, Portugal

* Correspondence: mveloso.inrb@gmail.com; Tel.: +351-21-446-3744 or +351-93-672-2540; Fax: +351-21-441-6011

Received: 2 November 2017; Accepted: 15 January 2018; Published: 18 January 2018

Abstract: The genetic diversity of "Gama" and "Bico de Corvo", local cultivars of olive tree (*Olea europaea*) from seven traditional orchards of Ficalho (Alentejo region, Portugal), was studied to characterize the local diversity and assess the level of *on farm* diversity. Two different analytical systems were used: endocarp morphological characteristics and genetic analysis by microsatellite markers (Simple Sequence Repeats or SSR). The seven screened *loci* were polymorphic and allowed the identification of 23 distinct SSR profiles within the 27 trees analyzed. A total of 52 different alleles were scored, with an average of 7.43 alleles/SSR locus, and considerable genetic diversity was found. Neighbor-Joining algorithm cluster analysis and principal co-ordinate analysis (PCoA) allowed for the identification of the genetic relationships between several accessions. The 27 *Olea* accessions were clearly separated into three different groups. SSR analysis was more precise than endocarp characterization in the classification of genetic diversity among the olive tree cultivars. The study shows reasonable olive tree diversity in Ficalho, indicating that these traditional orchards are important reservoirs of old minor cultivars and incubators of new genotypes.

Keywords: olive tree local cultivars; endocarp characterization; SSR genotyping; *on farm* conservation

1. Introduction

The olive tree preceded the Romans in the Iberian Peninsula and olive oil has always been of great significance for the whole of southern Europe. The recognition of olive oil dietetic properties has lead to an increase in its consumption worldwide. Portugal is the eighth-largest worldwide olive oil producer and according to the Olive Oil Council is the fourth largest producer in the European Union [1]. This accounts for about 7% of the country's agro-food exports that have been increasing exponentially since 2000, particularly of high quality extra virgin oil. The activity of olive oil press industry has likewise been increasing [2].

The domestication of the olive (*Olea europaea* subsp. *europaea* var. *europaea*) was a continuous, and complex process [3] with many crossings occurring between cultivated trees and local oleasters (*O. europaea* L. subsp. *europaea* var. *sylvestris* (Miller) Lehr) across the entire Mediterranean region. Portugal has a large genetic patrimony of such germplasm represented by many "old" local cultivars, some of which have restricted distribution [4]. Olive trees are grown throughout the country, mainly further inland, and represent 9.2% of the Portuguese Utilized Agriculture Area (UAA). About 60% of the oil is produced in the Alentejo region using two different cultivation systems, traditional farming systems (85% of the area and 35% of total olive oil production) and semi-intensive and intensive systems (15% of area and 65% of total olive oil production) [5].

Traditional orchards still account for 80% of the cultivated area [6], where "Cordovil de Serpa" and "Verdeal Alentejana" are the dominant cultivars in these traditional systems in the Alentejo region [7]. This cropping system is an important repository of genetic diversity as other local cultivars are always present [8], although in small numbers [9,10]. Such genetic variability contributes to the distinctness of the oil and determines crop resilience. The ability to adapt to changes in biotic and abiotic factors, such as pests, diseases, and climatic stresses, is important, particularly in areas where climate change may become more severe. So, the sustainability of such agroecosystems could be endangered by the emergence of modern olive growing systems that have poor genetic diversity.

Identification of existing genetic diversity is essential for olive germplasm management and preservation. This process should start with morphological studies of the olive cultivars, such as the endocarp characteristics [11,12], which serve to discriminate among trees. Methodologies involving molecular markers should complement the morphological characterization. For instance, microsatellites (or Simple Sequence Repeats—"SSRs") have been shown to be very useful in the characterization of olive germplasm [13–16], even for small areas of cultivation [17–20].

In the light of increased risk of genetic erosion of minor olive cultivars, the present study was carried out in Ficalho which is an important area within the traditional Alentejo olive oil producing region. The study aims to (i) characterize the old olive trees using endocarp parameters and SSR markers, (ii) investigate the genetic relationships between these olive trees and the dominant cultivars "Cordovil de Serpa" and "Verdeal Alentejana", and (iii) assess the level of *on-farm* genetic diversity.

2. Materials and Methods

The study was performed at seven small traditional orchards in Ficalho within the Alentejo region, Portugal (Figure 1). Approximately 87% of the olive trees from these orchards are "Verdeal Alentejana" and "Cordovil de Serpa" [10] and so, our study focused on minor cultivars. A total of 27 olive trees including two oleasters and one unnamed tree ("Desco 05") were studied, following the labelling used by farmers. Two traditional cultivars from other regions were also studied, "Maçanilha Algarvia" (from Algarve region, Portugal) and "Cornezuelo" (from Andalusia region, Spain) as shown in Table 1.

Table 1. Olive tree accessions (cultivars and wild) characterized in the study.

Olive Tree	Designations	Area of Cultivation
Local Cultivars	Bico de Corvo Carrasquenha Cordovil de Serpa Galega Gama Maçanilha Verdeal Alentejana	Traditional orchards, Ficalho
From other Regions	Cornezuelo Maçanilha Algarvia	Germplasm collection, Herdade da Abóboda *
Unnamed Tree	Desconhecida	Traditional orchards, Ficalho
Wild	Zambujeiro	Traditional orchards, Ficalho

* Farm from Ministry of Agriculture.

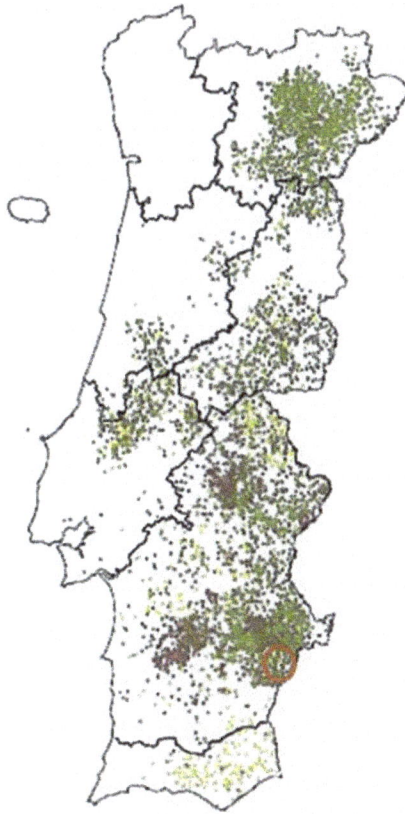

Figure 1. Olive tree distribution in Portugal, according to InstitutoNacional de Estatística (2011). The traditional orchards at Ficalho where the study was performed are delimited. ▨ 50 ha of olive tree with 60 trees/ha; ▨ 50 ha of olive tree with 61 to 100 trees/ha; ▨ 50 ha of olive tree with 101 to 300 trees/ha; ▨ 50 ha of olive tree with more than 300 trees/ha.

2.1. Morphological Study

Plant characterization is usually carried out using a large number of descriptors. In olives, descriptors based on endocarp characteristics can help determine the identity of a cultivar [12,21]. Forty fruits from each tree were collected during two seasons (2012 and 2013). Five endocarp characteristics were evaluated as follows: weight (g), length (mm), diameter (mm), and number and distribution of vascular bundles, using established methodologies [22]. Endocarp images were captured using a stereoscopic microscope Leica Wild MZ8 equipped with a Leica DC200 camera linked to a Leica IM50 image management software.

2.2. DNA Extraction, SSR Markers, PCR Amplification and Fragment Sizing

DNA was isolated from 100 mg of fresh young leaves ground in liquid nitrogen using the DNeasy Plant Mini Kit (QIAGEN GmbH, Hilden, Germany) according to the manufacturer's protocol. DNA quality was checked on 0.8% agarose gel, and the DNA concentration was estimated using a NanoDrop ND2000 spectrophotometer (Thermo Scientific, Massachusetts, MA, USA).

For olive tree genotyping, nine nuclear SSRs were used, selected among those available in the literature [16] and previously proven to be suitable for the characterization and identification of olive varieties as follows: OeUA-DCA04, OeUA-DCA05, OeUA-DCA13, OeUA-DCA14, OeUA-DCA16, OeUA-DCA18 [13], GAPU-71B, GAPU103-A [14] and EMO90 [15]. PCR was conducted in a final volume of 25 µL containing 25 ng of DNA, 10 mM Tris-HCl pH 8.0, 1.5 mM $MgCl_2$, 0.2 mM dNTPs, 0.25 µM forward primer fluorescently labelled with WellRED dyes at the $5'$-end and unlabelled reverse primers, and 0.05 units of Taq DNA polymerase (Invitrogen, Carlsbad, CA, USA). Capillary electrophoresis was performed to separate the PCR products using the CEQ 8000 Genetic Analysis System (Beckman Coulter Inc., Brea, CA, USA). The sizes of the amplified products were determined based on an internal standard included with each sample. Data Analysis was performed using the CEQ 8000 Fragment Analysis software, version 9.0, according to the manufacture's recommendations (Beckman Coulter Inc., Brea, CA, USA). Sizes of SSR fragments were automatically calculated using the CEQ 8000 Genetic Analysis System. The information obtained was used to study the genetic diversity of the selected olive trees and to determine their relatedness.

2.3. Data Analysis

The endocarp morphological characteristics were analyzed for their means and standard errors. Principal Component Analysis (PCA) was performed to examine the morphological variation and to identify the most relevant characteristics in order to distinguish different accessions. The program NTSYS-pc, version 2.1 was used in all the statistical multivariate analysis [23]. The statistical analysis of SSR data was performed using Microchecker software v2.2.3 [24] for the detection of null alleles, stuttering and allele dropout, and FSTAT [25] for genetic diversity parameters (Polymorphism Information Content, allelic richness, private allele number, and heterozygosity). The genetic distance between each pair of individuals was calculated as one minus the proportion of shared alleles across all *loci* [26] using the MICROSAT program package [27].

To establish the genetic relationships existing between several accessions, two methods were used as follows: Neighbor-Joining Algorithm cluster analysis and the principal co-ordinate analysis (PCoA). The Neighbor-Joining algorithm, as implemented in the DARwin software package, version 6.0.12 [28], was based on a dissimilarity matrix, and the reliability of the tree topology was assessed via bootstrapping over 1000 replicates. Regarding the PCoA, the distance matrix was performed based on the proportion of shared alleles calculated from SSR markers following NTSYS—pc v.2.1 [23].

3. Results and Discussion

Two different analytical systems were used, endocarp morphological characteristization and genetic analysis with microsatellite markers, in order to identify the germplasm and to study the diversity of the olive trees in Ficalho.

The endocarp images and characteristics for the different accessions are shown in Figure 2 and Supplementary Table S1. We found that the surface roughness and the number of vascular bundles over the endocarp surface were characteristics with little variation in keeping with observations from the Spanish Extremadura region [19] as well as from the Italian Tuscany region [29]. However, our data revealed important differences and variability in other endocarp characteristics. The PCA analysis of the data allowed for the discrimination of the olive trees, with the first two principal components explaining 77.27% of the total variation (48.37% and 28.90% for the first (I) and the second (II) principal components, respectively). The I principal component is controlled by weight, length and diameter of the endocarps and the II principal component by the number and distribution of vascular bundles on the endocarp surface (Figure 3). According to the projections of the 27 accessions onto the plane defined by the I and the II principal components, the endocarps with higher diameter are on the right (e.g., "Bico 32" and "Maçanilha Algarvia–Maca Al"), and those with low diameter are on the left (e.g., "Zambujeiro 01–Zambu 01" and "Galega–Galeg 01"). On the other hand, on the upper side of the plane are endocarps with a lower number of vascular bundles (e.g., "Zambujeiro 17–Zambu

17") and on the lower side are endocarps with higher surface roughness (e.g., "Maçanilha 16–Maca 16" and "Maçanilha Algarvia–Maca Al") (Figure 3A and Supplementary Table S1). The oleasters ("Zambu 01" and "Zambu 17"), which have a typical wild aspect, are quite distant from all the other accessions, although quite distinct from each other. It was not possible to conclude if they are true oleasters or feral forms. The unnamed tree ("Desco 05") is close to the traditional cultivar "Galega–Galeg 01" (Figure 3) and could have resulted from hybridization with a wild olive tree (Figure 3A). The "Maçanilha Algarvia–Maca Al" which is not grown in Ficalho, is in Figure 3A far from the local varieties. Considering the remaining traditional local varieties, it is observed that the majority are in the center of the plane defined by the I and the II principal components. Two "Verdeal Alentejana" trees ("Verde 10" and "Verde 15") are distant from the central group due to the surface roughness (Supplementary Table S1). "Bico de Corvo 32" ("Bico 32") is also different from all the other "Bico de Corvo" trees due to its heavier weight and longer diameter.

Figure 2. Endocarp images captured using a stereoscope Leica Wild. (**A**) "Zambu 17"; (**B**) "Zambu 01"; (**C**) "Galega 01"; (**D**) "Maca Al"; (**E**) "Maca 16"; (**F**) "Gama 32"; (**G**) "Cordo 01"; (**H**) "Verde 01"; (**I**) "Verde 15"; (**J**) "Bico 04"; (**K**) "Carra 01"; (**L**) "Maca 12".

The characterization of the accessions was further performed by SSR analysis. Using the microsatellite markers for the genetic analysis out of the nine SSR *loci* selected for the study, two (OeUA-DCA13 and EMO90) failed to amplify or yield monomorphic fragments and therefore, were not used. The seven remaining *loci* were polymorphic and allowed the identification of 23 distinct SSR profiles within the population of 27 olive trees (Supplementary Table S2). A total of 52 different alleles were scored, with an average of 7.43 alleles/SSR locus, ranging from 5 (for OeUA-DCA05 and GAPU71-B) to 14 alleles (for OeUA-DCA16), however, the average number of effective alleles was 3.352. Table 2 indicates that OeUA-DCA16 and OeUA-DCA18 *loci* had the highest number of unique alleles.

When evaluating the genetic diversity, it was found that the expected heterozygosity (He) was greater than 0.500 for the majority of the *loci*; the highest value (He 0.819) was observed with OeUA-DCA16 *locus*. The exception was the OeUA-DCA05 *locus*, with an He of 0.177. In regards to observed heterozygosity (Ho), the values were less than He except for those of OeUA-DCA05 (Ho 0.185) and GAPU71-B *loci* (Ho 1.000). The information obtained from the Polymorphism Information Content (PIC) value also indicated that the discriminatory power was quite good except for that of the OeUA-DCA05 *locus* (Table 2). Other studies of Portuguese cultivars [30,31] have also reported that OeUA-DCA05 and OeUa-DCA16 *loci* presented the lowest and the highest informative values, respectively. The number of alleles, the He and the PIC values per *locus* observed in our study are comparable to those reported by [16,19,20,32] for Mediterranean olive trees.

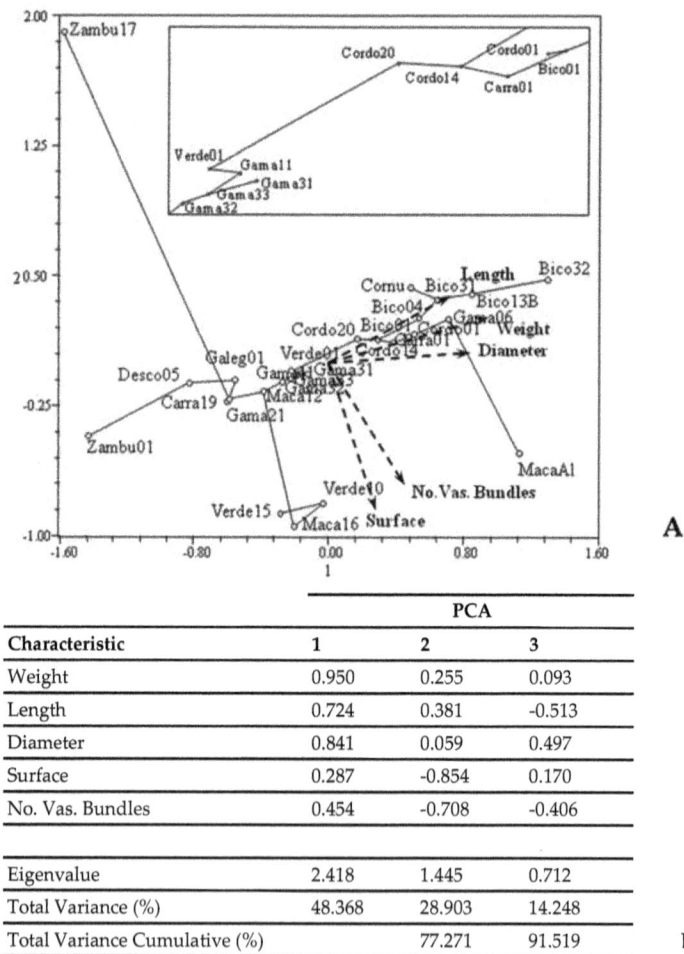

Characteristic	PCA		
	1	2	3
Weight	0.950	0.255	0.093
Length	0.724	0.381	-0.513
Diameter	0.841	0.059	0.497
Surface	0.287	-0.854	0.170
No. Vas. Bundles	0.454	-0.708	-0.406
Eigenvalue	2.418	1.445	0.712
Total Variance (%)	48.368	28.903	14.248
Total Variance Cumulative (%)		77.271	91.519

Figure 3. Principal Component Analysis (PCA) of the endocarp morphological characteristics. Projections of the olive tree endocarps onto the plane defined by the I (48.37%) and the II (28.90%) principal components with superimposition of the minimum spanning tree and eigenvectors. The variables correspond to weight, length, diameter and number and distribution of vascular bundles over the endocarps (**A**). Eigenvalues and total variance (%) describe the variation of the five endocarp characteristics analyzed (**B**).

Table 2. Genetic diversity of 27 *Olea europaea* accessions as assessed using seven SSRs *loci*. N—number of accessions; Na—number of alleles; Ne—effective number of alleles; Npa—number of unique alleles; Ho—observed heterozygosity; He—expected heterozygosity; PIC—polymorphism information content.

Locus	N	Na	Ne	Npa	Ho	He	PIC
OeUA-DCA04	27	6.000	2.666	1	0.222	0.578	0.538
OeUA-DCA05	27	5.000	1.211	3	0.185	0.177	0.170
OeUA-DCA14	27	7.000	2.285	1	0.667	0.671	0.618
OeUA-DCA16	27	14.000	5.080	6	0.667	0.819	0.782
OeUA-DCA18	27	9.000	3.636	4	0.630	0.739	0.684
GAPU71-B	27	5.000	4.599	-	1.000	0.797	0.747
GAPU103-A	27	6.000	3.984	1	0.704	0.753	0.693

Our data indicate that the olive tree is a highly polymorphic species, which confirms previous reports for Portuguese cultivars [30,31,33,34] and those for other Mediterranean olive trees [19,20,35–37].

To study the genetic relationships among the 23 different olive genotypes, an UPGMA dendrogram based on the proportion of shared alleles and a Neighbor-Joining Tree were constructed (Figure 4A,B). The UPGMA dendrogram showed a clear separation of cultivars from the wild olive trees. In the Neighbor-Joining Tree, the 27 *Olea* accessions were clearly separated into three different groups. The first group includes the wild trees ("Zambu 01" and "Zambu 17"), the "Galega–Galeg 01"cultivar, the unnamed tree ("Desco 05") and the cultivars not grown in Ficalho ("Maçanilha Algarvia–Maca Al" and "Cornezuelo–Cornu"). Concerning the two wild trees, it was not possible, at the molecular level, to disclose if they were truly wild or oleasters because they did not display close relationships with any of the studied varieties. The local cultivars are distributed in the second and third groups, the latter exclusively composed of "Gama". These two clusters are well supported by bootstrapping analysis, particularly in the case of "Gama" (Figure 4B). Two cases of identical plants were found: "Verde 01" and "Verde 15", on one side, and "Gama 11", "Gama 21" and "Gama 33", on the other side. The genetic uniformity of "Gama 11", "Gama 21" and "Gama 33" suggests that they may result from the clonal propagation of the same genotype. "Gama 32" differs in one *locus* (OeUA-DCA16) from the other "Gama" trees belonging to the same cluster (Figures 4 and 5). So, it probably resulted from a clonal mutation, although without phenotypic expression. Clonal variation in traditional olive varieties has been referred to in other studies [38]. Small genetic differences, probably due to somatic point mutations, were also reported in ancient olive trees [39] regardless of whether the phenotype was expressed [12]. The degree of correlation existing between somatic point mutations and phenotype variation remains to be clarified [39].

Three cases of mislabeling were also identified. One of them was "Gama 06" and "Bico 01", which are probably neither "Gama" nor "Bico de Corvo" trees. They have the same genotype and are in the first group due to the presence of the allele 141 (locus OeUA-DCA 04) that was found in the wild tree "Zambu 17". "Gama 31" has a different genotype than that of the "Gama" trees and has the allele 141 found in "Zambu 17" (Supplementary Table S2). "Bico 04", was also found to be different from the "Bico de Corvo" trees and very similar to "Cordovil 20", differing from it only by one different allele and is likely to be a "Cordovil" tree.

The unique allele 141 detected in the local cultivars "Bico 01", "Gama 06", "Gama 31" and in the oleaster "Zambu 17" (Supplementary Table S2) may suggest that these cultivars are the result of local tree domestication. This was similarly suggested for Moroccan trees [40,41] that some of the cultivated olive trees could result from local tree domestication. A similar conclusion can be made for the unnamed tree "Desco 05". Unique alleles at the OeUA-DCA05 and OeUA-DCA14 *loci* of wild material and alleles from local cultivars (GAPU71-B and GAPU103-A *loci*) suggest that "Desco 05" is autochthonous in origin. Earlier studies with Sardinian cultivated olive trees also suggested that local wild trees were involved in the domestication process [32].

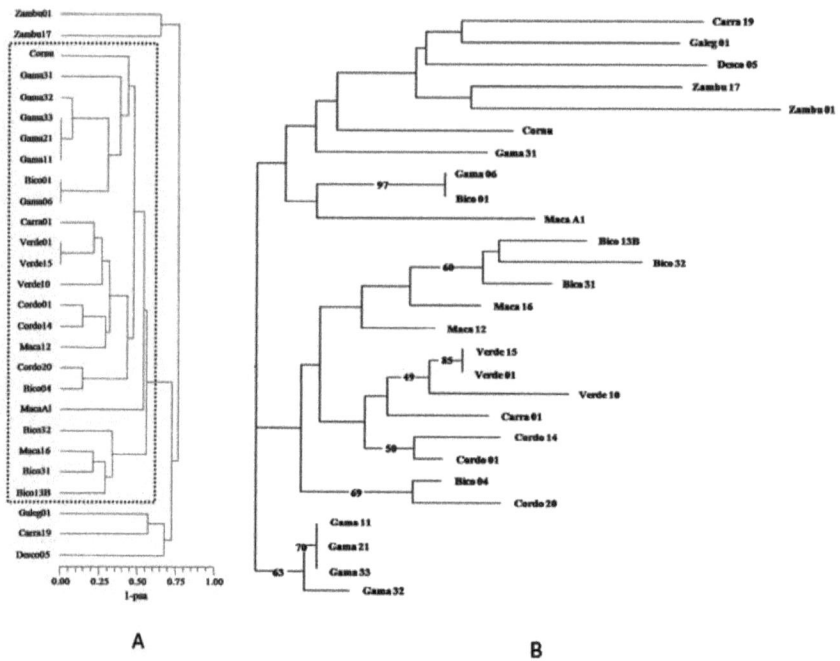

Figure 4. UPGMA (**A**) and Neighbor-Joining tree (**B**) of the 27 *Olea europaea* trees. All the cultivated olive trees are inside the dotted area (**A**). The tree construction was inferred from the bootstrapped dissimilarities, and a bootstrap value was given to each edge that indicates the occurrence frequency of this edge in the bootstrapped trees. Bootstrap values greater than 50 are indicated (**B**).

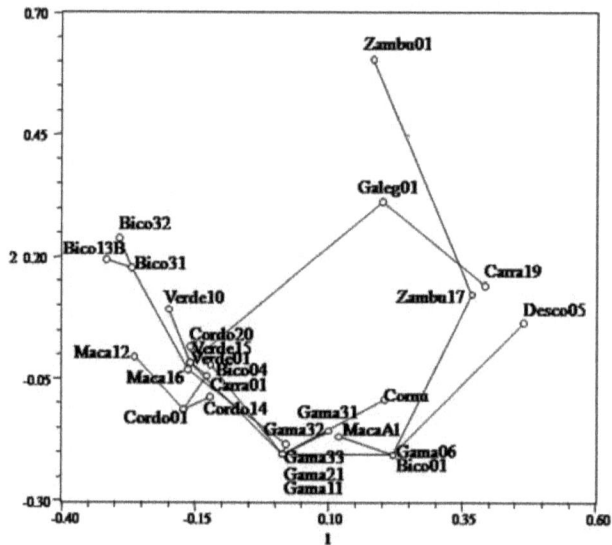

Figure 5. Principal co-ordinate analysis (PCoA) of the 27 *Olea europaea* trees based on the proportion-of-shared-alleles distance matrix. In all, 59.4% of the genetic variation was explained by the first two axes. The minimum length spanning tree was superimposed.

When performing the principal co-ordinate analysis (PCoA), it was found that considerable variation was explained by the first three axes (72.15%). The plane of the first two main PCoA axes accounted for 59.4% of the total genetic variation (Table 3).

Table 3. Percentage of variation explained by the first three axes as a result of a Principal co-ordinate analysis (PCoA) based on the proportion of shared alleles distance matrix for the 27 olive trees.

Axes	1	2	3
%	39.78	19.63	12.74
Cum %	39.78	59.41	72.15

The PCoA results are similar to those obtained by the UPGMA and the Neighbor-Joining algorithm. The wild trees ("Zambu 01" and "Zambu 17"), the unnamed tree ("Desco 05") and the cultivars not grown in Ficalho ("Maçanilha Algarvia–Maca Al" and "Cornezuelo–Cornu") are outliers. The local cultivars are distributed in three groups. "Bico de Corvo" ("Bico 13B", "Bico 31" and "Bico 32") and "Gama" ("Gama 11", "Gama 21", "Gama 32" and "Gama 33") constitute two separate groups. The remaining local varieties are clustered in the third group (Figure 5).

A comparison of the results obtained by endocarp characteristics and by means of SSR analysis showed that although both methods discriminate the 27 accessions in a similar way, some differences were observed as occurred for Italian cultivars from the Campanian region [42].The main differences relate to the "Verdeal" cultivar ("Verde 10" and "Verde 15") and the synonymy found between "Gama 21" and "Carra 19" using endocarp analysis which was not seen using SSR analysis (Figures 3–5). In summary, our data showed that the Ficalho area still has reasonable olive tree diversity. Such traditional orchards are important reservoirs of old minor cultivars and potential incubators of new genotypes. The genetic diversity observed in traditional farms is of great value and has been studied, for instance, in Morocco [40,41] and in Spain [39]. In Ficalho, there is a high probability that the observed genetic diversity of the traditional orchards is the result of mixed cultivars planting, of local trees originating from seed germination and of trees resulting from natural crosses between oleasters and domesticated trees. This was observed in the mountain region of Morocco, where apart from "Picholine marocaine", several other local varieties were found [40].

Keeping crop evolution on the farms contributes to the generation of a diversity of adaptive combinations of genes and traits in response to changing environmental conditions [43]. There is, presently, an increased risk of genetic erosion of olive germplasm due to the technological improvement in olive cultivation applied in the new commercial orchards [12,39].

It is recognized nowadays that *on farm* conservation is an important strategy of crop genetic resource conservation [38,44]. All over the Mediterranean region, traditional agroecosystems are of great importance as incubators of crop diversity [32,45,46] and as ecological infrastructures in general, providing essential ecological services. Portugal is very rich in traditional crop diversity, and a few studies [47–50] have demonstrated the existence of unexplored genetic resources. At Ficalho, the farmers have been developing an *on farm* conservation project by promoting the selection and utilization of local germplasm, and it is common to explore unnamed trees as a consequence of their agronomic characteristics. The *on farm* conservation is particularly important considering that, during the last several years, plant material from other provenances has been introduced, and some traditional orchards have been replaced by intensive systems in order to obtain higher yields. Some local producers are willing to adopt organic farming in these traditional orchards, thus producing higher quality products, in order to compensate for the financial loss associated to low yields and saving these orchards from being replaced. By promoting cultivar diversity, farmers benefit from the different fruiting times between cultivars, which allows staggering of field work such as harvest and olive oil mill work. They also benefit from the different tolerance/susceptibility to pests and diseases between cultivars [8,51,52], characteristics that may be used in breeding programs. Cultivar diversity also produces a diversity in flavors of the olives produced by a combination of cultivars.

4. Conclusions

Both methodologies used in this work showed that high genetic diversity still exists in the traditional orchards of Ficalho. Some cases of mislabeling were detected and it was found that several trees of "Gama" and "Verdeal Alentejana" are two distinct clonally propagated accessions. It was also found that the poorly represented cultivars "Gama" and "Bico de Corvo" are genetically distinct from all the other studied accessions, including the major local cultivars "Verdeal Alentejana" and "Cordovil de Serpa". "Gama", which is known to produce high quality oil, at a high yield, is specific to the Ficalho area and is a separate cluster from all the other accessions. Our work highlights the need to preserve minor cultivars in traditional orchards which can be reservoirs with great potential in the future.

Supplementary Materials: The following are available on line at www.mdpi.com/1424-2818/10/1/5/s1, Table S1: Endocarp characteristics of the 27 olive trees: Weight, Length, Diameter, Surface Roughness and Number of Vascular Bundles for the 27 olive trees. The values are mean values from 40 fruits from each tree, Table S2: SSR screening of 27 olive trees using 7 *loci*. The alleles sizes (in bp) are shown for each locus.

Acknowledgments: This work was supported by the Programme PRODER/SP3/Leader, project "Olival Tradicional". The authors are grateful to Ficalho's farmers, most particularly to Batista, Bento Sargento, Gemas and Lucas. We also thank Celina Matos for help of in endocarp characterization on a few accessions and Alexandra Seabra Pinto for her useful suggestions. Special thanks are due to Maria Costa Ferreira for checking the English writing.

Author Contributions: Maria Manuela Veloso designed the project. Maria Manuela Veloso and Cândido Pinto-Ricardo collected the plant material (fruits and leaves). Maria Manuela Veloso performed the endocarp characterization. Maria Cristina Simões-Costaperformed the DNA isolation and SSR analysis. Maria Manuela Veloso, Joana B. Guimarães and Luís C. Carneiro conceived and designed the statistical analysis. Maria Manuela Veloso, Célia Mateus and Cândido Pinto-Ricardo drafted the manuscript. Pedro Fevereiro critically revised the manuscript. All authors read and approved the final manuscript.

Conflicts of Interest: The authors declare no conflict of interest.

References

1. International Olive Council (IOC). *Market Newsletter*; International Olive Council: Madrid, Spain, 2017; Volume 121, p. 6.

2. Instituto Nacional de Estatística (INE). *EstatísticasAgrícolas 2016*, 1st ed.; Instituto Nacional de Estatística: Lisboa, Portugal, 2017; p. 171.

3. Breton, C.; Pinatel, C.; Médail, F.; Bonhomme, F.; Bervillé, A. Comparison between classical and Baysean methods to investigate the history of olive cultivars using SSR-polymorphisms. *Plant Sci.* **2008**, *175*, 524–532. [CrossRef]

4. Moreira, P.M.R.M.; Veloso, M.M. Landraces inventory for Portugal. In *European Landraces: On Farm Conservation, Management and Use*, 1st ed.; Vetelainen, M., Negri, V., Maxted, N., Eds.; Bioversity Technical Bulletin No. 15; Bioversity International: Rome, Italy, 2009; pp. 124–136, ISBN 978-92-9043-805-2.

5. Matos, M. As oliviculturas nacionais: Uma nova realidade em Portugal. In *Olival Tradicional: Contextos, Realidades e Sustentabilidade*, 1st ed.; Rota do Guadiana, Ed.; Rota do Guadiana: Serpa, Portugal, 2014; pp. 47–53, ISBN 978-989-98484-1-2.

6. Duarte, F.; Jones, N.; Fleskens, L. Traditional olive orchards on sloping land: Sustainability or abandonment? *J. Environ. Manag.* **2008**, *89*, 86–98. [CrossRef] [PubMed]

7. Gemas, V.J.; Rijo-Johansen, M.J.; Tenreiro, R.; Fevereiro, P. Inter and intra-varietal analysis of three *Olea europaea* L. cultivars using RAPD technique. *J. Hortic. Sci. Biotechnol.* **2000**, *75*, 312–319. [CrossRef]

8. Cidraes, F.G. Estudo das variedades de oliveira do Baixo Alentejo, Região de Serpa. In *Brigada Técnica da XVI Região Beja*; Série II, Numero 5; Direcção Geral dos Serviços Agrícolas: Beja, Portugal, 1939.

9. Veloso, M.M. Os agroecosistemas tradicionais na conservação da diversidadegenética da oliveira (*Olea europaea*) em Vila Verde de Ficalho. In *OlivalTradicional: Contextos, Realidades e Sustentabilidade*, 1st ed.; Rota do Guadiana, Ed.; Rota do Guadiana: Serpa, Portugal, 2014; pp. 147–153. ISBN 978-989-98484-1-2.

10. Veloso, M.M.; Reis, P.; Machado, D. *On farm* conservation of the olive tree (*Olea europaea*) landraces at Vila Verde de Ficalho (Portugal). *Landraces* **2015**, *3*, 16–17.

11. Fendri, M.; Trujillo, I.; Trigui, A.; Rodriguez-Garcia, I.M.; Ramirez, J.D.A. Simple Sequence Repeat identification and endocarp characterization of olive accessions in a Tunisian germplasm collection. *Hortscience* **2010**, *45*, 1429–1436.

12. Trujillo, I.; Ojeda, M.A.; Urdiroz, N.M.; Potter, D.; Barranco, D.; Rallo, L.; Diez, C.M. Identification of the worlwide olive germplasm bank of Córdoba (Spain) using SSR and morphological markers. *Tree Genet. Genomes* **2014**, *10*, 141–155. [CrossRef]

13. Sefc, K.M.; Lopes, M.S.; Mendonça, D.; Rodrigues Dos Santos, M.; Da Câmara Machado, M.L.; Da Câmara Machado, A. Identification of microsatellite loci in olive (*Olea europaea* L.) and their characterization in Italian and Iberian olive trees. *Mol. Ecol.* **2000**, *9*, 1171–1173. [CrossRef] [PubMed]

14. Carriero, F.; Fontanazza, G.; Cellini, F.; Giorio, G. Identification of simple sequence repeats (SSRs) in olive (*Olea europaea* L.). *Theor. Appl. Genet.* **2002**, *104*, 301–307. [CrossRef] [PubMed]

15. De la Rosa, R.; James, C.; Tobutt, K.R. Isolation and characterization of polymorphic microsatellite in Olive (*Olea europaea* L.) and their transferability to other genera in the Oleaceae. *Mol. Ecol. Notes* **2002**, *2*, 265–267. [CrossRef]

16. Baldoni, L.; Cultrera, N.G.; Mariotti, R.; Ricciolini, C.; Arcioni, S.; Vendramin, G.; Buonamici, A.; Porceddu, A.; Sarri, V.; Ojda, M.; et al. Consensus list of microsatellite markers for olive genotyping. *Mol. Breed.* **2009**, *24*, 213–231. [CrossRef]

17. Poljuha, D.; Sladonja, B.; Setic, E.; Milotic, A.; Bandelj, D.; Jakse, J.; Javornik, B. DNA fingerprinting of olive varieties in Istria (Croatia) by microsatellite markers. *Sci. Hortic.* **2008**, *115*, 223–230. [CrossRef]

18. Bracci, T.; Sebastiani, L.; Busconi, M.; Fogher, C.; Belaj, A.; Trujillo, I. SSR markers reveal the uniqueness of olive cultivars from the Italian region of Liguria. *Sci. Hortic.* **2009**, *122*, 209–215. [CrossRef]

19. Delgado-Martinez, F.J.; Amaya, I.; Sánchez-Sevilla, J.F.; Gomez-Jimenez, M.C. Microsatellite marker-based identification and genetic relationships of olive cultivars from the Extremadura region of Spain. *Genet. Mol. Res.* **2012**, *11*, 918–923. [CrossRef] [PubMed]

20. Roubos, K.; Moustakas, M.; Aravanopoulos, F.A. Molecular identification of greek olive (*Olea europaea*) cultivars based on microsatellite loci. *Genet. Mol. Res.* **2010**, *9*, 1865–1876. [CrossRef] [PubMed]

21. Bari, A.; Martin, A.; Boulouha, B.; Gonzalez-Andujar, J.L.; Barranco, D.; Ayad, G.; Padulosi, S. Use of fractals and moments to describe olive cultivars. *J. Agric. Sci.* **2003**, *141*, 63–71. [CrossRef]

22. Navero, D.B.; Cimato, A.; Fiorino, P.; Romero, L.R.; Touzani, A.; Castañeda, C.; Serafin, F.; Trujillo, I. *World Catalogue of Olive Varieties*, 1st ed.; International Olive Oil Council: Madrid, Spain, 2000; pp. 15–21. ISBN 84-931663-3-2.

23. Rohlf, J.F. *NTSYS-pc: Numerical Taxonomy and Multivariate Analysis System, Version 2.1*; Exeter Software: Setauket, NY, USA, 2000.

24. Van Oosterhout, C.; Hutchinson, W.F.; Wills, D.P.M.; Shipley, P. Micro-checker: Software for identifying and correcting genotyping errors in microsatellite data. *Mol. Ecol. Notes* **2004**, *4*, 535–538. [CrossRef]

25. Goudet, J. FSTAT (version 1.2): A computer program to calculate F-statistics. *J. Hered.* **1995**, *86*, 485–486. [CrossRef]

26. Nei, M.; Li, W.H. Mathematical model for studying genetic variation in terms of restriction endonucleases. *Proc. Natl. Acad. Sci. USA* **1979**, *76*, 5269–5273. [CrossRef] [PubMed]

27. Minch, E.; Ruiz-Linares, A.; Goldstein, D.; Feldman, M.; Cavalli-Sforza, L.L. *MICROSAT: A Computer Program for Calculating Various Statistics on Microsatellite Allele Data (Version 1.5d)*; Stanford University: Stanford, CA, USA, 1997. Available online: http://hpgl.stanford.edu/projects/microsat (accessed on 15 April 2016).

28. Perrier, X.; Jacquemoud-Collet, J.P.D. ARwin Software. 2006. Available online: http://darwin.cirad.fr/ (accessed on 20 June 2016).

29. Cantini, C.; Cimato, A.; Sani, G. Morphological evaluation of olive germplasm present in Tuscany region. *Euphytica* **1999**, *109*, 173–181. [CrossRef]

30. Lopes, S.M.; Mendonça, D.; Sefc, K.M.; Gil, S.F.; Câmara-Machado, A. Genetic evidence of intra cultivar variability within Iberian olive cultivars. *HortScience* **2004**, *39*, 1562–1565.

31. Gomes, S.; Martins-Lopes, P.; Lopes, J.; Guedes-Pinto, H. Assessing genetic diversity in *Oleaeuropaea* L. using ISSR and SSR markers. *Plant Mol. Biol. Report.* **2009**, *27*, 365–373. [CrossRef]

32. Erre, P.; Chessa, I.; Muñoz-Diez, C.; Belaj, A. Genetic diversity and relationships between wild and cultivated olives in Sardinia as assessed by SSR markers. *Genet. Resour. Crop Evol.* **2010**, *57*, 41–54. [CrossRef]

33. Gemas, V.J.V.; Almadanim, M.C.; Tenreiro, R.; Martins, A.; Fevereiro, P. Genetic diversity in the Olive tree (*Olea europaea* L. subsp. *europaea*) cultivated in Portugal revealed by RAPD and ISSR markers. *Genet. Resour. Crop Evol.* **2004**, *51*, 501–511. [CrossRef]

34. Martins-Lopes, P.; Lima-Brito, J.; Gomes, S.; Meirinhos, J.; Santos, L.; Guedes-Pinto, H. RAPD and ISSR molecular markers in *Olea europaea* L.: Genetic variability and molecular cultivar identification. *Genet. Resour. Crop. Evol.* **2007**, *54*, 117–128. [CrossRef]

35. Belaj, A.; Satovic, Z.; Rallo, L.; Trujillo, I. Genetic diversity and relationships in olive (*Olea europaea* L.) germplasm collections as determined by randomly amplified polymorphic DNA. *Theor. Appl. Genet.* **2002**, *105*, 638–644. [CrossRef] [PubMed]

36. Terzopoulos, P.J.; Kolano, B.; Bebeli, P.J.; Kaltsikes, P.J. Identification of *Olea europaea* L. cultivars using inter simple sequence repeat markers. *Sci. Hortic.* **2005**, *105*, 45–51. [CrossRef]

37. Belaj, A.; Muñoz-Diez, C.; Baldoni, L.; Satovic, Z.; Barranco, D. Genetic diversity and relationships of wild and cultivated olives at regional level in Spain. *Sci. Hortic.* **2010**, *124*, 323–330. [CrossRef]

38. Bakkali, A.; Haouane, H.; Hadiddou, A.; Oukabli, A.; Santoni, S.; Udupa, S.M.; Van Damme, P.; Khadari, B. Genetic diversity of on-farm selected olive trees in Moroccan traditional olive orchards. *Plant Genet. Resour.* **2012**, *11*, 97–105. [CrossRef]

39. Díez, C.M.; Trujillo, I.; Barrio, E.; Belaj, A.; Barranco, D.; Rallo, L. Centennial olive trees as a reservoir of genetic diversity. *Ann. Bot.* **2011**, *108*, 797–807. [CrossRef] [PubMed]

40. Ouazzani, N.; Lumaret, R.; Villemur, P. Genetic variation in the olive tree (*Olea europaea* L.) cultivated in Morocco. *Euphytica* **1996**, *91*, 9–20. [CrossRef]

41. Khadari, B.; Charafi, J.; Moukhli, A.; Ater, M. Substantial genetic diversity in cultivated Moroccan olive despite a single major cultivar: A paradoxical situation evidenced by the use of SSR loci. *Tree Genet. Genomes* **2008**, *4*, 213–221. [CrossRef]

42. Corrado, G.; La Mura, M.; Ambrosino, O.; Pugliano, G.; Varrichio, P.; Rao, R. Relationships of Campanian olive cultivars: Comparative analysis of molecular and phenotypic data. *Genome* **2009**, *52*, 692–700. [CrossRef] [PubMed]

43. Bellon, M.R.; Gotor, E.; Caracciolo, F. Conserving landraces and improving livelihoods: How to assess the success of *on-farm* conservation projects? *Int. J. Agric. Sustain.* **2015**, *13*, 167–182. [CrossRef]

44. Zhang, D.; Gardini, E.A.; Motilal, L.A.; Baligar, V.; Bailey, B.; Zuñiga-Cernades, L.; Arevalo-Arevalo, C.E.; Meinhardt, L. Dissecting genetic structure in farmer selections of *Theobroma cacao* in the Peruvian Amazon: Implications for *on farm* conservation and rehabilitation. *Trop. Plant Biol.* **2011**, *4*, 106–116. [CrossRef]

45. Baldoni, L.; Tosti, N.; Ricciolini, N.; Belaj, A.; Arcioni, S.; Pannelli, G.; Germana, M.A.; Mulas, M.; Porceddu, A. Genetic structure of wild and cultivated olives in the Central Mediterranean Basin. *Ann. Bot.* **2006**, *98*, 935–942. [CrossRef] [PubMed]

46. Achtak, H.; Ater, M.; Oukabli, A.; Santoni, S.; Kjellberg, F.; Khadari, B. Traditional agroecosystems as conservatories and incubators of cultivated plant varietal diversity: The case of fig (*Ficus carica* L.) in Morocco. *BMC Plant Biol.* **2010**, *10*, 28. [CrossRef] [PubMed]

47. Almandanim, M.C.; Baleiras-Couto, M.M.; Pereira, H.S.; Carneiro, L.C.; Fevereiro, P.; Eiras-Dias, J.E.; Morais-Cecílio, L.; Viegas, W.; Veloso, M.M. Genetic diversity of the grapevine (*Vitis vinifera* L.) cultivars most utilized for wine production in Portugal. *Vitis* **2007**, *46*, 116–119.

48. Queiroz, A.; Assunção, A.; Ramadas, I.; Viegas, W.; Veloso, M.M. Molecular characterization of Portuguese pear landraces (*Pyrus communis* L.) using SSR markers. *Sci. Hortic.* **2015**, *183*, 72–76. [CrossRef]

49. Monteiro, F.; Vidigal, P.; Barros, A.B.; Monteiro, A.; Oliveira, H.R.; Viegas, W. Genetic distinctiveness of rye in situ accessions from Portugal unveils a hotspot of unexplored genetic resources. *Front. Plant Sci.* **2016**, *7*, 1334. [CrossRef] [PubMed]

50. Leitão, S.T.; Dinis, M.; Veloso, M.M.; Satovic, Z.; Vaz-Patto, M.C. Establishing the bases for introducing the unexplored portuguese common bean germplasm into the breeding world. *Front. Plant Sci.* **2017**, *8*, 1296. [CrossRef] [PubMed]

Diversity **2018**, *10*, 5

51. Cordeiro, A.; Santos, M.L.; Morais, N.; Miranda, A. As variedades de oliveira, Portugal oleícola. In *O Grande Livro da Oliveira e do Azeite*, 1st ed.; Bohm, J., Godinho, C., Coelho, F., Eds.; Dinalivro: Lisboa, Portugal, 2013; pp. 188–220, ISBN 9789725766200.

52. Cordeiro, A.; Inês, C.; Morais, N. Principais cultivares de oliveira existentes em Portugal. In *Boas Práticas No Olival e No Lagar*, 1st ed.; InstitutoNacional de Investigação Agrária e Veterinária: Oeiras, Portugal, 2014; pp. 44–54, ISBN 978-972-579-041-0.

diversity

MDPI

Article

Patterns of Spontaneous Nucleotide Substitutions in Grape Processed Pseudogenes

Andrea Porceddu * (ID) and **Salvatore Camiolo** (ID)

Dipartimento di Agraria, University of Sassari, 07100 Sassari, Italy; scamiolo@uniss.it
* Correspondence: aporceddu@uniss.it; Tel.: +39-079-229-224

Received: 17 July 2017; Accepted: 9 October 2017; Published: 13 October 2017

Abstract: Pseudogenes are dead copies of genes. Owing to the absence of functional constraint, all nucleotide substitutions that occur in these sequences are selectively neutral, and thus represent the spontaneous pattern of substitution within a genome. Here, we analysed the patterns of nucleotide substitutions in *Vitis vinifera* processed pseudogenes. In total, 259 processed pseudogenes were used to compile two datasets of nucleotide substitutions. The ancestral states of polymorphic sites were determined based on either parsimony or site functional constraints. An overall tendency towards an increase in the pseudogene A:T content was suggested by all of the datasets analysed. Low association was seen between the patterns and rates of substitutions, and the compositional background of the region where the pseudogene was inserted. The flanking nucleotide significantly influenced the substitution rates. In particular, we noted that the transition of G→A was influenced by the presence of C at the contiguous 5′ end base. This finding is in agreement with the targeting of cytosine to methylation, and the consequent methyl-cytosine deamination. These data will be useful to interpret the roles of selection in shaping the genetic diversity of grape cultivars.

Keywords: processed pseudogenes; nucleotide substitution; transitions and tranversions; cytosine methylation and deamination; neutral sequence evolution

1. Introduction

Sequence diversity is generated by mutations that are transmitted across generations due to evolutionary forces. Understanding the patterns and frequencies of spontaneous mutations is very important, as these generate the molecular basis for gene and genome evolution [1–3]. The availability of sequence polymorphism data for a high number of individuals of one species can provide a wealth of information on the spectra and dynamics of nucleotide substitutions. However, most of these studies are conducted without any assessment of the selective constraints on the sites analysed, and thus it can be difficult to deduce how an identified substitution spectrum deviates from expectations constructed under the assumption of neutrality.

Several reports have demonstrated that the rate of nucleotide substitutions can be significantly influenced by the adjacent nucleotides, i.e., the probability that a nucleotide is substituted depends on the identity of its neighbouring nucleotides [4–6]. Datasets of nucleotide polymorphisms generated by next-generation sequencing usually do not report the identity of the neighbouring (unchanged) nucleotides, and thus information on the nucleotide context in which a mutation has occurred is not readily accessible.

A widely used approach to infer patterns and rates of nucleotide substitution at selectively unconstrained sites is based on the analysis of pseudogenes [7,8]. These are inactive copies of genes, and they are believed to experience very weak, if any, selective pressures [6,7,9]. Based on the mechanisms leading to their formation, pseudogenes are usually classified into two types: (i) duplicated (or non-processed), which are generated by genomic duplication and subsequent function disabling due to mutation; and (ii) processed (or retro-processed), which originate from genomic integration of a cDNA copy retro-transcribed from a spliced mRNA molecule [10,11]. The identification of either pseudogene type is generally based on their inferred intron–exon structure. Duplicated pseudogenes usually retain introns and other regulatory non-transcribed sequences, such as promoters and terminators, while processed pseudogenes are depleted of introns and of all untranscribed regulatory regions. The absence of regulatory elements means that processed pseudogenes are inactive, as at the time of their creation, and for this reason they were defined as "dead on arrival" (DOA) elements [8,12]. An important feature of DOA elements for diversity studies is that they represent a comprehensive catalogue of mutation events. In contrast, some mutations that occur in a non-processed pseudogene before its inactivation might potentially have been lost due to the action of purifying selection.

The determination of the spontaneous mutation patterns through an analysis of pseudogenes also poses some critical issues. Several lines of evidence have recently challenged the view that all DOA elements should evolve neutrally [9,13]. In this regard, an indication of the action of natural selection on a sequence can be obtained by studying the ratio of synonymous to non-synonymous substitutions. In theory, genes that evolve under purifying selection are expected to experience fewer non-synonymous than synonymous substitutions. On the contrary, synonymous and non-synonymous substitutions should be comparably frequent in pseudogenes, if these evolve neutrally.

As mutations are detected as changes in pairwise sequence comparisons, it is very difficult to tell in which of the compared sequences the mutations have occurred. Several possible solutions to this problem have been proposed [9]. The first and second base positions of codons generally determine the coding specificity. A change at these positions will be subjected to purifying selection in the functional locus, but not in the pseudogenes; thus, changes at these sites can be bona-fide attributed to the pseudogene [9]. An alternative solution is to use an orthologous sequence from a closely related species in combination with parsimony to infer the sequence that underwent the base change. In practice, given a polymorphic site, the substitution is attributed to the pseudogene only if the functional paralogous and orthologous sequences share the same nucleotide at that position [6,14,15].

The pattern of spontaneous mutation can vary across genomic regions, and therefore the genomic distribution of pseudogenes should be adequately scattered to represent the various genomic backgrounds. Analysis of spontaneous mutation patterns based on pseudogene–functional loci comparisons has been carried out for several mammalian systems [6,8,14], while there is little information available relating to plant genomes. Benovoy et al. [9] analysed 411 processed pseudogenes in *Arabidopsis thaliana* and reported that the spontaneous pattern of mutation in *A. thaliana* is different from that of mammals, but similar to that of *Drosophila*. Transitions were more abundant than expected by chance, but in contrast to other systems, they were less frequent than transversions. A study conducted by Ossowsky et al. [16] in resequenced *Arabidopsis* progenies, however, indicated that transitions were 2.5-fold more frequent than transversions. Two independent studies examined the variation dynamics of substitution rates in relation to regional and flanking nucleotides using datasets of single nucleotide polymorphisms that were constructed by resequencing of genomic portions of several individuals of *A. thaliana* and *Zea mays* (maize) [4,5]. Both of these studies reported significant correlation between mutation frequency and regional A + T content. However, the relationships pointed in different directions in these two species. There was a direct correlation between mutation frequency and A+T content in *A. thaliana*, whereas the opposite trend was reported in maize. These data demonstrated that some of the factors that influence the substitution rates can have different weights across different species.

The so-called methylation/deamination cycle that defines an increased rate of transition from cytosine to thymine due to rapid deamination of the methylated cytosine at CpG sites is one of the most prominent effects of nucleotide context on substitution rates in plant genomes [4,5,9]. Other context dependencies of the substitution rates have been reported in both *Arabidopsis* and maize, and in some cases, these effects have proved to be species-specific [4,5].

In the present study, we analysed the patterns of spontaneous base substitutions in *V. vinifera*, as detected through the analysis of 259 processed grape pseudogenes. These data demonstrate that the grape genome is almost at compositional equilibrium, and that transitions are more frequent than transversions. Some of the neighbouring nucleotide effects seen for other species are also confirmed, although very low regional effects on nucleotide substitutions emerged from this analysis of *V. vinifera*. Although our results are limited to a single grapevine line, and thus may need to be confirmed on other cultivars, we believe the findings reported here will be useful for comparisons with spectra of mutations deduced from re-sequencing studies with grape cultivars. These studies are expected to shed light on how natural selection impacts on sequence diversity.

2. Materials and Methods

2.1. Genomic Sequence

The *V. vinifera* genomic sequence, and the gene models and annotations, were downloaded from http://genomes.cribi.unipd.it/DATA/. As the reference genome sequence was determined from genomic DNA of cultivar Pinot 40024, all of the sequences used in this study should be referred to this cultivar [17]. Coding sequences, and intron and intergenic sequences were obtained using the gff2sequence software [18].

2.2. Pseudogene Identification

Processed pseudogenes were identified as described by Camiolo and Porceddu [19]. In brief, the sequences of the *V. vinifera* exons were extended (at both ends) with additional nucleotides (exon tails) from the neighbouring genomic regions [20]. As in Zheng and Gerstein (2006) [20], the length of these exon tails were 51, 52 or 53 nucleotides, which depended upon whether the exon started/ended with the first, second or third position of a codon. The tailed exon sequences were converted into amino-acid sequences and used as queries for tblastn searches against the *V. vinifera* reference genome sequence, which was previously masked in repetitive regions (Figure 1, Step A). The extra nucleotides are important for the separation of duplicated from processed pseudogenes, as they allow the queries to align to the pseudo-intron–exon boundaries of duplicated pseudogenes.

All of the tblastn hits with identity >40% and e-value <10^{-10} were kept for further analysis [20]. Adjacent hits that were identified by query sequences derived from adjacent exons of a functional locus were assembled in a pseudogene model (Figure 1, Steps B, C). We controlled for the distance between the assembled hits before incorporation into the model, by imposing their genomic distance as shorter by 500 nucleotides (base pairs; bp) than the distance between the matching exons added. This tolerance was set to take into account possible insertions within the identified pseudogenes [20], and the 500 bp was derived by a trial and error process, to reduce the risk of predicting pseudogene mosaics derived from independent pseudogenisation events. If present, the genomic region between the two hits was considered as a putative pseudo-intron, and the hits were assembled in a model with (eventually) an intervening (pseudo-)intron between the two (pseudo-)exons [21].

As paralogous loci identify (homologous) pseudogenes in overlapping chromosomal regions, pseudogene models that overlapped by ≥20% of their length were clustered together. Pseudogene models that overlapped by <20% of their length were assigned to different clusters, as these might represent either pseudogenes generated by nested insertions of homologous sequences, or pseudogene models generated by loci containing repetitive regions. The clusters identified at this step were defined as 'pseudogenic regions' (Figure 1, Steps C, D). For each pseudogenic region, the pseudogene model-functional locus pairs with the highest scores and the lowest e-values were selected for further analysis. The functional loci selected at this stage were referred to as the pater locus of the pseudogene, although they should be considered as a pseudogene sibling (i.e., functional paralogues).

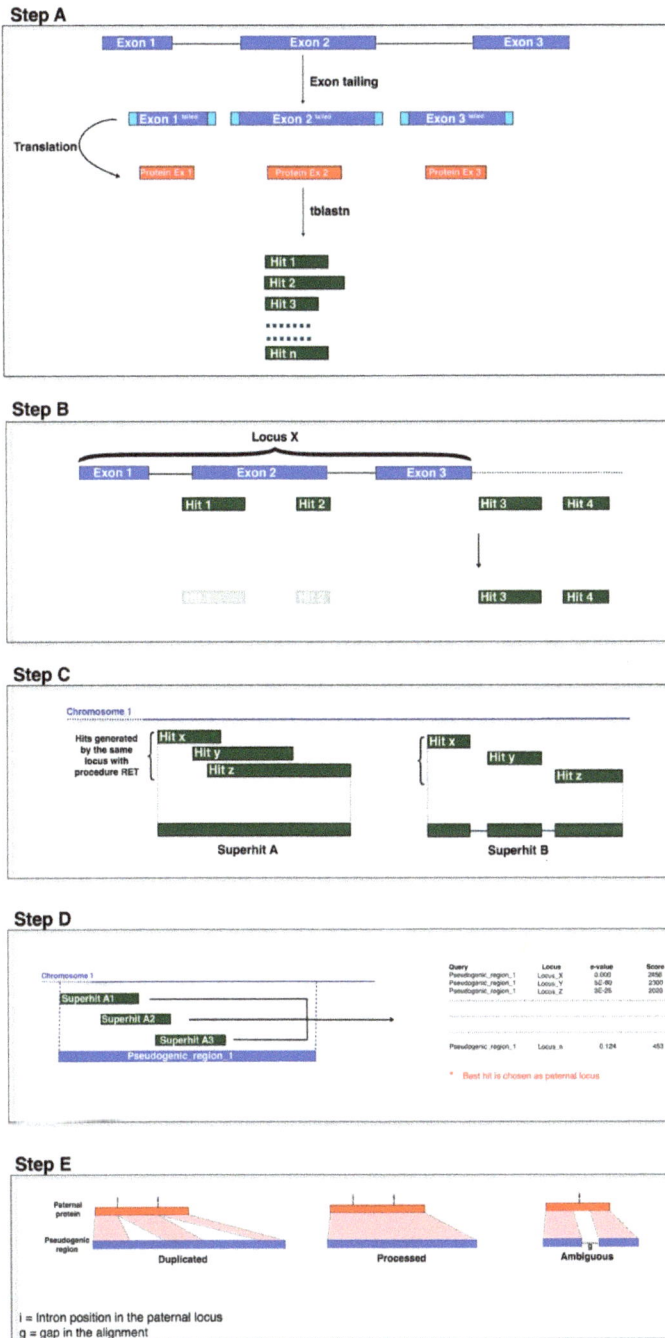

Figure 1. The pipeline outline. The pseudogenes were identified by sequence homology to functional exons using tblastn. Step **A**: Query preparation and tblastn homology search. Step **B**: Hit filtering. Step **C**: Hit merging to generate pseudogenic models. Step **D**: Pater locus assignment. Step **E**: Pseudogene classification.

The pseudogene models were classified as processed, non-processed and fragmented, based on the deduced exon–intron structure, e.g., the pseudogene sequences were aligned to the protein sequence of their pater locus and the deduced intron–exon structure was determined using the Genewise algorithm [22]. Here, blastn was run between each pseudogene model sequence and the introns of the pater locus, as reported by Khelifi et al. [23] (i.e., length > 50, no similarity threshold, e-value > 10^{-5}). This step is not shown in Figure 1. Pseudogene models with introns inferred at analogous positions of the exon–intron structure of the pater locus were classified as non-processed. Pseudogenes formed by at least two hits, that matched two exons without predicted pseudo-introns, and that showed no homology to the intron sequence of the pater locus in blastn searches were classified as processed and were kept for further sequence analyses (Figure 1, Step E).

Finally, as a pseudogene sequence can be generated by genomic duplication of a pre-existing pseudogene, we studied the phylogeny of pseudogenes in relation to the correspondent functional loci. The pseudogenes and functional loci were grouped into clusters using the CD-Hit software [24], with a sequence similarity threshold of 60% [24]. For each cluster of sequences, a neighbour-joining tree was constructed using ClustalW [25]. Only the processed pseudogenes that showed the highest homologies to the functional locus were inferred to be derived directly from retropositions of functional loci (i.e., primary pseudogenes), and these were kept for further analysis.

2.3. Ka/Ks Rate Determination

The number of synonymous substitutions per synonymous site (*Ks*) and of non-synonymous substitutions per non-synonymous site (*Ka*) were deduced using the method of Li et al. [26,27]. The *Ka* and *Ks* rates were calculated for pairwise alignments of collinear functional paralogues. The collinear paralogues were identified using the MCScanX software, with the default settings [28]. The *Ka* and *Ks* values were also calculated for alignments that involved the pseudogenes and their pater loci. The estimates obtained for the alignment involving the pseudogenes were slightly biased (e.g., underestimated) because the functional gene sequence is expected to evolve at a lower rate than the pseudogenes.

2.4. Analysis of Nucleotide Substitutions

The pseudogene-predicted amino-acid sequences were obtained by aligning the functional protein encoded by the pater locus and the pseudogene sequence using Genewise [22]. The pseudo-proteins were aligned to protein sequences that were encoded by functional proteins, using MUSCLE with the default settings [29]. The aligned (pseudo-)protein was then converted to a sequence of codons using a perl script. The substitutions between complimentary nucleotide pairs were joined and represented with a unique code. For example, the two complimentary events of A substituted for G (A→G), and hence T substituted for C (T→C), or vice versa, were joined and represented as A:T→G:C.

The directions of the nucleotide substitutions were inferred using the following two approaches.

2.4.1. The Non-Synonymous Dataset

Following Benovoy et al. [9], the substitutions at the first and second codon positions were assumed to have occurred in the pseudogene. This assumption is supported by the consideration that most mutations at these positions will change the codon coding specificity and hence will be filtered by purifying selection in the functional locus.

2.4.2. The Parsimony-Based Dataset

The substitutions were inferred as those that occurred in the pseudogene by applying parsimony to multiple alignments between the pseudogene, the pater locus, and an orthologous sequence from *A. thaliana*. In brief, the alignment positions that featured the same nucleotide in the pater locus and in its orthologues but showed a different nucleotide in the pseudogene were considered for further analysis [6].

3. Results

3.1. Identification of Primary Processed Pseudogenes

Processed pseudogenes were identified in *V. vinifera* through the homology search approach outlined in Figure 1 [19]. As processed pseudogenes are generated by genomic integration of a cDNA copy that is retro-transcribed from a spliced transcript, they are believed to be intronless. The pseudogene structure was inferred through the study of the pseudogene–pater locus alignments at intron–exon junctions, and by analysis of the sequence homology between the pseudogenes and the pater locus intron sequences.

In practice, a pseudogene was classified as processed (or retro-processed) if (i) it was identified by at least two matches to two consecutive exons of an intron-containing locus; (ii) the alignment spanned the exon–intron position; and (iii) no sequence homology to paralogous intron sequences was detected. Other diagnostic features showed low predictive power, such as for the presence of adenine-rich regions at the 3' end, or target site duplication, and thus these were not considered. For an example of a pseudogene–pater locus alignment, see Supplementary Figure S1.

In theory, an intronless pseudogene might originate from genomic duplication of a pre-existing processed pseudogene sequence. These duplicated pseudogene sequences will share the nucleotide changes that occurred in the common ancestor of these pseudogenes (i.e., from its generation by retroposition until genomic duplication) and thus provide (partially) redundant information for mutation analysis. To avoid such redundancy, the pseudogenes and functional paralogues were grouped into clusters based on sequence similarities. For each cluster, a phylogenetic tree was constructed, and only the pseudogene sequences with functional paralogues as their closest homologues (i.e., primary pseudogenes) were retained for further analysis resulting in 259 primary processed pseudogenes (Supplementary Table S1).

3.2. Compositional Features of Functional Genes and Pseudogenes

The distributions of the G + C content of the coding sequences (Figure 2) or at the synonymous codon positions (Supplementary Figure S2, G3 + C3) were unimodal, with means of 44% and 50% (see Figure 2). The introns (Supplementary Figure S2) and intergenic sequences (Figure 2) showed lower G + C content, with means of 33% and 32%, respectively. The *V. vinifera* pseudogenes showed a mean G + C content of 42% (Figure 2), and a mean G + C content at the third codon position of 48% (Supplementary Figure S2). These data were extrapolated from the reference genome sequence [17].

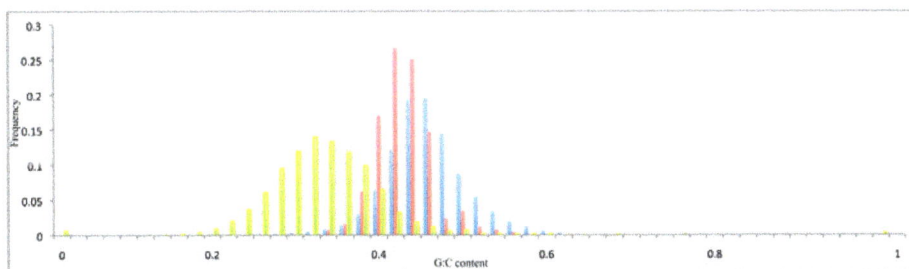

Figure 2. Relative G + C content distribution of the pseudogenes (red), the coding sequences of the functional genes (blue), and the intergenic sequences (yellow).

3.3. Pseudogene Evolutionary Rates

As pseudogenes are non-functional, they are expected to be under very weak selective constraints. Under this assumption, the non-synonymous (Ka) and synonymous (Ks) substitution rates should be equal, and hence the Ka/Ks ratio should approach unity [1,7]. Although all of the pseudogenes used in this study appeared to be disabled, their sequence evolutionary rates were analysed, as several studies have demonstrated that pseudogenes can undergo exaptation for other functions [9,13].

As most amino-acid changes (i.e., non-synonymous substitutions) are selected against in coding sequences, functional genes are expected to have significantly lower Ka/Ks ratio than pseudogenes. As illustrated in Figure 3, the Ka/Ks ratio of the pseudogenes was shifted significantly towards higher values compared to that for the functional genes ($p < 0.01$). However, the pseudogene Ka/Ks ratio did not peak at the expected value of 1.0. Note that the pseudogene sequences were compared to their 'sibling' functional sequences, and not to their true ancestral pater locus. This unavoidable approximation may have inflated the rate estimations [30].

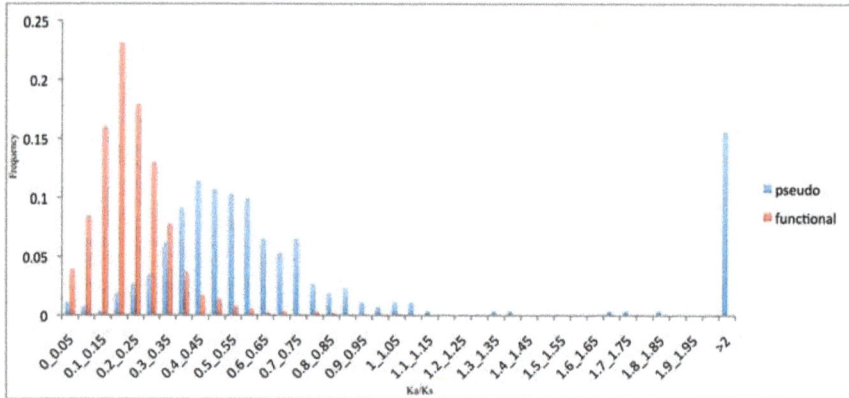

Figure 3. Pseudogene evolutionary rates. Frequency distributions of the Ka/Ks ratios of the functional genes (red) and pseudogenes (blue).

3.4. Transitions Are More Freequent Than Transversions

Two datasets of substitutions were analysed. The non-synonymous dataset (see Methods) comprised nucleotide substitutions at the first and second codon base positions. Generally, substitutions at these positions change the coding specificity of the codon, and are thus counter-selected in the functional locus. On the other hand, the parsimony-based dataset (see Methods) comprised the substitutions identified as single nucleotide variations with respect to the consensus nucleotide between the pater locus and orthologous locus at the corresponding position.

For each nucleotide type (i.e., adenine, guanine, cytosine, thymine), Table 1 provides information on the number of positions inferred to be unchanged in the pseudogene, the positions featuring substitutions in the pseudogene, and those missing in the aligned pseudogene sequence. The substitutions were classified as transversions if a purine was changed for a pyrimidine (or vice versa), or as transitions if they involved only purines or pyrimidines. Cytosine and guanine in CpG dinucleotides were treated separately.

Cytosine and guanine showed the highest frequencies of transitions (Table 1). In particular, cytosine and guanine that were part of the CpG dinucleotides were very frequently substituted by transitions. The CpG dinucleotide is a common target for cytosine methylation, and methylated cytosine is frequently converted to thymine by deamination. These data confirmed the high frequencies of transitions at CpG sites. However, high frequencies of transitions were also seen for non-CpG dinucleotides, which suggested that the methylation/deamination cycle alone cannot explain the high prevalence of transitional substitutions in these grape pseudogenes. The overall patterns detected in the two datasets were highly similar (see Supplementary Table S2 for an analogous summary of the non-synonymous dataset).

Table 1. Summary of nucleotide substitutions in the grape processed pseudogenes from the parsimony-based dataset.

Nucleotide Type	Group	Total (n)	Unchanged (n (%Total))	Deletions (n (%Total))	Substitutions [a]			
					Total	Transitions (n (%Substitutions))	Transversions (n (%Substitutions))	Transitions: Transversions Ratio
A	All	13,546	12,655 (93.42)	198 (1.46)	693 (5.12)	421 (60.75)	272 (39.25)	1.55
G	All	11,928	10,476 (87.83)	172 (1.44)	1280 (10.73)	994 (77.65)	286 (22.34)	3.48
	CpG	429	343 (79.95)	6 (1.39)	80 (18.65)	63 (78.75)	17 (21.25)	3.71
	Non-CpG	11,499	10,133 (88.12)	166 (1.44)	1200 (10.44)	931 (77.58)	269 (22.42)	3.46
C	All	8618	7516 (87.21)	102 (1.18)	1000 (11.60)	786 (78.60)	214 (21.40)	3.67
	CpG	429	336 (78.32)	6 (1.39)	87 (20.28)	81 (93.10)	6 (6.90)	13.50
	Non-CpG	8189	7180 (87.68)	96 (1.34)	913 (11.15)	705 (77.21)	208 (22.78)	3.39
T	All	13,835	12,955 (93.64)	158 (1.14)	722 (5.22)	446 (61.77)	276 (38.23)	1.62

[a] Transversion, purine changed for pyrimidine (or vice versa); transitions and substitutions that involve only purines or pyrimidines.

3.5. Effects of Compositional Background on Nucleotide Substitution

It has been reported that nucleotide substitution rates can vary among compositionally different genomic regions. To gain insight into this phenomenon (i.e., regional effects), the pseudogenes were divided into four groups based on the G + C content of the flanking genomic regions (30,000 bp at both sides of each pseudogene). In all of these pseudogene groups, most of the substitutions occurred by transitions with an overall balance towards a decrease in G:C content (see Figure 4a).

The substitution rates between the nucleotides were also investigated. Following the approach of Zhang and Gerstein [6], the substitution rate was considered to comprise two rates: one that measures the frequency of the change of one nucleotide to another (i.e., substitution rate); and the other that reports the relative proportions of the changes normalised to the identified substitutions (i.e., substitution preferentiality). For example, if we are studying the substitution of nucleotide j for i, we calculate the substitution rate as the ratio between the changes from j to i divided by the total number of nucleotides j in the pseudogene at the time of its integration. This index describes how often the j→i substitution occurs in a sequence. The index of substitution can be preferentiality obtained by normalising the substitutions j→i for the total number of nucleotides j that have been substituted; in other words, the number of j→i changes can be divided by the total number of substitutions that have involved the j nucleotide in the pseudogene. Substitutions between complementary nucleotide pairs were added together. For example, to study the A:T→G:C substitutions, we added up the adenines that were substituted by guanines, and the thymines that were substituted by cytosines. Figure 4a reports the substitution rates (for the parsimony-based dataset) between the nucleotides, and Figure 4b reports the substitution preferentialities (for the parsimony-based dataset). The analogous data for the substitution rates and preferentiality calculated for the non-synonymous dataset are reported in Supplementary Figure S3a,b.

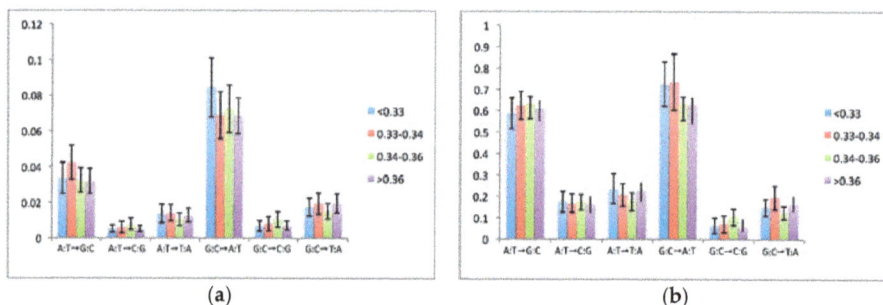

(a) (b)

Figure 4. Substitution patterns between the nucleotide pairs. (**a**) Substitution rates as normalised by the numbers of nucleotides of each type. Confidence intervals (95%) are given. (**b**) Substitutions rate (preferentiality) as normalised by the proportions of nucleotides that have mutated to another type. Reported values are intended as percentages. Each bar represents the sum of the substitution rates for complementary nucleotide pairs. As a way of example, A:T→G:C derives from the analysis of substitutions A→G and T→C.

The transitions of G:C→A:T were more frequent than the reverse, of A:T→G:C, which is in agreement with the higher rates of substitutions of G:C for A:T (Figure 4a). Such a bias towards an A:T increase was also observed for the transversions, as the G:C→T:A transversions were more frequent than the reverse, A:T→C:G. The rates of the G:C→A:T transitions were associated with the regional G:C content, with higher rates of these transitions in the groups with lower regional G:C content. Such an effect was, however, less evident for the non-synonymous dataset (see Supplementary Figure S3a).

A different picture emerged in the analysis of the substitution preferentiality (Figure 4b). The differences in the preferentiality rates among the two transitional types were less imbalanced in favour of the substitutions, indicating an A:T increase. Also, the effects of the regional G:C composition on the substitution preferentiality were less pronounced.

3.6. Neigbouring Effects on Nucleotide Substitutions

It has been proposed that nucleotide substitution rates are not only under regional effects, but also local effects, e.g., the effects of the nucleotides adjacent to the site that is mutated [4–6]. This factor is referred to as a neighbouring effect on substitutions. In analogy with the analysis on mononucleotides, the dinucleotide substitution rate was comprised of two factors: the substitution rate, and the preferentiality rate [6]. The dinucleotide substitution rate was calculated by dividing the number of a given dinucleotide substitution by the total number of occurrences of the dinucleotide in question in the pseudogene sequence at the time of its formation [6]. The dinucleotide preferentiality rate was obtained by normalising the dinucleotide substitutions for the total number of dinucleotides that had mutated in the pseudogene since its formation. To study the effects of adjacent nucleotides, it is essential that the change in question did not involve the nucleotide at the 5′ of the substituted position. This corresponds to an analysis of the alignment in units of two positions and considering only the units that differ in terms of the 3′ position [6]. The direction of the substitution was established following the parsimony approach. Due to the complementarity nature of DNA here, we needed to only consider the effects of the 5′ nucleotide. In other words, the effects of cytosine on thymine for the dinucleotide TpC are assumed to be equivalent to the effects of guanine on adenine for the complementary strand [20].

The neighbouring effects on the substitution rates are illustrated in the heatmap of Figure 5. The rows represent the 12 possible substitutions, while the colour intensities of the cells define the substitution rates. For example, if we consider the transition of A→G, it is evident that this occurs more frequently when the adenine is flanked at the 5′ by a thymine. The transition of G→A showed the highest neighbouring effect, with the highest substitution rates when the guanine was flanked by a cytosine in a CpG dinucleotide (Figure 5a). However, no neighbouring effect was seen on the substitution preferentially for these transitions (Figure 5b). We also noted a high rate of transition of C→T, especially when the cytosine was preceded by a guanine.

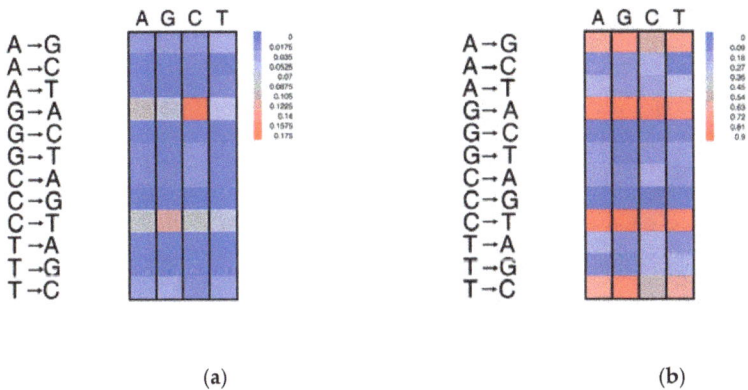

(a)   (b)

Figure 5. Neighbouring effects on the nucleotide substitution patterns. The dinucleotides are grouped on the basis of their first (5′) nucleotide. (**a**) Substitution rates normalised on the number of nucleotides of each type. (**b**) Proportions of substitutions normalised by the number of each type of substituted nucleotide type. The columns report the neighbouring nucleotide at the 5′ of the substituted nucleotide. The rows report the substitution type.

Porceddu et al. [31] compared the abundances of the dinucleotides in plant genes with the expectations based on the DNA (mono)nucleotide compositions. The TA dinucleotide was under-represented in both coding and non-coding genomic regions. For both A→G and A→T substitutions, TpA showed a higher substitution rate than the other nucleotides. However, this conclusion did not hold for the A→C substitutions.

4. Discussion

We have used processed pseudogene sequences here to infer the patterns of spontaneous mutations in the *V. vinifera* genome. As pseudogenes appear not to be subjected to functional constraints, all of their mutations (i.e., nucleotide substitutions) would be selectively neutral and become fixed with equal probability [7,8]. The rate of substitutions from G:C→A:T were higher than the reverse of A:T→G:C. This tendency towards A:T enrichment of pseudogenes was seen for both transitions and transversions. Taking into account the A:T→G:C substitution and the reverse rate, we calculated the equilibrium frequency as about 70 to 72% for A + T, which is a little higher than the average of 67 to 68% for introns and intergenic sequence.

Ossowsky et al. [32] calculated that the expected A+T equilibrium frequency for the *A. thaliana* genome would be 85%, which is higher than the 68% of the whole *A. thaliana* genome sequence. Based on the analysis of single nucleotide polymorphisms in maize transcript sequences from a panel of varieties, Morton et al. [5] calculated an expected A + T frequency of 62% for the maize genome. The average A + T relative content of the maize genome is 55% (for non-coding regions), which suggests that maize is far from reaching its compositional equilibrium [5].

The present data indicate that the grape genome is closer to compositional equilibrium than maize and *A. thaliana*. Whether this difference is due to any ascertainment bias of our pseudogene datasets or to other forces acting in the opposite direction (i.e., lower incidence of the G:C biased gene conversion in *V. vinifera*, as compared to other plant species) deserves further studies.

These data indicate that transitions are 3.5-fold more frequent than transversions in grape pseudogenes. Benovoy et al. [9] reported that in *A. thaliana* processed pseudogenes, although the transitions were more abundant than expected by chance, they were less frequent than transversions. In contrast, Ossowsky et al. [32] reported a transitions/transversions ratio of about 2.5 in 10 fully resequenced *A. thaliana* progenies. Studies conducted by Morton et al. [4] on 90 fully resequenced *A. thaliana* lines confirmed this preponderance of transition over transversion in introns and intergenic regions of this species. Similar findings have also been reported for several vertebrate genomes [6], and these have converged on the indication that the deamination of methylcytosine is the main determinant of the high transition rate. However, it was recently reported in the grasshopper *Podisma psedestris* that the transition and transversion rates were similar if the mutations that involved CpG sites were excluded from the computations [33]. To determine whether this conclusion holds for *V. vinifera*, we separately analysed the cytosines and guanines that were part of the CpG dinucleotides. Indeed, the CpG sites were subjected to very high rates of transitions, although there was also an excess of transitions over transversions even for cytosine and guanine in the non-CpG dinucleotides. Such a finding indicates that the methylation/deamination cycle alone cannot explain the high transition/transversion ratio in these grape pseudogenes. A similar finding was reported in *Homo sapiens* processed pseudogenes [6], and also by Ossowsky et al. [32] in a study that compared the patterns of methylated cytosines in an *A. thaliana* individual and the substitutions of 10 fully resequenced progenies.

Zhang et al. [6,34] suggested that the rates of spontaneous mutations can vary among regions. This finding implies that the A:T equilibrium frequency might vary between genomic regions, and it raises questions about the relationships between the genomic distribution of pseudogenes and the sampling errors for the rate estimates. To address these questions, we analysed the substitution rates in compositionally different genomic backgrounds. The differences between these 'regional' substitution rates were very small, and they defined a clear trend only in the parsimony-based dataset.

This finding is in sharp contrast with the data of Zhang and coworkers [6] in *H. sapiens*, where their rates of substitutions were clearly associated to compositional differences of the genomic regions. It is worth recalling that in contrast to *V. vinifera*, the *H. sapiens* genome is organised compositionally in highly differentiated regions, termed isochores [35]. Studies conducted in both *Arabidopsis* and maize have, instead, demonstrated significant regional context effects on substitution rates, with an evident association of the regional A:T content and substitutions replacing cytosine or guanine for adenine or thymine [4,5]. It is important to mention that these studies analysed the substitution patterns of non-coding sites, and the regional compositional context was determined based on the sequence of the locus scored for the mutations. The processed pseudogenes originated from retropositions of loci that mapped elsewhere in the genome. We speculate that the compositional differences between the sites and their pater locus genomic positions influenced our detection of substitution regional effects in *V. vinifera*. We acknowledge also that due to the limited size of the dataset used here, it might have lower power for studies on (weak) regional effects. For these reasons, we urge caution in drawing conclusions based solely on the findings presented in the present study.

Through this study, we investigated the substitution rates as determined by two main factors. The nucleotide substitution rate is a measure of the nucleotide stability, i.e., the probability of a given nucleotide to be changed for another nucleotide. The substitution preferentiality (or the proportion of substitutions) measured the probability of a given type of change over the other possible changes. Neither of these measures of the substitution rates showed strong dependence on regional genomic background. On the contrary, there were evident neighbouring effects of some substitutions. For example, the presence of cytosine at the $5'$ end of guanine had evident effects on the probability of guanine to be substituted by a transitional event. However, we did not find any strong neighbouring effects on the preferentiality rates, which indicated that, in this case, the neighbouring nucleotide influenced the vulnerability to mutation, but did not influence the direction of the substitution. Another interesting case was for nucleotides ending with adenine. The presence of thymine at the $5'$ targeted the adenine for mutation, which is in agreement with the under-representation of TA in plant genomes.

The advent of next-generation sequencing and the availability of a reference genome will soon allow the generation of sequence data from the wide grape germplasm [36,37]. The spontaneous mutation patterns of nucleotide substitutions presented here can be used as a reference for studies that are aimed at understanding how selection and other evolutionary forces shape the sequence diversity of grape cultivars.

Supplementary Materials: The following are available online at www.mdpi.com/1424-2818/9/4/45/s1, Figure S1: Pseudogene–pater locus alignment, Figure S2: GC3 content of pseudogenes and functional genes, Figure S3: Substitution patterns between nucleotide pairs (non-synonymous dataset), Table S1: List of the analysed processed pseudogenes.

Acknowledgments: We acknowledge the *Regione Autonoma della Sardegna* for grant LR7 CRP-79000.

Author Contributions: Andrea Porceddu and Salvatore Camiolo conceived, designed and performed the experiments, and analysed the data; Andrea Porceddu wrote the paper.

Conflicts of Interest: The authors declare no conflict of interest.

References

1. Fitch, W.M. Evidence suggesting a non-random character to nucleotide replacements in naturally occurring mutations. *J. Mol. Biol.* **1967**, *26*, 499–507. [CrossRef]
2. Kimura, M. Estimation of evolutionary distances between homologous nucleotide sequences. *Proc. Natl Acad. Sci. USA* **1981**, *78*, 454–458. [CrossRef] [PubMed]
3. Grantham, R. Amino-acid difference formula to help explain protein evolution. *Science* **1974**, *185*, 862–864. [CrossRef] [PubMed]
4. Morton, B.R.; Dar, V.-U.-N.; Wright, S.I. Analysis of site frequency spectra from Arabidopsis with context-dependent corrections for ancestral misinference. *Plant. Physiol.* **2009**, *149*, 616–624. [CrossRef] [PubMed]

5. Morton, B.R.; Bi, I.V.; McMullen, M.D.; Gaut, B.S. Variation in mutation dynamics across the maize genome as a function of regional and flanking base composition. *Genetics* **2006**, *172*, 569–577. [CrossRef] [PubMed]

6. Zhang, Z.; Gerstein, M. Patterns of nucleotide substitution, insertion and deletion in the human genome inferred from pseudogenes. *Nucleic Acids Res.* **2003**, *31*, 5338–5348. [CrossRef] [PubMed]

7. Gojobori, T.; Li, W.H.; Graur, D. Patterns of nucleotide substitution in pseudogenes and functional genes. *J. Mol. Evol.* **1982**, *18*, 360–369. [CrossRef] [PubMed]

8. Petrov, D.A.; Hartl, D.L. Patterns of nucleotide substitution in *Drosophila* and mammalian genomes. *Proc. Natl. Acad. Sci. USA* **1999**, *96*, 1475–1479. [CrossRef] [PubMed]

9. Benovoy, D.; Drouin, G. Processed pseudogenes, processed genes, and spontaneous mutations in the *Arabidopsis* genome. *J. Mol. Evol.* **2006**, *62*, 511–522. [CrossRef] [PubMed]

10. Vanin, E.F. Processed pseudogenes. Characteristics and evolution. *Biochim. Biophys. Acta* **1984**, *782*, 231–241. [CrossRef]

11. Esnault, C.; Maestre, J.; Heidmann, T. Human LINE retrotransposons generate processed pseudogenes. *Nat. Genet.* **2000**, *24*, 363–367. [PubMed]

12. Tutar, Y. Pseudogenes. *Comp. Funct. Genom.* **2012**. [CrossRef] [PubMed]

13. Wen, Y.-Z.; Zheng, L.-L.; Qu, L.-H.; Ayala, F.J.; Lun, Z.-R. Pseudogenes are not pseudo any more. *RNA Biol.* **2012**, *9*, 27–32. [CrossRef] [PubMed]

14. Ophir, R.; Graur, D. Patterns and rates of indel evolution in processed pseudogenes from humans and murids. *Gene* **1997**, *205*, 191–202. [CrossRef]

15. Mitchell, A.; Graur, D. Inferring the pattern of spontaneous mutation from the pattern of substitution in unitary pseudogenes of *Mycobacterium leprae* and a comparison of mutation patterns among distantly related organisms. *J. Mol. Evol.* **2005**, *61*, 795–803. [CrossRef] [PubMed]

16. Ossowski, S.; Schneeberger, K.; Clark, R.M.; Lanz, C.; Warthmann, N.; Weigel, D. Sequencing of natural strains of *Arabidopsis thaliana* with short reads. *Genome Res.* **2008**, *18*, 2024–2033. [CrossRef] [PubMed]

17. Jaillon, O.; Aury, J.-M.; Noel, B.; Policriti, A.; Clepet, C.; Casagrande, A.; Choisne, N.; Aubourg, S.; Vitulo, N.; Jubin, C.; et al. The grapevine genome sequence suggests ancestral hexaploidization in major angiosperm phyla. *Nature* **2007**, *449*, 463–467. [CrossRef] [PubMed]

18. Camiolo, S.; Porceddu, A. gff2sequence, a new user friendly tool for the generation of genomic sequences. *BioData Min.* **2013**, *6*, 15. [CrossRef] [PubMed]

19. Camiolo, S.; Porceddu, A. Identification of Pseudogenes in *Brachipodium dystachion*. In *Methods in Molecular Biology*; Humana Press: New York, NY, USA, 2018; Volume 1667, pp. 1–15.

20. Zheng, D.; Gerstein, M.B. A computational approach for identifying pseudogenes in the ENCODE regions. *Genome Biol.* **2006**, *7* (Suppl. 1), S13. [CrossRef] [PubMed]

21. Zhang, Z.; Carriero, N.; Zheng, D.; Karro, J.; Harrison, P.M.; Gerstein, M. PseudoPipe: An automated pseudogene identification pipeline. *Bioinformatics* **2006**, *22*, 1437–1439. [CrossRef] [PubMed]

22. Birney, E.; Clamp, M.; Durbin, R. GeneWise and Genomewise. *Genome Res.* **2004**, *14*, 988–995. [CrossRef] [PubMed]

23. Khelifi, A.; Adel, K.; Duret, L.; Laurent, D.; Mouchiroud, D.; Dominique, M. HOPPSIGEN: A database of human and mouse processed pseudogenes. *Nucleic Acids Res.* **2005**, *33*, D59–D66. [PubMed]

24. Huang, Y.; Niu, B.; Gao, Y.; Fu, L.; Li, W. CD-HIT Suite: A web server for clustering and comparing biological sequences. *Bioinformatics* **2010**, *26*, 680–682. [CrossRef] [PubMed]

25. Thompson, J.D.; Higgins, D.G.; Gibson, T.J. CLUSTAL W: Improving the sensitivity of progressive multiple sequence alignment through sequence weighting, position-specific gap penalties and weight matrix choice. *Nucleic Acids Res.* **1994**, *22*, 4673–4680. [CrossRef] [PubMed]

26. Yang, Z. PAML 4: Phylogenetic analysis by maximum likelihood. *Mol. Biol. Evol.* **2007**, *24*, 1586–1591. [CrossRef] [PubMed]

27. Li, W.H. Unbiased estimation of the rates of synonymous and nonsynonymous substitution. *J. Mol. Evol.* **1993**, *36*, 96–99. [CrossRef] [PubMed]

28. Wang, Y.; Tang, H.; Debarry, J.D.; Tan, X.; Li, J.; Wang, X.; Lee, T.-H.; Jin, H.; Marler, B.; Guo, H.; et al. MCScanX: A toolkit for detection and evolutionary analysis of gene synteny and collinearity. *Nucleic Acids Res.* **2012**, *40*, e49. [CrossRef] [PubMed]

29. Edgar, R.C. MUSCLE: Multiple sequence alignment with high accuracy and high throughput. *Nucleic Acids Res.* **2004**, *32*, 1792–1797. [CrossRef] [PubMed]

30. Thibaud-Nissen, F.; Ouyang, S.; Buell, C.R. Identification and characterization of pseudogenes in the rice gene complement. *BMC Genom.* **2009**, *10*, 317. [CrossRef] [PubMed]

31. Porceddu, A.; Camiolo, S. Spatial analyses of mono, di and trinucleotide trends in plant genes. *PLoS ONE* **2011**, *6*, e22855. [CrossRef] [PubMed]

32. Ossowski, S.; Schneeberger, K.; Lucas-Lledó, J.I.; Warthmann, N.; Clark, R.M.; Shaw, R.G.; Weigel, D.; Lynch, M. The rate and molecular spectrum of spontaneous mutations in *Arabidopsis thaliana*. *Science* **2010**, *327*, 92–94. [CrossRef] [PubMed]

33. Keller, I.; Bensasson, D.; Nichols, R.A. Transition-transversion bias is not universal: A counter example from grasshopper pseudogenes. *PLoS Genet.* **2007**, *3*, e22. [CrossRef] [PubMed]

34. Zhang, Z.D.; Frankish, A.; Hunt, T.; Harrow, J.; Gerstein, M. Identification and analysis of unitary pseudogenes: Historic and contemporary gene losses in humans and other primates. *Genome Biol.* **2010**, *11*, R26. [CrossRef] [PubMed]

35. Costantini, M.; Clay, O.; Auletta, F.; Bernardi, G. An isochore map of human chromosomes. *Genome Res.* **2006**, *16*, 536–541. [CrossRef] [PubMed]

36. Cardone, M.F.; D'Addabbo, P.; Alkan, C.; Bergamini, C.; Catacchio, C.R.; Anaclerio, F.; Chiatante, G.; Marra, A.; Giannuzzi, G.; Perniola, R.; et al. Inter-varietal structural variation in grapevine genomes. *Plant. J.* **2016**, *88*, 648–661. [CrossRef] [PubMed]

37. Mercenaro, L.; Nieddu, G.; Porceddu, A.; Pezzotti, M.; Camiolo, S. Sequence polymorphisms and structural variations among four grapevine (*Vitis vinifera* L.) cultivars representing Sardinian agriculture. *Front. Plant. Sci.* **2017**, *8*, 1279. [CrossRef] [PubMed]

diversity

MDPI

Article

Venetian Local Corn (*Zea mays* L.) Germplasm: Disclosing the Genetic Anatomy of Old Landraces Suited for Typical Cornmeal Mush Production

Fabio Palumbo [1], Giulio Galla [1], Liliam Martínez-Bello [2] and Gianni Barcaccia [1,*] (iD)

[1]   Department of Agronomy, Food, Natural Resources, Animals and Environment, University of Padova, 35020 Legnaro PD, Italy; fabio.palumbo.1@phd.unipd.it (F.P.); giulio.galla@unipd.it (G.G.)

[2]   Department of Crop and Soil Sciences, University of Georgia, Athens, GA 30602, USA; limabel@uga.edu

*   Correspondence: gianni.barcaccia@unipd.it; Tel.: +39-049-827-2814

Received: 14 July 2017; Accepted: 13 August 2017; Published: 16 August 2017

Abstract: Due to growing concern for the genetic erosion of local varieties, four of the main corn landraces historically grown in Veneto (Italy)—Sponcio, Marano, Biancoperla and Rosso Piave—were characterized in this work. A total of 197 phenotypically representative plants collected from field populations were genotyped at 10 SSR marker loci, which were regularly distributed across the 10 genetic linkage groups and were previously characterized for high polymorphism information content (PIC), on average equal to 0.5. The population structure analysis based on this marker set revealed that 144 individuals could be assigned with strong ancestry association (>90%) to four distinct clusters, corresponding to the landraces used in this study. The remaining 53 individuals, mainly from Sponcio and Marano, showed admixed ancestry. Among all possible pairwise comparisons of individual plants, these two landraces exhibited the highest mean genetic similarity (approximately 67%), as graphically confirmed through ordination analyses based on PCoA centroids and UPGMA trees. Our findings support the hypothesis of direct gene flow between Sponcio and Marano, likely promoted by the geographical proximity of these two landraces and their overlapping cultivation areas. Conversely, consistent with its production mainly confined to the eastern area of the region, Rosso Piave scored the lowest genetic similarity (<59%) to the other three landraces and firmly grouped (with average membership of 89%) in a separate cluster, forming a molecularly distinguishable gene pool. The elite inbred B73 used as tester line scored very low estimates of genetic similarity (on average <45%) with all the landraces. Finally, although Biancoperla was represented at K = 4 by a single subgroup with individual memberships higher than 80% in almost all cases (57 of 62), when analyzed with an additional level of population structure for K = 6, it appeared to be entirely (100%) constituted by individuals with admixed ancestry. This suggests that the current population could be the result of repeated hybridization events between the two accessions currently bred in Veneto. The genetic characterization of these heritage landraces should prove very useful for monitoring and preventing further genetic erosion and genetic introgression, thus preserving their gene pools, phenotypic identities and qualitative traits for the future.

Keywords: microsatellite; genetic erosion; local varieties; maize; barley; SSR; biodiversity; Veneto region

---

1. Introduction

The concepts of genetic erosion and conservation of plant genetic resources are rooted in the first decade of the twentieth century. Since then, several authors have warned of the consequences of the reduction of genetic variability in crop species mainly due to the dramatic loss of traditional landraces [1–3]. A landrace is an ancient population of a cultivated crop plant that has become adapted to the local conditions and to the agronomic practices of farmers. Most frequently, landraces are characterized by high diversity and thus provide a valuable source for potentially useful traits and an irreplaceable bank of co-adapted genotypes [4]. In practical terms, genetic diversity allows farmers and plant breeders to adapt a crop to heterogeneous and changing environments by, for example, providing it with resistance to pests and diseases [5].

Since modern, highly productive cultivars are irreversibly replacing many traditional varieties, the first priority is to arrest this loss of genetic diversity. Over the last decade, the rediscovery of local and traditional food products in the market has strengthened interest in local varieties. A fascinating case study is represented by "polenta", a traditional dish of the Italian cuisine, and by the four main corn landraces grown in Veneto (Italy) used for its production: Sponcio, Marano, Biancoperla and Rosso Piave. In the last few years, the demand of "polenta' from local varieties, has shown a steady increase due to the deeper attention that consumers pay to the autochthonous, locally cultivated crops, usually grown according to low-input agronomic practices, and to their consciousness towards the current dualism existing between conventional and novel foods [6]. In 2016 the total production of maize in Italy was approximately 6.84 million tons [7], and even if the amount destined to the human consumption is very small (few percentage points), the total market value of this product is estimated in millions of euro.

Assessing the genetic diversity and genetic structure of landraces could help to limit genetic erosion as well as to conserve landraces [8]. Several studies have been performed worldwide to assess the genetic diversity of local landraces of corn using molecular markers [9–12], but as far as we know, only one has been devoted to local varieties of Italian corn [13]. The four main corn landraces grown in Veneto (Italy) and examined in this work, namely, Sponcio, Marano, Biancoperla and Rosso Piave, represent a case study.

Sponcio is an ancient corn variety grown by a consortium of 20 farmers in a small plot that covers approximately 13 hectares in the area of the Val Belluna, specifically in the towns of Feltre, Cesiomaggiore and Santa Giustina [14]. This landrace, distinguishable by its sharp kernels, seems to have been known since the sixteenth century under variants of the name, but the first concrete documentation of its existence is a nineteenth century manuscript [15]. By the 1950s, the production of Sponcio had been reduced, and it was confined to marginal areas. Thanks to a few farmers and millers, the original germplasm was carefully preserved and later used to restart the current production, according to strict sustainable and environmentally friendly regulations determining the stages of cultivation, drying, grinding and packing. The yellow flour is the main ingredient of *"polenta"*, one of the most typical products of the Belluno cuisine.

An article dated 1939 reports that in 1890 Antonio Fioretti, a farmer from Marano Vicentino (Veneto, Italy), crossed two local varieties, Nostranino and Pignoletto d'Oro, and called this new hybrid Marano [16]. Although Marano was particularly esteemed during the 1970s in Veneto and Friuli Venezia Giulia and widely employed to produce new hybrids (e.g., ITALO 225, ITALO 260 and ITALO 270) and pure lines (Cinquantino San Fermo and Cinquantino Bianchi), the cultivation of this local variety was progressively abandoned and replaced by more productive lines. Currently, it survives only in the area of Val Leogra, specifically in the towns of Marano Vicentino, Malo, Schio, San Vito di Leguzzano, Torrebelvicino, Valli del Pasubio, Santorso and Piovene Rocchette [17]. The *"polenta maranelo"*, a typical dish of this area, is produced starting from the orange flour of this landrace.

A book published at the end of seventeenth century reports that a white *"sorgoturco"* [dialectal word referring to corn] was widespread in Veneto at that time, and that white variety probably represents the ancestor of the current Biancoperla [18]. Several documented sources state that this landrace, which owes its name to the vitreous and pearly white color of its kernels, was widely grown (>50,000 hectares) in the eastern part of Veneto and in Friuli Venezia Giulia in the first half of the twentieth century [19]. As with the aforementioned landraces, this local variety was progressively replaced by more profitable corn varieties from the USA immediately after the Second World War. Currently, thanks to a consortium of approximately 13 member producers promoting its conservation, this variety survives on less than 50 hectares in some rural areas of Vicenza, Treviso and the northern part of Padua district. It is strongly appreciated for the production of white *"polenta"* [20].

Little is known about the fourth landrace, Rosso Piave. Miniscalco [19] reports that, unlike the other three local varieties, this landrace was rarely grown even in the past, since its color permanently soiled mills. Today, it is grown mainly in the Venice area in the towns of Musile di Piave, Fossalta di Piave, Noventa di Piave and San Donà di Piave. Its peculiar burgundy color, which also characterizes its *"polenta"*—the main derivative product of this landrace—comes from the presence of anthocyanins that have been recently recognized as compounds able to reduce the risk of myocardial infarction [21].

In this study, the genetic diversity of the four main corn landraces in Veneto was assessed by means of simple sequence repeat (SSR) markers. The assembled molecular data were used to evaluate their population genetic structure and their genetic relationships. The characterization of these old local varieties supports a more general discussion of possibilities for avoiding genetic erosion, promoting and safeguarding local populations, thereby maintaining stable seed yields, and preserving phenotype and qualitative identity.

2. Materials and Methods

2.1. Plant Material and Genomic DNA Isolation

Four different Venetian Institutes for Agricultural Research kindly donated the corn samples used in the present study. The germplasm collection conserved in each institute was originally constituted combining hundreds of kernels from as many ears selected on the basis of their morphology. Marano seeds were provided by the "N. Strampelli Institute" (Lonigo, VI, Italy), whereas Sponcio seeds were obtained from the "A. Della Lucia Institute" (Feltre, BL, Italy). The "D. Sartor Institute" (Castelfranco Veneto, TV, Italy) and Veneto Agricoltura (Legnaro, PD, Italy) supplied Biancoperla and Rosso Piave seeds, respectively.

For germination, 40 to 70 seeds of each variety were placed in Petri dishes on two layers of filter paper moistened with water. After fifteen days of incubation, a total of 197 seedlings (64 Marano, 32 Sponcio, 62 Biancoperla and 39 Rosso Piave) were collected and used for the analyses described below. The elite public inbred line B73, initially developed from the Iowa Stiff Stalk Synthetic (BSSS) populations and commonly used for the development of heterotic F1 hybrids, was chosen as reference corn germplasm accession in order to test the relatedness and distinctiveness of the four local varieties with and from modern varieties.

Then, 100 mg of fresh leaf tissue was used to isolate genomic DNA using a DNeasy plant kit (Qiagen, Valencia, CA, USA), following the procedure provided by the suppliers. Electrophoresis on an 0.8% agarose/1× TAE gel containing 1× Sybr Safe DNA stain (Life Technologies, Carlsbad, CA, USA) allowed estimation of the integrity of extracted DNA samples. The purity and quantity of DNA extracts were evaluated with a NanoDrop 2000c UV-Vis spectrophotometer (Thermo Scientific, Pittsburgh, PA, USA).

2.2. Analysis of SSR Markers

PCR amplifications were performed using the M13-tailed SSR method described by Schuelke [22], with some minor modification. Briefly, the amplification procedure is based on a three-primer system consisting of a specific SSR-targeting forward primer with a 5′-M13 tail, a specific SSR-targeting reverse primer and an M13-labelled primer (5′-TTGTAAAACGACGGCCAGT-3′). The set of 10 SSR marker loci investigated in this study was obtained from Register et al. [23] and, based on the highest Polymorphic Information Content (PIC) values, one SSR marker per linkage group was selected (Table 1).

Table 1. List of SSR loci selected from Register et al. [23] for use in this study. For each microsatellite locus, linkage group, locus ID, motif, amplicon size in bp, forward and reverse primer sequences used to amplify the microsatellite region, melting temperature and polymorphism information content (PIC) coefficient related to the previously mentioned study are shown.

Linkage Group	Locus ID	Motif	Size (bp)	Primer	Tm (°C)	PIC
1	phi056	GCC	103–112	M13-ACGCCCAGATCTGTTCCTTCTC ATGGCGGCAGGCCGATTGTT	63	0.67
2	phi127	AGAC	129–145	M13-ATATGCATTGCCTGGAACTGGAAGGA AATTCAAACACGCCTCCCGAGTGT	62	0.70
3	phi073	CAG	107–116	M13-TTACTCCTATCCACTGCGGCCTGGAC GCGGCATCCCGTACAGCTTCAGA	69	0.65
4	phi076	GAGCGG	182–192	M13-TTCTTCCGCGGCTTCAATTTGACC GCATCAGGACCCGCAGAGTC	61	0.65
5	phi024	CCT	183–195	ACTGTTCCACCAAACCAAGCCGAGA M13-AGTAGGGGTTGGGGATCTCCTCC	69	0.69
6	phi031	GTAC	174–177	M13-GCAACAGGTTACATGAGCTGACGA CCAGCGTGCTGTTCCAGTAGTT	66	0.57
7	phi057	GCC	211–215	M13-CTCATCAGTGCCGTCGTCCAT CAGTCGCAAGAAACCGTTGCC	66	0.61
8	umc1075	ATTGC	156–166	M13-GAGAGATGACAGACACATCCTTGG ACATTTATGATACCGGGAGTTGGA	57	0.69
9	phi016	GGT	173–176	M13-TTCCATCATTGATCCGGGTGTCG AAGGAGCAACATCCCATCCAGGAA	60	0.52
10	phi084	GAA	174–178	M13-AGAAGGAATCCGATCCATCCAAGC CACCCGTACTTGAGGAAAACCC	59	0.49

The PCR reaction consisted of a 20 µL final volume containing $1\times$ NH$_4$ Reaction Buffer, 3 mM MgCl$_2$, 1 IU of BioTAQ™ DNA polymerase (Bioline, London, UK), 0.25 mM each dNTPs, 0.25 µM tailed forward primer, 0.75 µM reverse primer, 0.5 µM M13-labelled primer (Invitrogen, Carlsbad, CA, USA), 20 ng of DNA, and dH$_2$O up to the final volume. Amplifications were performed in a 96-well plate using a 9600 thermal cycler (Applied Biosystems, Carlsbad, CA, USA). The following thermal conditions were used: 5 min at 94 °C for the initial denaturing; 5 cycles at 94 °C for 30 s, at 62 °C for 30 s decreasing by 0.8 °C with each cycle, and at 72 °C for 45 s; and 35 cycles at 94 °C for 30 s, 58 °C for 30 s and 72 °C for 45 s. A final extension at 72 °C for 10 min terminated the reaction to fill in any protruding ends of the newly synthesized strands. Capillary electrophoresis with an ABI PRISM 3130xl Genetic Analyzer, adopting LIZ500 as molecular weight standard, was used to assess the PCR products. The size of each peak was determined using Peak Scanner software 1.0 (Applied Biosystems).

2.3. Marker Data Analysis

PIC values were calculated with PICcalc software [24] to estimate the marker allele variation in microsatellite loci in the 197 corn individuals. GenAlEX v. 6.5 [25] and POPGENE v. 1.32 [26] software were used to estimate the number of observed alleles (N_a), number of effective alleles (N_e), Shannon's information index of genetic diversity (I), observed (H_o) and expected (H_e) heterozygosity according to Nei [27]. The presence of private alleles in each population and the occurrence of locally common alleles, defining as "locally common" those alleles with a frequency higher than 5% in a local population and occurring in less than 25% of all populations examined [28], were also considered.

F-statistics were calculated according to Wright [29] to investigate the variance of heterozygosity in our population at different levels of population structure (i.e., individual, subpopulation and population levels). Inbreeding coefficients were computed to measure the deficiency (positive values) or excess (negative values) of heterozygotes for each assessed microsatellite marker and to assess hierarchical organization of sample individuals. Similarly, inbreeding coefficients were calculated at multilocus level in order to estimate the genetic effect of total population subdivision as proportional reduction in overall heterozygosity due to variation in SSR allele frequencies among different subpopulations [30]. Finally, gene flow (N_m) was calculated from F_{ST}.

Genetic similarity (GS) was calculated in all possible pairwise comparisons of individuals by applying the simple matching coefficient [31]. A principal coordinates analysis (PCoA) was applied to compute the first two principal components of the similarity data matrix. All analyses and calculations were conducted using NTSYS-pc v. 2.21q [31]. In addition, pairwise GS values were also used to construct a dendrogram, using the unweighted pair-group arithmetic average (UPGMA) method and PAST software v. 3.14 [32] with 1000 bootstrap repetitions.

The genetic structure of the four landraces was modeled using a Bayesian clustering algorithm implemented in STRUCTURE v. 2.2 [33]. Since no prior knowledge about the origin of the populations under study was available, the "admixture model" was used and then a "correlated allele frequencies model" was selected, because it guarantees that a previously undetected correlation will be identified, without affecting the results if no such correlation exists [34]. Ten replicate simulations were conducted for each value of K, with the number of founding groups ranging from 2 to 22, using a burn-in of 2×10^5 and a final run of 10^6 MCMC steps. The method described by Evanno et al. [35] was used to evaluate the most likely estimation of K. Estimates of membership were plotted as a histogram using an Excel spreadsheet.

3. Results

3.1. Descriptive Statistics of SSR Marker Loci

All SSR loci were determined to be polymorphic (Table 2). PIC values were considered to estimate the ability of each locus to discriminate among different genotypes and the selected SSR loci scored a mean PIC of 0.50, with a minimum of 0.32 (phi084) and a maximum of 0.71 (umc1075). A total of 36 marker alleles were detected across the four populations with an average number of observed alleles (N_a) of 3.6, ranging from 2 (phi084, phi031 and phi016) to 6 (phi127). Moreover, the effective number of alleles (N_e) per locus varied from 1.68 (phi084) to 4.02 (umc1075), as reported in Table 2.

SSR loci were highly polymorphic within each landrace, except for phi084, which was monomorphic in the Marano landrace for a 177 bp marker allele. The same allele was also the most common one overall, being detected in 141 out 197 samples (71.68%, Supplementary Data S1). Six alleles of the 36 were private to single populations. Specifically, Sponcio showed two private alleles, one at the phi057 locus and one at the phi056 locus, while Rosso Piave showed four different allelic variants never detected in the other landraces, three at the phi127 locus and one at the phi024 locus (Supplementary Data S1).

Over all SSR loci, the observed (H_o) and expected (H_e) heterozygosity estimates were, on average, equal to 0.43 ± 0.12 and 0.58 ± 0.12, respectively (Table 2). The same parameters calculated within each landrace were, on average, equal to 0.43 ± 0.04 and 0.48 ± 0.04, respectively (Table 2). The Shannon's index (I) was used to characterize population diversity and it was found to be, on average, equal to 0.97 ± 0.30 over all loci and 0.80 ± 0.09 within landraces. The inbreeding coefficient (F_{IS}) had an average value of 0.08 ± 0.04 for SSR loci. Finally, F_{IT} and F_{ST} were both positive and, on average, equal to 0.25 ± 0.03 and 0.18 ± 0.03, respectively, while the gene flow (N_m) was equal to 1.62 ± 0.40.

The same F-statistics were then applied to each landrace to assess the genetic effects of total population subdivision as proportional reduction in overall heterozygosity due to variation in SSR allele frequencies among landraces (Table 2). Overall, Wright's inbreeding coefficients F_{IS} and F_{IT} scored positive values, revealing a general deficiency of heterozygotes across individual accessions and landraces. As displayed in Table 2, reduction of heterozygosity was marked for the two landraces Rosso Piave ($F_{IS} = 0.16$) and Biancoperla ($F_{IS} = 0.15$), whereas it was minimal for the two landraces Sponcio ($F_{IS} = 0.06$) and Marano ($F_{IS} = 0.02$). Interestingly, the variation observed in our estimates of

F_{IT} was much lower, as this parameter was on average equal to 0.25, ranging from 0.21 (Sponcio and Marano) to 0.35 (Biancoperla). As displayed in Table 2, F_{ST} was, on average, equal to 0.17, ranging from 0.08 (Rosso Piave) to 0.24 (Biancoperla, Table 2). Altogether, these data suggest that the proportion of genetic variation found among landraces was relatively low (17% on average) and to some extent variable across landraces (8% to 24%).

Table 2. Genetic parameters with respect to the SSR markers and to the four landraces object of this study. Average number of observed alleles (N_a), effective number of alleles (N_e) per locus, polymorphism information content (PIC), estimates of Shannon's information index of genetic diversity (I), observed heterozygosity (H_o) and unbiased Nei's genetic diversity equivalent to the expected heterozygosity (H_e) are shown. Wright's inbreeding coefficients F_{IS}, F_{IT} and F_{ST} and gene flow (N_m) estimates are also indicated.

Locus	N_a	N_e	PIC	I	H_o	H_e	F_{IS}	F_{IT}	F_{ST}	N_m
phi024	5.00	3.54	0.66	1.32	0.59	0.72	0.07	0.19	0.13	1.69
phi127	6.00	2.24	0.47	0.95	0.43	0.55	0.12	0.22	0.11	1.97
phi084	2.00	1.68	0.32	0.60	0.29	0.41	−0.03	0.26	0.28	0.65
phi076	3.00	2.95	0.59	1.09	0.48	0.66	0.14	0.24	0.12	1.84
phi031	2.00	1.75	0.34	0.62	0.39	0.43	0.04	0.09	0.05	4.87
phi057	4.00	2.72	0.55	1.07	0.48	0.63	0.01	0.27	0.26	0.70
phi056	4.00	1.86	0.38	0.77	0.29	0.46	−0.08	0.37	0.42	0.34
phi073	3.00	2.94	0.59	1.09	0.55	0.66	0.01	0.17	0.17	1.24
phi016	2.00	1.00	0.37	0.69	0.250	0.50	0.34	0.47	0.19	1.04
umc1075	5.00	4.02	0.71	1.47	0.53	0.75	0.16	0.26	0.12	1.84
All loci	3.60	2.57	0.50	0.97	0.43	0.58	0.08	0.25	0.18	1.62
St. dev.	1.43	0.80	0.14	0.30	0.12	0.12	0.04	0.03	0.03	0.40
Sponcio	3.00	2.09	na	0.80	0.46	0.49	0.06	0.21	0.16	1.34
Marano	2.50	2.08	na	0.74	0.46	0.47	0.02	0.21	0.19	1.04
Biancoperla	2.50	1.90	na	0.69	0.37	0.44	0.15	0.35	0.24	0.78
Rosso Piave	3.30	2.30	na	0.89	0.45	0.53	0.16	0.23	0.08	3.01
All landraces	2.82	2.09	na	0.80	0.43	0.48	0.10	0.25	0.17	1.54
St. dev.	0.39	0.16	na	0.09	0.04	0.04	0.07	0.07	0.07	1.00

3.2. Genetic Diversity and Cluster Analysis

Within the genetic similarity (GS) matrix calculated for all possible pairwise comparisons among the 197 individuals' DNA genotypes, Rohlf's index ranged from 0.29 (between MAR_128 and RSM_25) to 0.97 (between MAR_129 and SPO_32). When calculated within each landrace, this index varied, on average, from 0.70 (±0.11) within the Rosso Piave population to 0.78 (±0.08) within the Biancoperla population. In pairwise comparisons between landraces, Marano and Biancoperla populations showed the lowest average value (0.59 ± 0.07), while Marano and Sponcio populations exhibited the highest one (0.67 ± 0.08, Supplementary Data S1).

Principal coordinate analysis (PCoA) showed that most of the individual DNA samples were clustered into four major subgroups (Figure 1).

The first principal coordinate accounted for 14.7% of the total variation and clearly separated Marano from Biancoperla, whereas the second principal coordinate accounted for 7.4% of the total variation and separated Sponcio from Rosso Piave. Biancoperla and Rosso Piave were firmly grouped in two distinct clusters and only few individuals, namely, RSM_5, RSM_12, RSM_15, RSM_25 and RSM_26, were partially separated from the rest of their landrace. This finding is also supported by low mean genetic similarity values (always lower than 69.0%) calculated through pairwise comparisons between these five samples and the Rosso Piave collection as a whole. The PCoA analysis further underlines the existence of some overlaps between the Marano and Sponcio clusters, a scenario that is mirrored by the high mean similarity value (67.2%) calculated at all loci between these two populations.

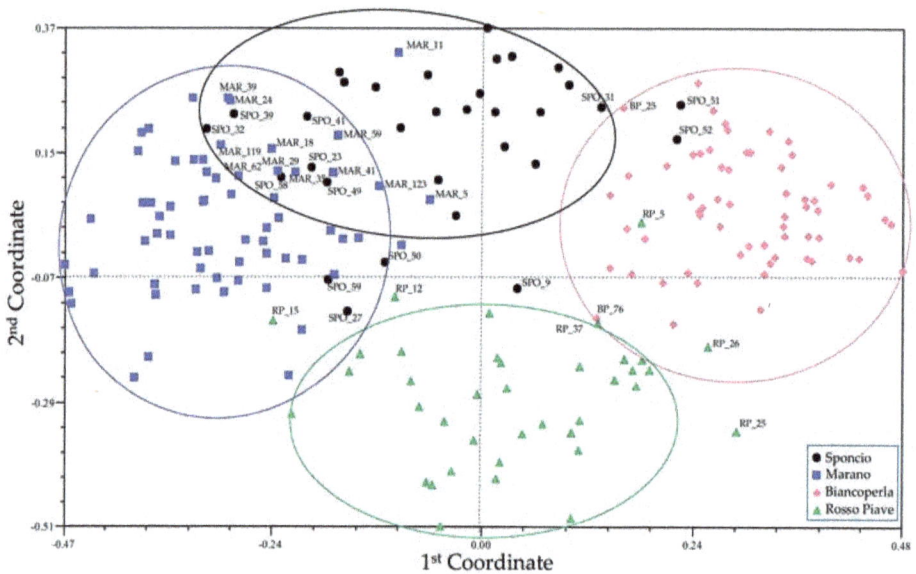

Figure 1. Two-dimensional centroids derived from genetic similarity estimates computed among the 197 accessions in all possible pairwise comparisons using the SSR marker data set. Only the names of those genotypes with unclear membership to one of the four subgroups are reported.

The UPGMA tree revealed a marked genetic differentiation of the four local varieties. Using this analysis, it was possible to distinguish three main clusters of individuals that were firmly supported by a bootstrap value of 100% (Figure 2). The inbred line B73 was grouped separately from all individuals representing the four Venetian landraces (data not shown). In particular, genetic similarity estimates of the inbred line B73 with respect to all individual genotypes, measured in all possible pairwise combinations, were much lower than the average values computed among individuals either within landraces (ranging from 0.70 to 0.78) or between landraces (varying from 0.59 to 0.67). As a matter of fact, the mean genetic similarity of this inbred line used as tester with each of the four landraces was equal to 0.43 ± 0.04 (with Marano), 0.44 ± 0.05 (with Rosso Piave) and 0.45 ± 0.06 (with both Sponcio and Biancoperla).

The largest group consisted of two subgroups with bootstrap support of 76%, one including approximately 50% of the Sponcio population and the other one including the Marano population and the remaining Sponcio individuals. The second cluster included the entire Biancoperla population, which further split into two subgroups (bootstrap support of 52%). Finally, the third group represented most of the Rosso Piave individuals.

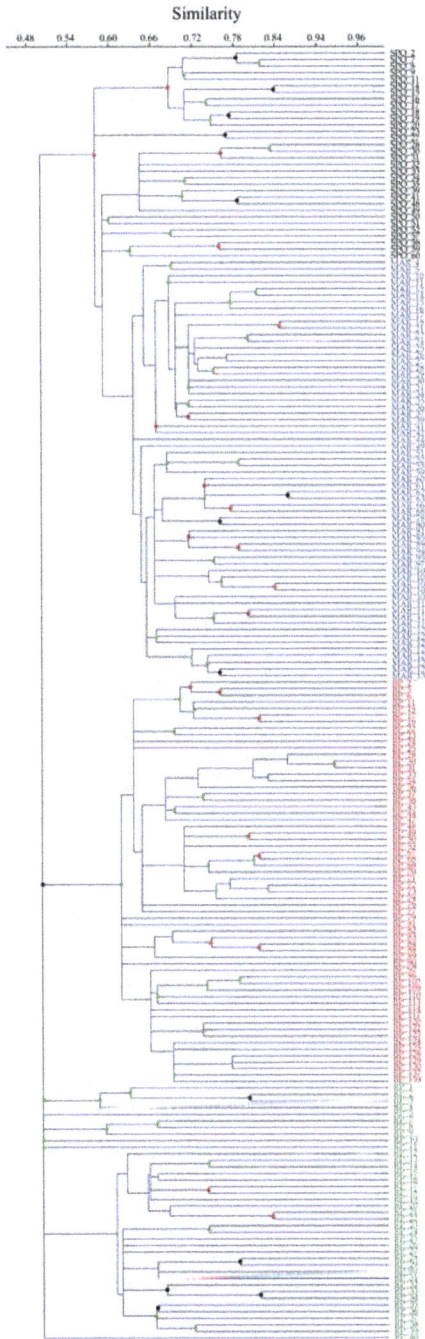

Figure 2. Constrained UPGMA tree of genetic similarity estimates computed among pairwise comparisons of corn accessions using the SSR marker data set, with nodes supported by bootstrap values. Black circle: bootstrap values ≥ 90%; red circle: 70% ≤ bootstrap values < 90%; green circle: 50% ≤ bootstrap values < 70%. The color scheme for the text is the same as for the symbols described in Figure 1 (black = Sponcio, blue = Marano, magenta = Biancoperla and green = Rosso Piave).

3.3. Genetic Structure Analysis

STRUCTURE v2.2 [33] was used to investigate the genetic structure of the corn core collection. Following the procedure of Evanno et al. [35], a clear maximum for ΔK value at K = 4 was found (ΔK = 548.79, Figure 3).

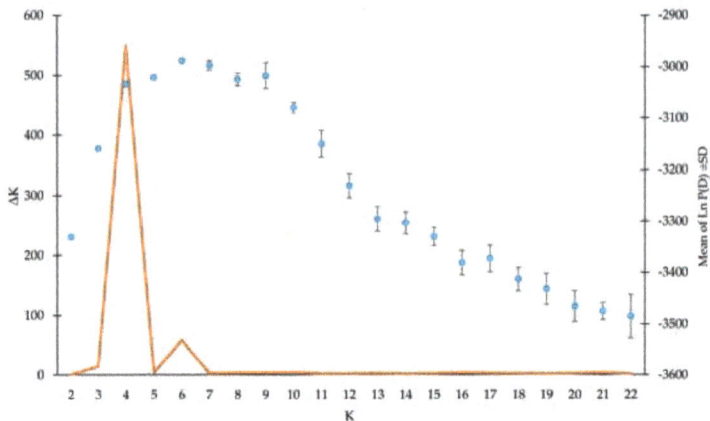

Figure 3. Definition of the number of ancestral corn populations based on the SSR marker dataset. Mean LnP(D) ± SD over 10 runs is a function of K, as L'(K) = ΔLnP(D). Mean ΔK is calculated as |L''(K)|/(SD(L(K)) following Evanno et al. [35]. ΔK values are represented by the orange line, while the blue points indicate the mean LnP(D) ± SD values.

Since the ancestral population size K = 4 also corresponds to the number of local varieties used in this study, it was considered the best estimate of the current population structure (Figure 4, panel a). The 197 corn samples were grouped into four genetically distinct clusters. In this graphical representation, each genotype is plotted as a vertical histogram divided into K = 4 colored segments representing the estimated membership in each hypothesized ancestral genotype. The clustering of genotypes revealed that 144 of 197 samples showed strong ancestry association (>90%). Almost all individuals from Biancoperla (94%) and Rosso Piave (90%) showed an individual membership to their respective founding groups higher than 80% while most of the admixed genotypes (<80% membership to a single ancestral genotype) were from Sponcio and Marano. Specifically, these two landraces included a substantial number of genotypes with admixed ancestry, namely, 11 genotypes for the variety from Val Belluna (32%) and 9 for the one from Marano Vicentino (15%). Most of the admixed genotypes (Figure 4, panel a) originated in the overlapping region between these two clusters in the PCoA analysis (Figure 1, MAR_5, MAR_11, MAR_29, MAR_59, SPO_27, SPO_32, SPO_39, SPO_49 and SPO_58, SPO_59).

The second largest ΔK, at K = 6 (ΔK = 58.58, Figure 3), revealed an additional level of population structure and allowed the clustering of all investigated genotypes into six additional subgroups. In this interpretative framework, all the Marano samples and most of the Sponcio population (91%) were organized into three main clusters. The first one included 21 individuals from Sponcio (membership > 50%), a second one grouped 39 Marano genotypes (membership > 50%) and a third cluster comprised most of the Sponcio and Marano samples that showed admixed ancestry at K = 4 (Figure 4, panel b). As already found for K = 4, Rosso Piave continued to cluster apart, but all the individuals belonging to the Biancoperla landrace showed admixed ancestry from two different clusters (Figure 4, panel b).

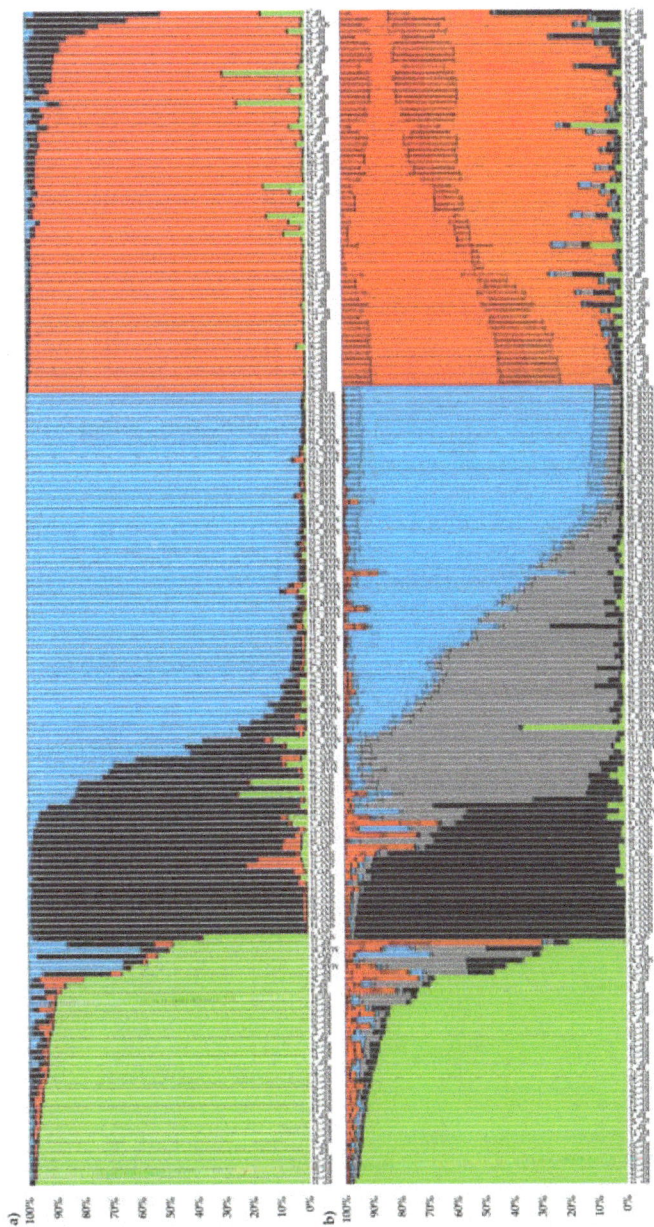

Figure 4. Population genetic structure of the four main corn landraces in Veneto (N = 197) as estimated by STRUCTURE using the SSR marker data set. Each sample is represented by a vertical histogram partitioned into K = 4 (panel **a**) or K = 6 (panel **b**) colored segments that represent the estimated membership. The proportion of ancestry (%) is reported on the ordinate axis and the identification number of each accession is reported below each histogram. The color scheme for the figure is the same as for the symbols and the text described in Figures 1 and 2, respectively. Green = Rosso Piave, black = Sponcio, blue = Marano and magenta = Biancoperla. For K = 6 two shades of red are used for the two clusters of Biancoperla and the third new cluster between Sponcio and Marano is marked in grey.

4. Discussion

The Sponcio, Marano, Biancoperla and, to a lesser extent, Rosso Piave landraces of corn were abundantly grown in the past and characterized the Veneto region (Italy) for centuries, as reported by several documents [15,16,19]. Nevertheless, since the twentieth century, they have been progressively abandoned and replaced by more productive lines. Currently, they survive only in a few hectares, and extinction is becoming a real threat. To the best of our knowledge, this work represents the first attempt to describe the genetic diversity and structure of these four varieties.

For the purposes of this analysis, 10 microsatellite markers equally distributed into 10 linkage groups (see Table 1) were chosen from Register et al. [23] on the basis of their high PIC. The PIC values calculated for this SSR marker dataset (see Table 2) were slightly lower than those reported by Register et al. [23], probably because in the original work PIC values were obtained from an analysis of over 500 genotypes largely representative of the whole North American germplasm. Moreover, we cannot exclude that the different methods adopted to run and screen PCR products could have influenced the detection of allelic variants. According to Botstein et al. [36], five of the selected SSR markers (phi024, phi076, phi057, phi073 and umc1075) would be considered highly informative (PIC > 0.5), while the other loci would be considered informative (0.25 < PIC < 0.5, see Table 2). Interestingly, there was no direct correlation between population size and number of observed alleles (N_a) or number of effective alleles (N_e). In fact, the two populations numerically less represented, Sponcio (N = 32) and Rosso Piave (N = 39), showed the highest N_a (N_a = 3.00 and N_a = 3.30, respectively) and N_e values (N_e = 2.09 and N_e = 2.30, respectively). The number of effective alleles in a population is estimated from the gene diversity (i.e., N_e = $1/(1 - H_e)$), and denotes the number of equally frequent alleles necessary to achieve a given level of gene diversity. The finding that the number of effective alleles indirectly correlates with the size of the assessed populations is consistent with a reduction of gene diversity within these populations (see Figure 1). Furthermore, the observation that the difference between the number of observed alleles (N_a) and number of effective alleles (N_e) is higher in Sponcio and Rosso Piave could indicate the presence of several low-frequency alleles in these landraces. Roughly speaking, without considering allele frequencies at this level of interpretation, a high number of observed alleles theoretically produces many genetically possible genotypes and, thus, high genetic diversity within the population. Accordingly, we observed that mean percentages of genetic similarity scored within Rosso Piave (69.98%) and Sponcio (74.71%) were lower than those calculated within Marano and Biancoperla (75.40% and 78.12%, respectively). Of the 36 alleles, six appeared to be private to specific populations (Sponcio and Rosso Piave). SSR private alleles are recognized as an efficient food traceability tool since they can be assigned unambiguously to a specific variety. Recently, a molecular system entirely based on private alleles was developed by Palumbo et al. [37] in order to verify the genetic authenticity of food products deriving from an Italian barley landrace. Unfortunately, all six private alleles observed in this study were present at very low frequencies (<0.05%), so they could not be used, even in combination, for the same purpose. The large number of polymorphisms and the presence of both rare allele and alleles unshared with the inbred B73, potentially confirm that these four landraces could represent a valuable source of genetic variants and unique germplasm traits [13].

The fact that the overall mean observed heterozygosity (H_o = 0.43 ± 0.12) for all loci was lower than expected (H_e = 0.58 ± 0.12) suggests an excess of homozygosity in the core collection. This is further supported by the positive values of the individual inbreeding coefficients (F_{is} = 0.08 ± 0.04 and F_{it} = 0.25 ± 0.03, Table 2). The observed heterozygosity calculated within the four landraces was, on average, equal to 0.43 (±0.04), consistent with the allogamous reproductive system of corn and with that reported in other works focused on corn landraces [11,12]. As found in the only other Italian work available on corn landraces [13], a deficiency of heterozygotes was observed for each local variety.

We are confident that the deficiency of heterozygosity is not correlated with the size of the assessed populations: the lowest value (H_o = 0.37 ± 0.20) was recorded for Biancoperla, one of the two most numerically represented landraces (N = 62) and vice versa the smallest group (Sponcio, N = 32) showed the highest value (H_o = 0.47 ± 0.15). Moreover, we rule out the possibility that the cause of low levels of heterozygosis is ascribable to the limited number of plants from which the seeds sampled and analyzed in this study were originally selected by the institutes. In fact, germplasm collections were constituted combining hundreds of kernel corns from as many ears, carefully avoiding seeds from the same plant and ear. More likely, the repeated crosses of genetically similar individuals played a crucial role in the homozygosity excess showed by the loci investigated [38]. This could be the case when farmers select, every year, very small seed stocks based on an "ear ideotype", applying a strong selective pressure [13]. More in details, the traditional selection carried out annually by farmers is oriented to maintain (i) the distinctive morphological traits of the landrace, (ii) the peculiar qualitative characteristics of kernels used for "polenta", and (iii) the level of distinctiveness even when the pollen source is not controlled [6]. Biancoperla also scored the highest mean value of similarity (78.12%), providing a reasonable connection between the low level of heterozygosity and high genetic similarity calculated within each local variety.

Overall, Wright's inbreeding coefficients F_{IS} and F_{IT} scored positive values, confirming a general deficiency of heterozygotes across individual accessions and landraces. Based on our marker set, the reduction of heterozygosity was higher for the two landraces Rosso Piave and Biancoperla, while it was relatively low for the two landraces Sponcio and Marano. Interestingly, F_{IT} estimates did not mirror the variation observed for F_{IS}, as the four cultivars displayed very similar values for this parameter. Estimates of F_{ST} varied considerably among the four landraces, indicating unbalanced contributions of the investigated populations to the total assayed genetic variation. Accordingly, our estimates of inbreeding coefficients suggest that these landraces are characterized by a relatively low degree of genetic differentiation, with approximately 17% of the genetic variation found among landraces (average $F_{ST} = 0.17$) and approximately 83% of the total genetic variation expressed within landraces.

Based on a pairwise comparison among varieties, Sponcio and Marano landraces exhibited the highest mean genetic similarity estimates (on average, 67.23%), as graphically confirmed through ordination analyses based on the definition of PCoA centroids (see Figure 1) and construction of the UPGMA tree (see Figure 2). Our findings support the hypothesis of marked gene flow between these two landraces, which could have been promoted by their geographical proximity and recently overlapping cultivation areas as a clear-cut distribution in the Veneto region has been progressively lost. To further corroborate this hypothesis, Marano and Sponcio revealed genetically differentiated populations for K = 4 (see Figure 4, panel a) and subpopulations grouping individuals with admixed ancestry for K = 6 (see Figure 4, panel b). Further analysis will be needed, and combining genetic data with phenotypic observations will help determine whether genotypes with admixed ancestry (K = 4) also share the morphological characteristics of both landraces. As already reported in [10], the inbred line B73 showed very low average estimates of genetic similarity (<45%) with all the landrace populations.

Rosso Piave, whose production is mainly confined to the extreme east of the Veneto region, was the landrace showing the lowest mean genetic similarity values in pairwise comparisons with the other three landraces. Consistent with these results, this landrace grouped in a cluster apart from the other landraces for both K = 4 and K = 6 (see Figure 4).

Although for K = 4 Biancoperla was represented by a unique group with individual memberships almost always (57 of 62 samples) higher than 80% (see Figure 4, panel a), the clustering of individuals for K = 6 revealed that this variety was totally (100%) constituted by admixed individuals and each individual showed a variable percentage of membership to both clusters (see Figure 4, panel b). Since two main accessions of Biancoperla are currently bred in Veneto (ITA0340323 and ITA0340324), which differ in plant size, spike length and kernel color [39], based on STRUCTURE results, we speculate that the current Biancoperla population could be the result of repeated events of hybridization and/or introgression between these two accessions.

Conservation of the genetic resources in the agro-ecosystem in which they have evolved (in situ conservation) is now being more widely considered as complementary to strategies based on gene banks (ex situ conservation, [6]). By taking advantage of the molecular markers and population genetics data here presented, an attempt to increase the genetic purity and to improve the genetic stability of these very old landraces could be made. For each population, only individuals characterized by the highest within-population genetic similarity and ancestry estimates could be maintained and multiplied by open pollination in isolated fields in order to yield farmer's seed stocks.

Supplementary Materials: The following are available online at www.mdpi.com/1424-2818/9/3/32/s1, Data S1: Statistics related to the frequency of all marker alleles, including private alleles, sorted by populations and samples, and genetic similarity matrix of all possible pair-wise comparisons among individuals and mean genetic similarity estimates among populations.

Acknowledgments: This work was supported by the project 'PROGRAMMA BIO.NET, rete regionale per la conservazione e caratterizzazione della biodiversità di interesse agrario-Gruppo di lavoro cerealicolo' funded by Programma di Sviluppo Rurale per il Veneto 2007–2013. The authors wish to thank Flavio Da Ronch for technical assistance with DNA extraction and PCR amplification experiments.

Author Contributions: G.B. and G.G. conceived and designed the experiments; F.P. and G.G performed the experiments; F.P. and L.M.-B analyzed the data; G.B. contributed reagents, materials and analysis tools; all authors contributed to data interpretation and manuscript preparation.

Conflicts of Interest: The authors declare no conflict of interest.

References

1. Baur, E. Die Bedeutung der primitiven Kulturrassen und der wilden Verwandten unserer Kulturpflanzen fur die Pflanzenzuchtung. *Jahrb. DLG* **1914**, *21*, 104–110.
2. Frankel, O.; Bennett, E. *Genetic Resources in Plants—Their Exploration and Conservation (IBP Handbook No. 11)*; Blackwell Science Ltd.: London, UK, 1970.
3. Harlan, J. Our vanishing genetic resources. *Science* **1975**, *188*, 618–621. [CrossRef] [PubMed]
4. Brush, S. In situ conservation of landraces in centers of crop diversity. *Crop Sci.* **1995**, *35*, 346–354. [CrossRef]
5. Bellon, M.R. Conceptualizing interventions to support on-farm genetic resource conservation. *World Dev.* **2004**, *32*, 159–172. [CrossRef]
6. Lucchin, M.; Barcaccia, G.; Parrini, P. Characterization of a flint maize (*Zea mays* L. convar. *mays*) Italian landrace: I. Morpho-phenological and agronomic traits. *Genet. Resour. Crop Evol.* **2003**, *50*, 315–327. [CrossRef]
7. Eurostat: Crop Statistics. Available online: http://ec.europa.eu/eurostat/web/agriculture/data/database?p_p_id=NavTreeportletprod_WAR_NavTreeportletprod_INSTANCE_ff6jlD0oti4U&p_p_lifecycle=0&p_p_state=normal&p_p_mode=view&p_p_col_id=column-2&p_p_col_pos=1&p_p_col_count=2 (accessed on 31 July 2017).
8. Shanbao, Q.; Yuhua, W.; Tingzhao, R.; Kecheng, Y.; Shibin, G.; Guangtang, P. Effective improvement of genetic variation in maize lines derived from R08xDonor backrosses by SSRs. *Biotechnology* **2009**, *8*, 358–364. [CrossRef]
9. Cömertpay, G.; Baloch, F.S.; Kilian, B.; Ülger, A.C.; Özkan, H. Diversity assessment of Turkish maize landraces based on fluorescent labelled SSR markers. *Plant Mol. Biol. Report.* **2012**, *30*, 261–274. [CrossRef]
10. Pineda-Hidalgo, K.V.; Méndez-Marroquín, K.P.; Alvarez, E.V.; Chávez-Ontiveros, J.; Sánchez-Peña, P.; Garzón-Tiznado, J.A.; Vega-García, M.O.; López-Valenzuela, J.A. Microsatellite-based genetic diversity among accessions of maize landraces from sinaloa in México. *Hereditas* **2013**, *150*, 53–59. [CrossRef] [PubMed]
11. Qi-Lun, Y.; Ping, F.; Ke-Cheng, K.; Guang-Tang, P. Genetic diversity based on SSR markers in maize (*Zea mays* L.) landraces from Wuling mountain region in China. *J. Genet.* **2008**, *87*, 287–291. [CrossRef] [PubMed]
12. Oppong, A.; Bedoya, C.A.; Ewool, M.B.; Asante, M.D.; Thomson, R.N.; Adu-Dapaah, H.; Lamptey, J.N.; Ofori, K.; Offei, S.K.; Warburton, M.L. Bulk genetic characterization of Ghanaian maize landraces using microsatellite markers. *Maydica* **2014**, *59*, 1–8.
13. Barcaccia, G.; Lucchin, M.; Parrini, P. Characterization of a flint maize (*Zea mays* var. indurata) Italian landrace, II. Genetic diversity and relatedness assessed by SSR and Inter-SSR molecular markers. *Genet. Resour. Crop Evol.* **2003**, *50*, 253–271. [CrossRef]

14. Mais Sponcio—Cooperativa Agricola La Fiorita. Available online: http://www.cooperativalafiorita.it/?scheda_prodotto+prodotti=mais_sponcio (accessed on 30 June 2017).

15. Perco, D. *Antonio Maresio Bazolle/Il Possidente Bellunes*; Tip. B. Bernardino: Feltre, Italy, 1986.

16. Zapparoli, T.V. Il Granoturco Marano. *L'Italia Agric.* **1939**, *76*, 155–159.

17. Zaccaria, L. Valutazione delle Variazioni dei Caratteri Morfo-Fisiologici Intervenute nel Corso Degli Anni Nella Varietà di Mais Marano Vicentino. Bachelor's Thesis, University of Padua, Padua, Italy, 31 Octorber 2012.

18. Bernardi, U.; Demattè, E. *Giacomo Agostinetti/Cento e Dieci ricordi che Formano il Buon Fattor di Villa*; Neri Pozza: Pordenone, Italy, 1998.

19. Miniscalco, V. *Il Granoturco*; Arti grafiche F.lli Cosarini: Pordenone, Italy, 1946.

20. Clemens, R.L.B. Keeping Farmers on the Land: Adding Value in Agriculture in the Veneto Region of Italy. *MATRIC Briefing Papers 6*. 2004. Available online: http://lib.dr.iastate.edu/matric_briefingpapers/6 (accessed on 13 August 2017).

21. Cassidy, A.; Mukamal, K.J.; Liu, L.; Franz, M.; Eliassen, A.H.; Rimm, E.B. High anthocyanin intake is associated with a reduced risk of myocardial infarction in young and middle-aged women. *Circulation* **2013**, *127*, 188–196. [CrossRef] [PubMed]

22. Schuelke, M. An economic method for the fluorescent labeling of PCR fragments A poor man's approach to genotyping for research and high-throughput diagnostics. *Nat. Biotechnol.* **2000**, *18*, 233–234. [CrossRef] [PubMed]

23. Register, J.I.; Sullivan, H.; Yun, Y.; Cook, D.; Vaske, D. A set of microsatellite markers of general utility in maize. *Maize Genet. Coop. News Lett.* **2001**, *75*, 31–34.

24. Nagy, S.; Poczai, P.; Cernák, I.; Gorji, A.M.; Hegedus, G.; Taller, J. PICcalc: An online program to calculate polymorphic information content for molecular genetic studies. *Biochem. Genet.* **2012**, *50*, 670–672. [CrossRef] [PubMed]

25. Peakall, R.; Smouse, P.E. GenALEx 6.5: Genetic analysis in Excel. Population genetic software for teaching and research-an update. *Bioinformatics* **2012**, *28*, 2537–2539. [CrossRef] [PubMed]

26. Yeh, F.C.; Yang, R.C.; Boyle, T.B.J.; Ye, Z.H.; Mao, J.X. *POPGENE, the User Friendly Shareware for Population Genetic Analysis*; Molecular Biology and Biotechnology Centre, University of Alberta: Edmonton, AB, Canada, 1997.

27. Nei, M. Analysis of gene diversity in subdivided populations. *Proc. Natl. Acad. Sci. USA* **1973**, *70*, 3321–3323. [CrossRef] [PubMed]

28. Van Zonneveld, M.; Scheldeman, X.; Escribano, P.; Viruel, M.A.; van Damme, P.; Garcia, W.; Tapia, C.; Romero, J.; Sigueñas, M.; Hormaza, J.I. Mapping genetic diversity of cherimoya (*Annona cherimola* Mill.): Application of spatial analysis for conservation and use of plant genetic resources. *PLoS ONE* **2012**, *7*, e29845. [CrossRef] [PubMed]

29. Wright, S. The interpretation of population structure by F-statistics with special regard to systems of mating. *Evolution* **1965**, *19*, 395–420. [CrossRef]

30. Barcaccia, G.; Felicetti, M.; Galla, G.; Capomaccio, S.; Cappelli, K.; Albertini, E.; Buttazzoni, L.; Pieramati, C.; Silvestrelli, M.; Verini Supplizi, A. Molecular analysis of genetic diversity, population structure and inbreeding level of the Italian Lipizzan horse. *Livest. Sci.* **2013**, *151*, 124–133. [CrossRef]

31. Rohlf, F.J. *NTSYS-pc: Numerical Taxonomy and Multivariate Analysis System, Version 2.1*; Applied Biostatistics Inc.: Port Jefferson, NY, USA, 2004.

32. Hammer, Ø.; Harper, D.A.T.; Ryan, P.D. PAST: PAleontological STatistics software package for education and data Analysis. *Palaeontol. Electron.* **2001**, *4*, 1–9.

33. Falush, D.; Stephens, M.; Pritchard, J.K. Inference of population structure using multilocus genotype data: Linked loci and correlated allele frequencies. *Genetics* **2003**, *164*, 1567–1587. [PubMed]

34. Porras-Hurtado, L.; Ruiz, Y.; Santos, C.; Phillips, C.; Carracedo, A.; Lareu, M.V. An overview of STRUCTURE: Applications, parameter settings, and supporting software. *Front. Genet.* **2013**, *4*, 98. [CrossRef] [PubMed]

35. Evanno, G.; Regnaut, S.; Goudet, J. Detecting the number of clusters of individuals using the software STRUCTURE: A simulation study. *Mol. Ecol.* **2005**, *14*, 2611–2620. [CrossRef] [PubMed]

36. Botstein, D.; White, R.L.; Skolnick, M.; Davis, R.W. Construction of a genetic linkage map in man using restriction fragment length polymorphisms. *Am. J. Hum. Genet.* **1980**, *32*, 314–331. [PubMed]

37. Palumbo, F.; Galla, G.; Barcaccia, G. Developing a molecular identification assay of old landraces for the genetic authentication of typical agro-food products: The case study of the barley 'Agordino'. *Food Technol. Biotechnol.* **2017**, *55*, 29–39. [CrossRef]
38. Russell, W.A. Genetic improvement of maize yields. *Adv. Agron.* **1991**, *46*, 245–298. [CrossRef]
39. Mais Biancoperla. Available online: http://www.legambientepadova.it/files/Scheda%20mais.pdf (accessed on 30 June 2017).

diversity

MDPI

Article

Putting Plant Genetic Diversity and Variability at Work for Breeding: Hybrid Rice Suitability in West Africa

Raafat El-Namaky [1], **Mamadou M. Bare Coulibaly** [2], **Maji Alhassan** [1,3], **Karim Traore** [1], **Francis Nwilene** [4], **Ibnou Dieng** [5], **Rodomiro Ortiz** [6,*] and **Baboucarr Manneh** [1]

[1] Africa Rice Center (Africa Rice) Sahel Station, B.P. 96 Saint Louis, Senegal; R.Elnamaky@cgiar.org (R.E.-N.); A.Maji@cgiar.org (M.A.); K.Traore@cgiar.org (K.T.); B.Manneh@cgiar.org (B.M.)
[2] Institute of Rural Economy (IER), B.P. 258 Bamako, Mali; mamadoumbare@yahoo.fr
[3] National Cereals Research Institute (NCRI), Badeggi, P.M.B. 8 Bida, Niger State, Nigeria
[4] Africa Rice Center, Ibadan, P.M.B. 5320 Oyo State, Nigeria; F.Nwilene@cgiar.org
[5] Africa Rice, 01 B.P. 2031 Cotonou, Benin; I.Dieng@cgiar.org
[6] Swedish University of Agricultural Sciences (SLU), Box 101, SE 23053 Alnarp, Sweden
*   Correspondence: rodomiro.ortiz@slu.se; Tel.: +46-725418386

Received: 17 April 2017; Accepted: 7 July 2017; Published: 10 July 2017

Abstract: Rice is a staple food in West Africa, where its demand keeps increasing due to population growth. Hence, there is an urgent need to identify high yielding rice cultivars that fulfill this demand locally. Rice hybrids are already known to significantly increase productivity. This study evaluated the potential of Asian hybrids with good adaptability to irrigated and rainfed lowland rice areas in Mali, Nigeria, and Senegal. There were 169 hybrids from China included in trials at target sites during 2009 and 2010. The genotype × environment interaction was highly significant ($p < 0.0001$) for grain yield indicating that the hybrids' and their respective cultivar checks' performance differed across locations. Two hybrids had the highest grain yield during 2010 in Mali, while in Nigeria, four hybrids in 2009 and one hybrid in 2010 had higher grain yield and matured earlier than the best local cultivar. The milling recovery, grain shape and cooking features of most hybrids had the quality preferred by West African consumers. Most of the hybrids were, however, susceptible to African rice gall midge (AfRGM) and *Rice Yellow Mottle Virus* (RMYV) isolate Ng40. About 60% of these hybrids were resistant to blast. Hybrids need to incorporate host plant resistant for AfRGM and RYMV to be grown in West Africa.

Keywords: *Oryza sativa*; adaptability; food security; genotype × environment interaction; grain yield; heterosis; hybrid vigor; quality; resilience; sustainability

---

1. Introduction

Rice (*Oryza sativa* L.) is the most important staple food crop in the world. About 3.5 billion people depend on rice globally, since this staple provides in excess of 20% of their daily calorie intake [1]. In Africa alone, where currently rice consumption is the most rapidly growing food source, about 30 million t more rice will be required by 2035, thus representing an increase of 130% in rice consumption from 2010. Nigeria would require almost one third of this additional rice [2]. More than 90% of West African rice farmers are smallholders (mostly women) who cultivate less than 1 ha and whose crop yields depend on rainfall. Crop production from these small plots is often insufficient to provide a reasonable household income for maintaining a minimum standard of living. These farmers manage complex farming systems, cultivating rice and other food crops based on its degree of importance as food and as a cash crop, unlike farmers in Asia, where rice is a crop mainly grown in lowland and irrigated agro-ecosystems. The major constrains of global rice production are drought, flood, heat, pathogens, pests, declining productivity in intensive rice production systems, low grain yield in some areas of the developing world, increasing production costs in the industrialized world, and rising public concern regarding sustainability of rice farming. In addition, mechanization, the high cost of irrigated rice production, as well as the poor management of uplands and rainfed lowlands are among the main challenges for producing rice in Africa [3,4].

Advances in rice production in Africa are at various developmental stages due to its relative importance to respective local economy. Hybrid rice seed technology is key for increasing its production and maintaining self-sufficiency and food security. Hybrid rice has been used in rice production for more than 40 years in Asia and North America, and more recently in Egypt because of its high grain yield potential. This yield advantage plus water and nitrogen use efficiency, and host plant resistance to pathogens and pests are the main determining factors for adapting hybrid seed technology in rice production. Heterosis improves grain yield and quality for many crops especially when facing a limited area for farming. Hybrid cultivars have been developed to take advantage of heterosis in the production of many field crops such as cotton, maize, oilseed rape, rice, sorghum, sunflower, and vegetables [5,6]. African farmers may boost rice production by using hybrids particularly in the largely unutilized lowland rice areas. Mali, Nigeria, and Senegal are willing to adopt this seed technology to increase their rice production.

Rice breeders in China led the developing and commercializing of rice hybrids that had 15 to 20% yield advantage [7] or at least 1 t ha^{-1} [8] over inbred cultivars. The new set of Green Super Rice (GSR) hybrids need less chemical inputs to increase grain yields than the old rice hybrids. Likewise, many of these GSR hybrids show adaptation to drought, thereby requiring less water and could be grown in rainfed agro-ecosystems. Nonetheless, their adaptability to African farming systems needs to be assessed across target areas, as well as the relative inputs they require vis-à-vis local cultivars. Farmers need to invest in both seeds and inputs for getting high grain yields when using rice hybrids. The increased grain yield of rice hybrids should pay for this investment and bring profitability to those using this seed technology.

There are various pathogens and pests affecting rice production in Africa. The African rice gall midge (AfRGM; *Orseolia oryzivora*; Diptera: *Cecidomyiidae*) is an important insect pest in irrigated lowland rice areas, causing 25 to 80% grain yield loss in West Africa [9]. AfRGM is an endemic pest to Africa, where was first reported in Sudan, and currently spreading throughout the continent. It can be found in 12 West African, two Central African, and five East and Southern African countries [10]. The insect pest causes 20 to 100% grain yield loss in the worst-affected areas. There are 16 quantitative trait loci (QTL) associated with host plant resistance to AfRGM, of which three are in [ITA306 × BW348-1], five in [ITA306 × TOS14519], and eight in [ITA306 × TOG7106] breeding populations. The major effect genomic region for AfRGM resistance was in the [ITA306 × TOS14519] population, which was at 111cM on chromosome 4 (qAfrGM4), had a LOD score of 60 and accounted for 34.1% of the total phenotypic variance [11]. Likewise, *Rice Yellow Mottle Virus* (RYMV, a Sobemo virus) is another major constraint to rice production in the continent [12] because it causes 17 to 100% grain yield loss according to both the infection date and time, and the cultivar host [13,14]). RYMV is highly infectious to rice, especially to Asian *indica* cultivars in lowland and irrigated agro-ecosystem. RYMV is prevalent in all major rice growing ecosystems of Africa [15]. Rice blast, caused by *Magnaporthe oryzae* (anamorph: *Pyricularia oryzae*) [16], is another serious disease affecting rice in temperate and tropical regions, including Africa [17,18].

The main purpose of this research was to determine the suitability of GSR hybrid cultivars bred in China at the irrigated lowland rice areas of Mali, Nigeria, and Senegal. We assessed their grain yield across suitable West African rice growing areas.

2. Materials and Methods

2.1. Plant Materials

The Chinese Academy of Agricultural Sciences provided two sets of GSR hybrids consisting of 122 and 47 F_1s. Africa Rice included eight inbred rice cultivars that were used as checks because farmers in Mali, Nigeria, and Senegal grow them widely.

2.2. Methods

The multi-location trials were conducted during 2009 and 2010. For each country, two trials were conducted in an augmented design layout [19] in a randomized complete block (RCB), with each cultivar plot comprising of five rows of 3 m length with a plant to row spacing of 20 × 20 cm. In Senegal, 122 hybrids and 13 checks were evaluated during the 2009 wet season (June–November) at Ndiaye (the Africa Rice Research Station near Saint-Louis), and 47 hybrids were evaluated with checks during the 2010 dry season (February–June) 2010 at the Research Station of L' Institut Sénégalais de Recherche Agricole (ISRA) in Fanaye. There are two seasons in Senegal, the rainy season from June to October, characterized by heat, humidity and storms, and the dry season from December to May, characterized by cool ocean breeze and dust from the Harmattan winds. A total of 122 hybrids and 8 checks were evaluated in Mali during the 2010 dry season (January–May) at N'Debougou in the Office du Niger, and 47 hybrids and checks were evaluated during 2010 wet season (June–November) at the Agriculture Research Station in Niono. The Office du Niger irrigation scheme is the largest in West Africa and located in the Segou region of Central Mali. Rainfall in the Office du Niger area ranges from 450 to 600 mm per year. The rainy season lasts three to five months, and the dry season is divided into a cool and a hot period. In Nigeria, 122 hybrids and checks were evaluated during the 2009 dry season (October–April) at the research field of the National Cereals Research Institute (NCRI) in Badeggi, and 47 hybrids evaluated with eight checks during the 2010 wet season (May–October) at Wushishi. Both Nigerian locations are in the Southern Guinea Savanna, a typical rainfed lowland agro-ecology with iron toxicity and high AfRGM severity. The trials used known cultural practices of each country. The fertilizer applications varied as follows: 115–36 kg N–K ha^{-1} in Mali, 80–40–40 kg N–P–K ha^{-1} in Nigeria, and 150–17.5–33 kg N–P–K ha^{-1} in Senegal. The data recorded in the field included plant height, panicles m^{-2}, spikelets plant^{-1}, 1000-grain weight and grain yield. Grain quality traits and milling properties of 18 promising hybrids were evaluated at Africa Rice Sahel Station, near Saint Louis, Senegal. The most popular cultivar Sahel 108, widely preferred by farmers and consumers, was used as grain quality check. The traits assessed were brown rice length, grain length and shape, percentage of total milling, alkali-spreading value (ASV), and gelatinization temperature (GT). A high and low GT indicates more or less energy necessary for cooking rice, respectively. All traits were measured according to the standard's evaluation system used by the International Rice Research Institute (IRRI, Manila, Philippines) [20].

2.3. Host Plant Resistance (HPR) to AfRGM, RYMV and Blast

The first set of 122 hybrids along with eight inbred lowland cultivars were evaluated for HPR to AfRGM at a paddy screenhouse in Ibadan, Nigeria during 2009. Two rice cultivars were used as checks: ITA 306 as susceptible and TOS14519 as resistant to AfRGM. Conventional "spreader rows" of the highly susceptible cultivar ITA 306 were around the GSR plants. Newly hatched larvae were deposited on "spreader" plants to increase infestation in the screenhouse [18]. Data were taken 45 and 70 days after transplanting (DAT) to estimate the percentage of tillers infested or damage by AfRGM. In the 2010 wet season, 47 hybrid lines and eight check cultivars were screened for resistance to AfRGM under natural infestation at Edozhigi in Niger State, Nigeria. The GSR hybrids and checks were laid out in an RCB with three replications. Data were taken at 42 and 63 DAT. Rice blast was evaluated in a trapping nursery as described by Sere, et al. [21] on blast hot spot at Ouedeme, Benin in 2009. Trapping nurseries consists of exposing cultivars with known resistance genes to natural inoculum. The reaction of each of the cultivars is an indication of the presence/activation (non-compatible reaction) or absence/inactivation (compatible reaction) of the corresponding avirulence genes without prejudging their association into distinct races. This trapping nursery was laid out in an RCB design. We scored disease severity weekly using IRRI's standard evaluation system [20]. Standard cultivars and near isogenic lines (NIL) were used as checks to define host plant resistance clusters to blast. A highly resistant rating was given to those hybrids similar to the cultivar check and a NIL possessing the gene *Pi9*, while for the resistant cluster the check was the NIL bearing the *Pi2* gene. The susceptible cluster included as checks five NIL with *Pi33*, *Pif*, *Pii*, *Pish*, and *Pi1b* genes. Hybrids were rated as highly susceptible when showing scores similar to a susceptible check and 22 NIL. There were 122 hybrids tested against RYMV through artificial inoculation of two strains (NG-01 and NG-40) in a screenhouse. We included two resistant (Gigante and TOG 5681) and one susceptible check in this screening. RYMV was evaluated 21 and 42 days after inoculation using the standard evaluation system with a 1–9 scale, in which 1 means lack of any symptom and 9 a completely damaged leave. The disease severity score (S) was calculated as $S = \{(n_1 \times 1 + n_3 \times 3 + n_5 \times 5 + n_7 \times 7 + n_9 \times 9) \times 100\}/\{(n_1 + n_3 + n_5 + n_7 + n_9) \times 9\}$, where n_1, n_3, n_5, n_7, and n_9 represent the number of plant scoring 1, 3, 5, 7, and 9, respectively [22]. The following hybrid clusters were defined for both strains according to their host plant resistance scores: (1) rated as the resistant checks, (2) moderately susceptible, and (3) rated as the susceptible checks.

2.4. Statistical Analysis

Data were subjected to an analysis of variance using SAS/STAT 9.2 (SAS Institute, Cary, NC, USA). We investigated the genotype × country interaction (G × E) using a SAS mixed model [23]. If a significant G × E was detected, we analyzed the performance of hybrids and their respective cultivar checks within each country because they performed differently across environments. A multiple comparison adjustment for the probability (*p*) values was then performed to test whether the adjusted means of the hybrids were significantly higher than the best check for each country at a significance level of $p \leq 0.05$.

3. Results

The G × E was highly significant ($p < 0.0001$) for grain yield in both sets of experiments, thereby indicating that hybrids and check cultivars performed differently across Mali, Nigeria, and Senegal. Hence, we analyzed separately each set of experiments within each country (Table 1).

Table 1. Combined analysis of variance for grain yield of two sets of hybrids and their cultivar checks across sites in Mali, Nigeria, and Senegal.

Source of Variation	DF [z]	MSE [y]	F Value	*p* Value
First set: 122 hybrids + 13 inbred				
Genotypes (G)	134	4,833,250	2.58	<0.001
Country (E)	2	723,984,958	386.79	<0.001
G × E	240	4,316,391	2.306	<0.001
Error	117	1,871,732		
Second set: 47 hybrids + 13 inbred				
Genotypes (G)	59	6,770,649.3	1.34	0.0990
Country (E)	2	120,454,097.3	23.91	<0.001
G × E	111	6,880,333.9	1.37	0.0496
Error	95	5,038,496.0		

[z] Degrees of freedom, [y] Mean square error.

3.1. Country Performance

Most of the hybrids showed on average 10 days earlier 50% flowering than check cultivars during the 2009 wet season at Saint Louis, Senegal (Table 2a). The best yielding check cultivar, Sahel 108, had the highest number of tillers (23), while hybrids, whose grain yield was above Sahel 108, had 13 to 15 tillers. The promising hybrids QYI, HanF1-39, and HS706 exhibited panicles whose lengths were 25, 24, and 29 cm, respectively, while the panicle of Sahel 108 measured on average 20 cm. There were not significant differences for average of plant height between hybrids and checks cultivars.

The average grain yield of hybrids (7 t ha^{-1}) and checks (6.5 t ha^{-1}) was not significantly different. Only hybrids QYI, HanF1-39, and HS706 had a significant grain yield advantage (38 to 43.5%) vis-à-vis Sahel 108. Most of the hybrids and checks had early flowering (89 and 92 days, respectively) during the 2010 wet season at Fanaye, Senegal (Table 2b). The hybrid CNY498 was the earliest flowering (82 days) and shortest (80 cm). There was no significant difference for average plant height between hybrids (101 cm) and checks (102 cm). The check cultivars showed a higher average number of tillers plant^{-1} and panicle length than the hybrids. The hybrid HS706 had the longest panicle (29 cm), and the hybrid XYR24 was the only one out-yielding significantly the check cultivar Sahel 108.

Table 2. Grain yield and other agronomic traits of promising rice hybrids and best cultivar check (C) at (**a**) Ndiaye (2009 wet season) and (**b**) Fanaye (2010 wet season), Senegal. The *p*-value indicates that hybrids' yield was significantly higher than Sahel 108. Standard errors (S.E.) given for each trait.

Cultivar	Days to 50% Flowering	Plant Height (cm)	Tiller Plant^{-1}	Panicle Length (cm)	Grain Yield (t ha^{-1})	Yield Advantage (%)	*p*-Value
		(a)	Ndiaye 2009 Wet Season				
Qy1	80	108	14	25	10.5	43.54	0.0285
HanF1-39	87	97	13	24	10.5	42.86	0.0307
HS706	87	108	15	29	10.1	37.96	0.0480
Sahel 108 (C)	88	86	23	20	7.4		
Hybrid mean	84	102	15	24	7.1	8.91	
Check mean	94	101	18	27	6.5		
Hybrid S.E.	4	5	3	2	1.5		
Check S.E.	2	2	1	1	0.6		
		(b)	Fanaye 2010 Wet Season				
XYR24	91	106	20	25	12.2	82.46	0.027
CXY727	93	121	20	25	10.7	59.97	0.077
NEY2123	92	79	20	23	10.6	59.37	0.078
CNY498	82	99	20	27	10.5	57.42	0.085
3LYR24	89	108	20	23	10.3	54.87	0.095
Sahel 108 (C)	87	82	25	23	6.7		
Hybrids mean	89	101	21	21	6.1	59.74	
Check mean	92	102	20	20	3.8		
Hybrid S.E.	4	7	2	3	2.0		
Check S.E.	3	14	2	2	0.9		

There were no significant differences for days to 50% flowering and panicle m^{-2} between hybrids and check cultivars grown during the 2010 dry season at N'Debougou in Mali (Table 3a). The average plant height for the hybrids was 108 cm and 111 cm for the check cultivars. The tallest hybrid, among those out-yielding the best check, was 49youR24 (122.5 cm). The hybrid 49youR24 had the highest number of spikelets per panicle (242), which was about 49% above the best check WITA 9. The average 1000-grain weight of this hybrid (27.5 g) was 23% higher than the average of the checks (23.5 g). Grain yield of the high-yielding promising hybrids ranged from 11.5 to 12.5 t ha $^{-1}$, but there were not significantly different that the best check cultivar WITA 9 (10.2 t ha^{-1}).

There was an early 50% flowering for both hybrids (88 days) and checks (95 days) during the 2010 wet season at Niono, Mali (Table 2b). XYR24 was the earliest (82 days) among the most promising hybrids due to their high grain yield, though most of these hybrids (and high yielding checks) were among the tallest in this trial. The best hybrids and check cultivar had a high number of spikelets per panicle (152–254). On average the hybrids had less panicles per m^2 (292) than the check cultivars (325), but the hybrid GXY803067 had the most panicles (347). The hybrids' grains weighed higher than the check cultivars; and the hybrid CYX2 had the largest grains. Only one hybrid showed a significantly higher grain yield (14.7 t ha^{-1}) than the best cultivar check NericaL-19 (10.6 t ha^{-1}).

Table 3. Grain yield and other agronomic traits of promising rice hybrids and best cultivar check (C) in 2010 at (**a**) N'Debougou, Mali dry season and (**b**) Niono, Mali wet season. The *p*-value indicates that hybrids' yield was significantly higher that WITA 9 and NERICA-L19. Standard errors (S.E.) given for each trait.

Cultivar	Days to 50% Flowering	Plant Height (cm)	Panicle (m^2)	Spikelets Panicle^{-1}	1000 Grain Weight (g)	Grain Yield (t ha^{-1})	Yield Advantage (%)	*p*-Value
(a) N'Debougou Dry Season								
HanF1-35	88	111	168	158	19.8	12.5	22.31	0.0938
HS33	94	108	228	126	27.2	12.0	17.42	0.1508
49youR24	96	122	288	242	27.5	11.5	12.52	0.2282
HanF1-22	95	117	175	187	24.5	11.5	12.52	0.2282
HanF1-30	98	105	214	212	28.5	11.5	12.52	0.2282
WITA 9 (C)	96	109	245	162	22.4	10.2	0.00	
Hybrid mean	96	108	238	164	25.5	6.3	8.58	
Check mean	97	111	237	142	23.5	5.8		
Hybrid S.E.	4	7	27	27	2.1	2.8		
Check S.E.	2	3	11	11	0.8	2.4		
(b) Niono Wet Season								
QS2	92	117	281	246	22.3	14.7	38.57	0.046
GSR-H-0007	89	133	291	152	26.4	12.0	12.89	0.291
CXY2	90	112	331	218	31.4	12.0	12.89	0.291
QS3	90	124	300	175	26.8	11.7	9.78	0.339
ZhongyouR24	85	125	271	254	27.9	11.3	6.59	0.389
GXY803067	89	125	347	183	29.9	11.3	6.59	0.389
NERICA-L19 (C)	92	143	255	209	25.2	10.6	0.00	
Hybrid mean	88	122	292	215	28.0	8.4	−4.99	
Check mean	95	119	325	182	25.3	8.8		
Hybrid S.E.	2	7	45	20	2.0	2.4		
Check S.E.	1	3	20	9	0.9	1.1		

The early sowing date affected days to 50% flowering, and most of the hybrids and check cultivars have their blooming after 100 days in the 2009 dry season at Badeggi (Table 4a). The average number of days to 50% flowering was 104 for the hybrids and 109 days for the check cultivars. Most of the promising hybrids were dwarf to semi-dwarf in plant height; i.e., not exceeding 100 cm. The most promising hybrids had more panicles per area than the best check WITA 4. Grain yield was on average very low for hybrids and checks (2.55 and 2.51 t ha^{-1}, respectively). The grain yield of the most promising hybrids ranged between 4.9 and 5.6 t ha^{-1}, which was at least 50% above the grain yield of the best check cultivar WITA 4. Most of the hybrids had early flowering (about two weeks of advantage than the average of the check cultivars) during the 2010 wet season at Wushishi. TY8022 was the earliest flowering (77 days) hybrid (Table 4b). Most hybrids and checks exhibited a tall plant height (112 and 119 cm on average, respectively). There were no significant differences for grain yield between most promising hybrids (ranging from 8.5 to 11.5 t ha^{-1}) and best check WITA 1 (8.5 t ha^{-1}).

3.2. Grain Quality

All hybrids and check cultivar had medium to long grain types (Table 5). Their grain shape was either medium or slender. The most promising hybrids exhibited milling recovery that was above the check Sahel 108. The hybrid ERHAO had high total milling recovery (70%), slender grain shape and low GT. The hybrids WZY2, CNY 549, and IL YOU Z2 showed the highest total milling recovery (above 70%). Only two promising hybrids had low GT while the others and Sahel 108 exhibited a medium GT.

Table 4. Grain yield and other agronomic traits of promising rice hybrids and best cultivar check (C) at (a) Badeggi 2009 dry season and (b) Wushishi 2010 wet season. The *p*-value indicates that the hybrids' yield was higher that WITA1 and WITA4. Standard errors (S.E.) given for each trait.

Cultivar	Days to 50% Flowering	Plant Height (cm)	Panicle (m^{-2})	Grain Yield $(t\ ha^{-1})$	Yield Advantage (%)	*p*-Value
			(a) Badeggi 2009 Dry Season			
HanF1-26	104	87	294	5.6	58.51	0.0267
XYR24	101	93	395	5.4	74.92	0.0078
3LYR24	112	98	325	5.1	66.25	0.0151
WZY2	109	73	293	4.9	51.08	0.0447
WITA 4 (C)	111	83	279	3.2	0.00	
Hybrid mean	104	93	248	2.6	1.59	
Check mean	109	88	301	2.5		
Hybrid S.E.	6	16	72	0.9		
Check S.E.	2	6	29	0.3		
			(b) Wushishi 2010 Wet Season			
GSR-H-0141	85	104	17	11.5	35.50	0.144
WZY2	84	114	11	9.2	9.08	0.392
GXY3169	92	126	12	9.2	8.84	0.395
Ilyou623	96	112	13	8.9	4.83	0.442
TY8022	77	114	12	8.7	2.12	0.475
Huayou7109	79	118	12	8.5	0.59	0.493
WITA 1 (C)	103	146	15	8.5	0.00	
Hybrid mean	84	117	11	6.8	17.12	
Check mean	99	119	15	5.8		
Hybrid S.E.	16	7	4	2.7		
Check S.E.	7	3	2	1.2		

Table 5. Grain traits and milling properties of promising rice hybrids and cultivar check (C) Sahel 108 after harvest in Ndjaye, Senegal during 2009 wet season.

Hybrid	Brown Rice Length (mm)	Grain Length	Grain Shape	Total Milling (%)	Alkali-Spreading Value	Gelanization Temperature (GT)
QY 1	6.3	Medium	Medium	68	5.67	Low
HAN F1-47	6.4	Medium	Medium	69	4.00	Intermediate
HS 706	6.2	Medium	Slender	70	4.58	Intermediate
HAN F1-39	6.3	Medium	Medium	66	3.67	Intermediate
HAN F1-36	6.8	Long	Slender	67	3.50	Intermediate
PLY 1108	6.1	Medium	Medium	70	5.67	Low
QY 3	6.4	Medium	Medium	66	4.00	Intermediate
ERHAO	6.1	Medium	Slender	70	6.67	Low
HAN F1-26	6.7	Long	Slender	69	3.50	Intermediate
HAN F1-31	6.6	Long	Medium	69	3.33	Intermediate
YX 9	6.9	Long	Medium	68	5.33	Intermediate
WZY 2	6.0	Medium	Medium	72	4.67	Intermediate
HAN F1-10	6.4	Medium	Medium	70	3.42	Intermediate
GXY–207	6.9	Long	Slender	69	3.50	Intermediate
CNY 549	6.1	Medium	Medium	72	3.50	Intermediate
HAN F1-9	6.2	Medium	Medium	70	4.42	Intermediate
IL YOU Z2	5.8	Medium	Medium	71	4.25	Intermediate
Sahel 108 (C)	6.1	Medium	Slender	65	3.33	Intermediate

3.3. Host Plant Resistance

Most of the 122 hybrids evaluated at a paddy screen house in Ibadan 2009 were highly susceptible to AfRGM. Only four hybrids (Han F1-21, YG 206, HS 33, and YG 17) were moderately susceptible to this insect pest (Table 6a), while the hybrid FFY66 showed partial resistance (5.9% tiller infestation) to AfRGM under natural infestation during the 2010 wet season at Edozhigi (Table 6b). Among hybrids,

FFY66 had the lowest mean plant damage and could be described as possessing an acceptable level of resistance to the pest. The hybrid FFY66 needs to be further evaluated at Edozhigi to confirm its stability and acceptability as possessing desirable or appreciable level of host plant resistance to AfRGM. The population structure of AfRGM in different rice farming systems in West Africa revealed that the biotype at Edozhigi is a virulent type—hence, the high susceptibility of the hybrids to this pest. The percent tiller infestation reduced at 63 days because of other parasites in the fields.

Table 6. Host plant resistance of most promising rice hybrids plus resistant and susceptible check cultivars to African rice gall midge after artificial infestation in screen house at (**a**) Ibadan 2009 wet season and (**b**) Edozhigi 2010 wet season.

Cultivar	Tiller Infestation (%) at 45 DAT [z]	Tiller Infestation (%) at 70 DAT	Mean Tiller Infestation (%)	Host plant Resistance Rating [y]
(a) Ibadan 2009 Wet Season				
Han F1-21	7.5	16.1	11.8	MS
YG 206	20.5	17.1	18.8	MS
HS 33	12.3	26.4	19.4	MS
YG 17	21.0	18.1	19.6	MS
ITA 306 (S)	32.1	54.0	43.1	HS
TOS14519 (R)	0.3	0.3	0.3	R
(b) Edozhigi 2010 Wet Season				
FFY66	5.9	4.4	5.3	MR
Hanyou3	10.0	4.5	7.2	MS
HuF1-8	11.2	3.4	7.3	MS
FFY366	7.7	8.5	8.1	MS
EYH2	11.9	5.5	8.7	MS
QS2	12.7	4.9	8.8	MS
Huhan7A/07ZR15	11.0	7.2	9.1	MS
HLYR24	10.9	8.9	9.9	MS
CXY6	14.8	5.8	10.3	MS
CXY2	15.4	5.4	10.4	MS
HuF1-18	16.5	4.4	10.4	MS
Huhan9A/07ZR15	16.6	4.4	10.5	MS
ITA 306 (S)	20.7	14.9	17.8	S
TOS 14519 (R)	1.0	0.9	1.0	R

[z] Days after transplanting, [y] R: resistant, MS: moderately susceptible, HS: highly susceptible (after [20]).

Table 7. Percentage of rice hybrids showing host plant resistance to *Rice Yellow Mottle Virus* strains Ng-01 and Ng-40.

Days after Inoculation	Isolate	Number of Lines Per Cluster	Host Plant Resistance [z]	Hybrid Resistance (%)
21	Ng-01	36	R	35.3
		0	MS	0.0
		66	S	64.7
	Ng-40	26	R	25.5
		31	MS	30.4
		45	S	44.1
42	Ng-01	5	R	4.9
		25	MS	24.5
		72	S	70.6
	Ng-40	0	R	0.0
		28	MS	27.5
		74	S	72.5

[z] R: resistant, MS: moderately susceptible.

The hybrids were in four clusters according to their host plant resistance to blast when comparing them to known resistance and susceptible standard cultivars and NIL. We noted that two NIL bearing the same resistant gene (IRBLzt-IR56/CO and C104PKT) were in two contrasting clusters. This result could be due to either a non-detected additional major gene or minor sensitive resistant genes [24,25]. The highly resistant cluster included 49 hybrids, while there were 28 hybrids showing partial resistance, 23 hybrids were susceptible, and the remaining 22 hybrids were rated as highly susceptible.

There were 23% and 32% of the hybrids showing host plant resistance to isolates Ng-40 and Ng-01 after two weeks from inoculation (Table 6). With the epidemic progression, none of the hybrids showed host plant resistance to Ng-40 as the resistant check, but 18% hybrids were moderately susceptible. There were five hybrids rated as resistant, as the resistant check to isolate Ng-01 (Table 7).

4. Discussion

Hybrid rice seed technology was developed in China more than 35 years ago. Many of the released rice hybrids have shown between 15% and 20% grain yield advantage over inbred cultivars in Asia, South America, and Egypt [26,27]. These results encouraged some to test hybrid rice in sub-Saharan Africa. Our previous research indicated that the grain yield of IRRI-bred hybrids was similar to that of check cultivars in Senegal [28]. This research shows, however, that there was a high yield potential of hybrid rice bred in China after testing them in Mali, Nigeria, and Senegal. The grain yield of the most promising hybrids was higher than that of the best local inbred check cultivar. The hybrids exhibited a wide range grain yield, which was affected by the testing location and growing season.

The genotype × environment interactions (G × E) were highly significant for both sets of the hybrids and their respective cultivar checks across the three countries; i.e., their performances were different across sites from country to country. There was not the same hybrid or check cultivar showing superiority over others across the sites used in these three countries. This result may be due the soil characteristics, fertilizer applications, and weather conditions in the sites used for testing. The results of this study exhibited the high potential of these hybrids under the optimum conditions (Mali and Senegal), while the yield affected by RYMV and AfRGM stresses (in Nigeria). This finding suggests developing different hybrids for each region. In West Africa, high rice grain yields are often associated with high solar radiation, high maximum temperature, intermediate air humidity, multiple split nitrogen (N) fertilizer applications, high frequency of weeding operations, the use of certified seeds, and well-leveled fields in the irrigated lowland system [29]. Local cultivars with host plant resistance to pathogens and pest may be used to develop new hybrids with high adaptability to stress-prone African sites.

The most promising hybrids across the testing locations had 20% more grain yield than the best check cultivars. This additional grain yield advantage may encourage some African farmers to grow hybrid rice cultivars. Testing sites in Mali and Senegal have irrigated areas with higher solar radiation than the Nigerian testing site [30], which could be optimum environments for rice hybrids to show their heterosis for grain yield. The most promising hybrids also showed a good plant type and other interesting attributes. For example, most of them had medium or slender grain shape and higher milling recovery than the preferred check cultivar. Likewise, two hybrids had low GT, thereby needing less energy inputs for their cooking.

A major shortcoming of most of the hybrids bred in China is their susceptibility to both AfRGM and RYMV. Host plant resistance to both of them needs to be bred or incorporated into the parental lines of the hybrid rice germplasm before their release to West African farmers. In order words, indigenous parental lines such as traditional African (*O. glaberrima*) and Asian (*O. sativa*) rice-derived cultivars grown already in West Africa are proven useful sources of resistance to AfRGM and RYMV. They should be used in developing hybrids for farmers in West Africa. The levels of resistance shown in *O. glaberrima* gene pool are better than those of found in *O. sativa*. TOG7106 and TOS14519 were rated highly stable and highly resistant to AfRGM [31]. Crossing should be made with donors that possess multiple tolerance to biophysical stresses, and off-sites should be used to select appropriate lines from segregating generations [32]. Recently the major QTL of AfRGM were located and validated [11], being the QTL with major effect (qAfRGM4) on chromosome 4. They could be easily incorporated into the parental lines of the promising hybrids through marker-aided backcrossing [33]. Knowledge on the distribution of virulent RYMV populations across different target sites will also help to deploy safely rice hybrids in sub-Saharan Africa. It was interesting to note that in excess of 60% of the hybrids were either partially resistant (with slight symptoms) or highly resistant (without any symptom) to blast. Moreover, multi-environment testing will be still necessary to assess whether their host plant resistance will remain across space and time. Microsatellite-aided screening for fertility restoration genes *Rf* may further facilitate hybrid development with high adaptability in West Africa [34]. Preliminary results are promising: some adapted breeding lines and cultivars had high grain yield per plant with cytoplasmic male sterility lines and may be used as fertility restorers.

Acknowledgments: We are grateful for the assistance given by Ngoni Jiri, the AGRITEX officer for Chiota, and for the assistance by the Chiota community during data collection. This work was supported by the International Foundation of Science (IFS) (grant C/4569-1); DAAD Fellowship (grant number A/10/03022) and the Climate Food and Farming (CLIFF) network under the CGIAR Research Program on Climate Change, Agriculture and Food Security (CCAFS).

Author Contributions: Raafat El-Namaky and Baboucarr Manneh conceived and designed the experiments; Raafat El-Namaky, Mamadou M. Bare Coulibaly, Maji Alhassan, Karim Traore, Francis Nwilene, Ibnou Dieng, and Baboucarr Manneh performed the experiments; Raafat El-Namaky, Karim Traore, Francis Nwilene, Ibnou Dieng, Baboucarr Manneh; and Rodomiro Ortiz analyzed the data and wrote the paper.

Conflicts of Interest: The authors declare no conflict of interest. The founding sponsors had no role in the design of the study; in the collection, analyses, or interpretation of data; in the writing of the manuscript, or in the decision to publish the results.

References

1. International Rice Research Institute (IRRI); Africa Rice; CIAT. *Global Rice Science Partnership (GRiSP)*; International Rice Research Institute: Los Baños, Philippines, 2010.

2. Seck, P.A.; Diagne, A.; Mohanty, S.; Wopereis, M.C.S. Crops that feed the world 7: Rice. *Food Sec.* **2012**, *4*, 7–24. [CrossRef]

3. Nguyen, N.V.; Ferrero, A. Meeting the challenges of global rice production. *Paddy Water Environ.* **2006**, *4*, 1–9. [CrossRef]

4. Balasubramanian, V.; Sie, M.; Hijmans, R.J.; Otsuka, K. Increasing rice production in sub-Saharan Africa: Challenges and opportunities. *Adv. Agron.* **2007**, *94*, 55–132. [CrossRef]

5. Bentolila, S.; Alfonso, A.A.; Hanson, M.R. A pentatricopeptide repeat-containing gene restores fertility to cytoplasmic male sterile plants. *Proc. Natl. Acad. Sci. USA* **2002**, *99*, 10887–10892. [CrossRef] [PubMed]

6. Bastawisi, A.O.; El-Mowafi, H.F.; Abo Yousef, M.I.; Draz, A.E.; Aidy, I.R.; Maximos, M.A.; Badawi, A.T. Hybrid Rice Research and Development in Egypt. In *Hybrid Rice for Food Security, Poverty Alleviation and Environmental Protection, Proceedings of the 4th International Symposium on Hybrid Rice, Hanoi, Vietnam, 14–17 May 2002*; Virmani, S.S., Mao, C.X., Hardy, B., Eds.; International Rice Research Institute: Los Baños, Philippines, 2003; pp. 257–263.

7. Yuan, L.P. Hybrid Rice Breeding in China. In *Advances in Hybrid Rice Technology, Proceedings of the 3rd International Symposium on Hybrid Rice, Hyderabad, India, 12–14 November 1996*; Virmani, S.S., Siddiq, E.A., Muralidharan, K., Eds.; International Rice Research Institute: Metro Manila, Philippines, 1988; pp. 592–607.

8. He, G.T.; Zhu, X.; Gu, H.Z.; Flinn, J.C. The use of hybrid rice technology: An economic evaluation. In *Hybrid Rice*; International Rice Research Institute: Los Baños, Philippines, 1988; pp. 229–241.

9. Nwilene, F.E.; Traore, A.K.; Asidi, A.N.; Sere, Y.; Onasany, A.; Abo, M.E. New records of insect vectors of *Rice Yellow Mottle Virus* (RYMV) in Côte d'Ivoire, West Africa. *J. Entomol.* **2009**, *6*, 198–206. [CrossRef]

10. Nacro, S.; Heinrichs, E.A.; Dakouo, D. Estimation of rice yield losses due to the African rice gall midge, *Orseolia oryzivora* Harris and Gagne. *Int. J. Pest Manag.* **1996**, *42*, 331–334. [CrossRef]

11. Yao, N.; Lee, C.R.; Semagn, K.; Sow, M.; Nwilene, F.; Kolade, O.; Bocco, R.; Oyetunji, O.; Mitchell-Olds, T.; Ndjiondjop, M.N. QTL mapping in three rice populations uncovers major genomic regions associated with African rice gall midge resistance. *PLoS ONE* **2016**, *11*, e0160749. [CrossRef] [PubMed]

12. Kouassi, N.K.; Guessan, P.N.; Albar, L.; Fauquet, C.M.; Brugidou, C. Distribution and characterization of *Rice Yellow Mottle Virus*: A threat to African farmers. *Plant Dis.* **2005**, *89*, 124–132. [CrossRef]

13. Joseph, A.; Olufolaji, D.B.; Nwilene, F.E.; Onasanya, A.; Omole, M.M. Effect of leaf age on *Rice Yellow Mottle Virus* severity and chlorophyll content with mechanical inoculation and vector transmission method. *Trends Appl. Sci. Res.* **2011**, *6*, 1345–1351. [CrossRef]

14. Onwughalu, J.T.; Abo, M.E.; Okoro, J.K.; Onasanya, A.; Sere, A. *Rice Yellow Mottle Virus* infection and reproductive losses on rice (*Oryza sativa* L.). *Trends Appl. Sci. Res.* **2011**, *6*, 182–189. [CrossRef]

15. Sere, Y.; Onasanya, A.; Nwilene, F.E.; Abo, M.E.; Akator, S.K. Potential of insect vector screening method for development of durable resistant cultivars to *Rice Yellow Mottle Virus* disease. *Int. J. Virol.* **2008**, *4*, 4–47. [CrossRef]

16. Couch, B.C.; Kohn, L.M. A multilocus gene genealogy concordant with host preference indicates segregation of a new species, *Magnaporthe oryzae*, from *M. grisea*. *Mycologia* **2002**, *94*, 683–693. [CrossRef] [PubMed]

17. Bonman, J.M.; Mackill, D.J. Durable resistance to rice blast disease. *Oryza* **1988**, *25*, 103–110.

18. Vera Cruz, C.M.; Kobayashi, N.; Fukuta, Y. Rice blast situation, research in progress, needs and priorities in 13 countries: Summary of results from a blast survey. *JIRCAS Work. Rep.* **2007**, *53*, 97–103.

19. Federer, W.T. Augmented (or hoonuiaku) designs. *Hawaii. Plant. Rec.* **1956**, *55*, 191–208.

20. IRRI. *Standard Evaluation System for Rice*, 4th ed.; International Rice Research Institute: Los Baños, Philippines, 1996. Available online: http://www.knowledgebank.irri.org/images/docs/rice-standard-evaluation-system.pdf (accessed on 9 July 2017).

21. Sere, Y.; Sy, A.A.; Sie, M.; Akator, S.K.; Onasanya, A.; Kabore, B.; Conde, C.K.; Traore, M.; Kiepe, P. *Importance of Varietal Improvement for Blast Disease Control in Africa*; JIRCAS Working Report 70; Japan International Research Centre for Agricultural Sciences: Tsukuba, Japan, 2011.

22. Finninsa, C. Relationship between common bacterial blight severity and bean yield loss in pure stand and bean-maize intercropping system. *Int. J. Pest Manag.* **2003**, *49*, 177–185. [CrossRef]

23. Littel, R.C.; George, A.M.; Walter, W.S.; Russell, D. *SAS System for Mixed Models*; SAS Institute Inc.: Cary, NC, USA, 1996.

24. Mekwatanakarn, P.; Kositratana, W.; Levy, M.; Zeigler, R.S. Pathotype and avirulence gene diversity of *Pyricularia grisea* in Thailand as determined by rice lines near-isogenic for major resistance genes. *Plant Dis.* **2000**, *84*, 60–70. [CrossRef]

25. Sere, Y.; Onasanya, A.; Afolabi, A.; Mignouna, H.D.; Akator, K. Genetic diversity of the blast fungus, *Magnaporthe grisea* (Hebert) Barr, in Burkina Faso. *Afr. J. Biotech.* **2007**, *6*, 2568–2577.

26. Cheng, S.H.; Zhuang, J.Y.; Fan, Y.Y.; Du, J.H.; Cao, L.Y. Progress in research and development on hybrid rice: A super-domesticate in China. *Ann. Bot.* **2007**, *100*, 959–966. [CrossRef] [PubMed]

27. El-Mowafi, H.F.; Bastawisi, A.O.; Abdekhalek, A.F.; Attia, K.A.; El-Namaky, R.A. Hybrid rice technology in Egypt. In *Accelerating Hybrid Rice Development, Proceedings of the 5th International Symposium on Hybrid Rice, Changsha, China, 12–14 September 2008*; Xie, F., Hardy, B., Eds.; International Rice Research Institute: Los Baños, Philippines, 2008; pp. 592–607.

28. Kanfany, G.; El-Namaky, R.; Ndiaye, K.; Traore, K.; Ortiz, R. Assessment of rice inbred lines and hybrids under low fertilizer levels in Senegal. *Sustainability* **2014**, *6*, 1153–1162. [CrossRef]

29. Abibou, N.; Becker, M.; Ewert, F.; Dieng, I.; Gaiser, T.; Tanaka, A.; Senthilkumar, K.; Rodenburg, J.; Johnson, J.M.; Akakpo, C.; et al. Variability and determinants of yields in rice production systems of West Africa. *Field Crops Res.* **2017**, *207*, 1–12. [CrossRef]

30. Dingkuhn, M.; Saw, A. Potential yield of irrigated rice in the Sahel. In *Irrigated Rice in the Sahel: Prospects for Sustainable Development*; Miezan, K.M., Wopereis, M.C.S., Dinkuhn, M., Deckers, J., Randolph, T.F., Eds.; West Africa Rice Development Association: Bouaké, Côte d'Ivoire, 1997; pp. 361–379.

31. Nwilene, F.E.; Williams, C.T.; Ukwungwu, M.N.; Dakouo, D.; Nacro, S.; Hamadoun, A.; Kamara, S.I.; Okhidievbie, O.; Abamu, F.J.; Adam, A. Reactions of differential rice genotypes to African rice gall midge in West Africa. *Int. J. Pest Manag.* **2002**, *48*, 195–201. [CrossRef]

32. Singh, B.N.; Williams, C.T.; Ukwungwu, M.N.; Maji, A.T. Breeding for resistance to African rice gall midge, *Orseolia oryzivora* Harris and Gagné. In *New Approaches to Gall Midge Resistance in Rice*; Bennett, J., Bentur, J.S., Pasalu, I.C., Krishnaiah., K., Eds.; International Rice Research Institute: Metro Manila, Philippines, 2004; pp. 121–130.

33. Himabindu, K.; Sundaram, R.M.; Neeraja, C.N.; Mishra, B.; Bentur, J.S. Flanking SSR markers for allelism test for the Asian rice gall midge (*Orseolia oryzae*) resistance genes. *Euphytica* **2007**, *157*, 267–279. [CrossRef]
34. El-Namaky, R.; Seedek, S.; Moukoumbi, Y.D.; Ortiz, R.; Manneh, B. Microsatellite-aided screening for fertility restorer genes (*Rf*) facilitates hybrid improvement. *Rice Sci.* **2016**, *23*, 160–164. [CrossRef]

diversity

MDPI

Article

Allelic Variants of Glutamine Synthetase and Glutamate Synthase Genes in a Collection of Durum Wheat and Association with Grain Protein Content

Domenica Nigro [1], Stefania Fortunato [1], Stefania Lucia Giove [2], Giacomo Mangini [1], Ines Yacoubi [3], Rosanna Simeone [1], Antonio Blanco [1] and Agata Gadaleta [2,*]

[1] Department of Soil, Plant & Food Sciences, Plant Breeding Section, University of Bari Aldo Moro, Via Amendola, 165/a, 70126 Bari BA, Italy; domenica.nigro@uniba.it (D.N.); stefania.fortunato26@gmail.com (S.F.); giacomo.mangini@uniba.it (G.M.); rosanna.simeone@uniba.it (R.S.); antonio.blanco@uniba.it (A.B.)

[2] Department of Agricultural & Environmental Science, University of Bari Aldo Moro, Via Amendola, 165/a, 70126 Bari BA, Italy; s.giove1@inwind.it

[3] Biotechnology and plant improvement laboratory, Biotechnology Centre of Sfax Tunisia, Kef 7119, Tunisia; ines.bouchrityaccoubi@cbs.rnrt.tn

[*] Correspondence: agata.gadaleta@uniba.it

Received: 6 October 2017; Accepted: 7 November 2017; Published: 16 November 2017

Abstract: Wheat is one of the most important crops grown worldwide. Despite the fact that it accounts for only 5% of the global wheat production, durum wheat (*Triticum turgidum* L. subsp. *durum*) is a commercially important tetraploid wheat species, which originated and diversified in the Mediterranean basin. In this work, the candidate gene approach has been applied in a collection of durum wheat genotypes; allelic variants of genes glutamine synthetase (*GS2*) and glutamate synthase (*GOGAT*) were screened and correlated with grain protein content (GPC). Natural populations and collections of germplasms are quite suitable for this approach, as molecular polymorphisms close to a locus with evident phenotypic effects may be closely associated with their character, providing a better physical resolution than genetic mapping using ad hoc constituted populations. A number of allelic variants were detected both for *GS2* and *GOGAT* genes, and regression analysis demonstrated that some variations are positively and significantly related to the GPC effect. Additionally, these genes map into homoeologous chromosome groups 2 and 3, where several authors have localized important quantitative trait loci (QTLs) for GPC. The information outlined in this work could be useful in breeding and marker-assisted selection programs.

Keywords: durum wheat; genetic diversity; grain protein content; glutamine synthetase; glutamate synthase

1. Introduction

Wheat, together with rice and maize, is one the most important cereal crops grown worldwide. Most of the cultivated cultivars and varieties belong to the hexaploid *Triticum aestivum* L. (genomes AABBDD, bread wheat) or to the tetraploid *T. turgidum* L. var. *durum* (genomes AABB, durum wheat), which are different in genome size (bread wheat also having a D genome), grain composition, and food end-use quality attributes. Several agronomic traits and composition aspects determine the final quality of grains.

Among them, grain protein content (GPC) contributes to the nutritional value and the baking properties of common wheat and to the pasta-making technology characteristics of durum wheat. GPC is a quantitative trait influenced by a complex genetic system and affected by environmental factors as well as by management practices. Indeed, one of the most pursued goals of breeders in the last decades has been the improving of grain protein concentration. However, because of the negative correlation between grain yield and GPC, simultaneous increases of both traits have been difficult to achieve [1,2]. Nitrogen fertilizers are today extensively used to increase both crop yield and protein content. The current agricultural system requires growers to optimize the use of nitrogen fertilizers to avoid pollution, while maintaining reasonable profit margins. Numerous studies allowed the identification of candidate genes that encode enzymes involved in nitrogen assimilation and recycling [3], and many of them co-localized with agronomic and physiological traits related to nitrogen metabolism [4,5]. Two genes resulted to be particularly important in the first step of ammonia assimilation: glutamine synthetase and glutamate synthase. These two enzymes work synergistically in a cycle known as GS-GOGAT shunt, involved in the first step of N metabolism and glutamate synthesis.

The glutamine synthetase (GS) enzyme has an essential role in the assimilation and re-assimilation of inorganic N. GS genes represent a gene family with three to five isoforms, depending on the species [6]. On the bases of phylogenetic studies and mapping data in wheat, 10 GS cDNA sequences were classified into four sub-families denominated *GS1* (a, b and c), *GSr* (1 and 2), *GSe* (1 and 2), active in cytosol, and *GS2* (a, b and c), localized in plastids [7,8]. Several studies have been carried out on various isoforms as candidates for improving nitrogen use efficiency (NUE) in wheat, and their relationships with GPC have been investigated [8–12]. Once glutamine is synthetized, another important enzyme is involved in the second step of the reaction to the synthesis of glutamate: the glutamate synthase (Glutamine-2-oxoglutarate amidotransferase). This enzyme is responsible for the transfer of the amide group of glutamine to 2-oxoglutarate, with the result of two yielded glutamate molecules, one of which is then available for aminoacid synthesis and the other of which returns to the GS-GOGAT cycle [13]., Gene regulation, as well asenzymes structure, have been reported in previous works [4,14]. Based on the electron donors, GOGAT exists in plants in two different isoforms: A ferredoxin (Fd)-dependent (EC 1.4.7.1) form and an NADH-dependent (EC 1.4.1.14) form, both of which are located in plastids but in two different types of tissues. The Fd-GOGAT enzyme is usually present in photosynthesizing tissues, while the NADH-GOGAT enzyme is the predominant form in non-photosynthesizing cells. These differences in tissue location and enzyme roles have been well studied in both rice and conifers [15,16]. Additionally, mutagenesis studies showed that mutated and/or silenced *GOGAT* not only reduced enzyme activity but also seemed to be involved in changes in aminoacid metabolism [17–20].

Only a few studies have reported gene isolation and sequencing of *GOGAT* genes in plants, probably due to their length and structural complexity. *GOGAT* genes genomic sequences have been reported for maize [21], tobacco [22], Arabidopsis [23] and barley [24] and partial sequences were also reported in bread wheat [25]. Recently, NADH and Fd-GOGAT gene structures, genomic sequences, gene localization, and involvement in GPC control have been reported in durum wheat [26,27].

Considering the importance and central role of these two genes in nitrogen metabolism, the main objectives of this study were to investigate the presence of allelic variants of both the plastidic *GS2* and the *GOGAT* genes in a collection of tetraploid wheat genotypes and to validate the relationships between variants and GPC.

2. Materials and Methods

2.1. Plant Material

A collection of 236 tetraploid wheat genotypes (*Triticum turgidum*), including wild and cultivated accessions of seven subspecies (*durum, turanicum, polonicum, turgidum, carthlicum, dicoccum* and *dicoccoides*), was screened for *GS2* and *GOGAT* allelic variants (List of accessions is reported in Table S3). The collection was bred in an experimental field of the University of Bari at Valenzano (Bari, Italy) in 2009 and 2010 and in Foggia (Italy) in 2009 in a randomized complete block design study with three replications. Each plot consisted of 1 m rows, 30 cm apart, and the seeding rate was 50 seeds per plot.

Genomic DNA was isolated from fresh leaves and subsequently purified using a method previously described [28] and subsequently purified via phenol-chloroform extraction.

2.2. PCR Condition and Sequencing

PCR reactions were performed in final volumes of 20 μL in BIORAD thermo cyclers. The reaction mixture contained each deoxynucleotide in 200 μM concentrations, each primer in 0.5 μM concentrations, 1× buffer, 0.02 U/μL Taq polymerase (Phusion High-Fidelity DNA Polymerase, Thermo Fisher Scientific, Waltham, MA, USA), and 50 ng of template DNA. Using Oligo Explorer software, a set of genome specific primer pairs were designed for *GS2* and *GOGAT* genes, as reported in Tables S1 and S2. PCR fragments were purified with a QIAquick PCR purification kit (Qiagen, Hilden, Germany) cloned into the pCR4-TOPO vector (Invitrogen, Cloning Kit, Thermo Fisher Scientific, Waltham, MA, USA) following the manufacturer's instructions, and subsequently sequenced (3500 Genetic Analyzer, Applied Biosystem, Foster City, CA, USA). Sequences alignments were carried out using ClustalOmega (http://www.ebi.ac.uk/Tools/msa/clustalo/) and CodonCode Aligner software (CodonCode Corporation, Centerville, MA, USA).

2.3. Digestion with CEL I and Revelation Fragments

In order to discover mutations within *GOGAT* gene sequences in the durum wheat collection previously described, single nucleotide polymorphisms (SNPs) were detected using the Surveyor nuclease kit (Transgenomic, Omaha, NE, USA). Heteroduplex formation, *Cel I* digestion and gel analysis were performed following a procedure previously reported [27].

2.4. Protein Content Quantification and Regression Analysis

Grain protein content (GPC) and yield components were evaluated in the durum wheat collection previously described grown in three different environments. GPC was assessed on 3 g of whole meal flour using a dual beam near infrared reflectance spectrophotometer (Zeutec Spectra Alyzer Premium, Zeutec Büchi, Rendsburg, Germany). Linear regression analysis between each allelic variant of both *GS* and *GOGAT* genes and GPC were carried out using MSTAT-C software developed by Freed et al. [29], Michigan State University.

3. Results and Discussion

3.1. Phenotypic Characterization for the Protein Content of a Collection of Tetraploid Wheat Genotypes

A collection of 236 tetraploid wheats genotypes, including durum cultivars, landraces, and wild accessions, has been characterized in terms of genetic diversity and population structure [30] and used for genome-wide association mapping of loci controlling some qualitative important traits, such as β-glucan content [31], carotenoid content [32] and phenolic acids [33]. GPC was determined in three replicated field experiments. Table 1 shows the summary data of protein content expressed as grams of protein on 100 g of dry weight of wholegrain. Variance analysis revealed significant differences between genotypes ($p < 0.001$).

Table 1. Mean, standard deviation (SD), ranges, and coefficient of variation (CV %) of grain protein content (GPC) (μg/g dry matter) in a tetraploid wheat collection evaluated in three different environments.

Environment	Foggia 2009	Valenzano 2009	Valenzano 2010
Mean	14.1	16.2	15.5
SD	1.55	2.01	2.21
Min	10.8	11.8	12.5
Max	20.4	22.9	23.5
CV (%)	10.9	12.4	14.3

The highest GPC mean value (16.2%) was found for the trial of Valenzano 2009; however, looking at the three trials, minimum and maximum values ranged from 10.8 and 23.5, respectively. Additionally, the coefficient of variation showed different values among the trials, ranging from 10.9 to 14.3. Variance analysis revealed significant differences in $p < 0.001$ between genotypes.

3.2. GS2 Allelic Variants and Relationship with GPC

Based on the genomic sequences of the two homoeologous genes *GS2-2A* and *GS2-2B* previous isolated [10], several pairs of specific genomic primers reported in Table S1 were drawn. These were amplified and screened for genomic variations in the entire collection of tetraploid wheat (see Materials and Methods).

Three major polymorphisms were found in the *GS2* gene. An insertion/deletion of a 239 bp MITE (Miniature Inverted-Repeat Transposable Element), located in the second intron of the 2A homoelogous chromosome, identifies the only allelic variant for the *GS2-2A* gene. Out of the analyzed 236 wheat genotypes, the intronic Miniature Inverted Transposable Element was found in 96 genotypes. On the other hand, two different polymorphisms were identified on the 2B homoelogous chromosome: a repeated 5 bp microsatellite (GATTA) and a 33 bp indel, respectively, located in the first and second introns of the *GS2-2B* gene. Figure 1 shows peaks corresponding to MITE deletion in the second intron of the *GS2-2A* gene (peak at 480 bp, and the 33 bp deletion/insertion of the *GS2-2B* gene (473 and 507 bp, respectively) [10].

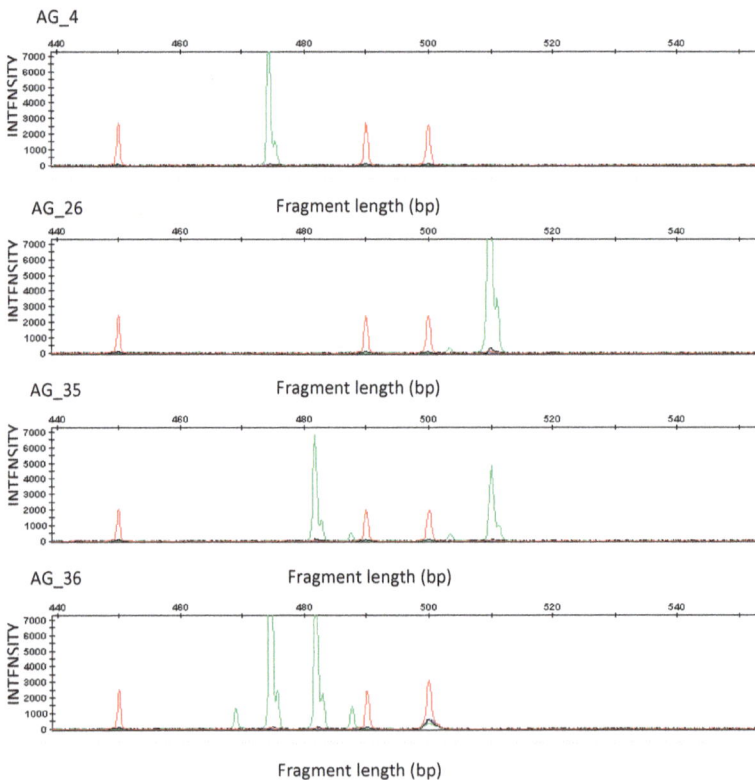

Figure 1. Polymorphisms detected in the *GS2* genes in genotypes of the durum collection. Peaks corresponding to MITE deletion in the second intron of the *GS2-2A* gene (peak at 480 bp) and the 33 bp deletion/insertion of the *GS2-2B* gene (473 and 507 bp, respectively) are shown.

We ran a regression analysis between the *GS2* allelic variants and the GPC in the described tetraploid wheat collection evaluated in three replicated field experiments. The regression analysis with the *GS2-A2* alleles showed that the amplicon of 480 bp absent of MITE is positively correlated with GPC in all three environments (Table 2). The stronger correlation was found in the Foggia 2009 environment ($p \geq 0.001$), while the lowest one, albeit still significant at $p \geq 0.05$, was found in Valenzano 2009. The phenotypic variation ranged from 4.1 to 9.2%. Although significant results were previously reported [10], no significant correlation was instead outlined by the regression analysis between GPC and the 33 bp indel of the *GS2-B2* gene, likely due to the fact that an unequal distribution of alleles occur in the collection of genotypes.

Table 2. Regression analysis between *GS2* allelic variants and GPC (% DW^{-1}) in a tetraploid wheat collection evaluated in three replicated field experiments.

Gene	Amplicons (bp)	Frequency	Environments								
			Foggia 2009			Valenzano 2009			Valenzano 2010		
			$\log_{10}(p)$	Effect	R^2	$\log_{10}(p)$	Effect	R^2	$-\log_{10}(p)$	Effect	R^2
GS2-A2	480/719	131–90	5.3 ***	0.92	9.2	2.6 *	0.83	4.1	3.7 **	1.11	6.1
GS2-B2	473/507	184–37	0.4	0.24	0.4	0.4	−0.33	0.4	0.1	0.13	0.0

*, **, and *** = significant at $p \geq 0.05$, $p \geq 0.01$, and $p \geq 0.001$, respectively, using the Bonferroni threshold ($p/12$) to control for multiple testing. R^2 = Phenotypic variation (%).

Glutamine synthetase plays a key role in the use of absorbed nitrogen, process of primary importance for plant growth. The absorption and utilization of nitrogen are closely related both to production and to the accumulation of grain proteins in various cereal species. Haplotype analysis of *GS2* gene and their association with NUE and yield-related traits was performed in bread wheat [34]. MITE was also found in Chinese bread wheat genotypes, and an interesting association with yield-related traits was also found.

Recently, a genome-wide association analysis with high-density SNP markers was conducted in order to identify genomic regions that may be associated with NUE traits in Great Plains hard winter wheat germplasm [35]. Interestingly, it was found that SNP markers on the long arm of a 2D chromosome were associated with NUE traits, in the same distal position where *GS2* genes map.

Other authors reported a colocalization between GS genes and GPC quantitative trait loci (QTLs) in a DHL population in three different environments [36]. A Meta-QTL for GY and GPC detection was carried out using three inter-connected doubled haploid populations grown in a large multi-environment trial network, identifying several genomic regions having GY and GPC [37]. Among them, they found 2A and 2D chromosomes carrying important QTLs, suggesting that genomic regions close to *GS2* genes are involved in GPC control.

3.3. GOGAT Allelic Variants and Relationship with GPC

Along with glutamine synthetase, glutamate synthase (GOGAT) forms an enzymatic complex considered to be one of the bottlenecks of the early stages of nitrogen metabolism, in particular as regards the absorption, assimilation, and amelioration of ammonia nitrogen in the plant. As previously reported [26,27], *GOGAT* genes have a complex intronic-exon structure: *Fd-GOGAT* is comprised of 33 exons and has a size of about 15 kb, while *NADH-GOGAT* is comprised of 22 exons separated by 21 introns and has a size of about 10 kb. In order to find polymorphisms more easily in such long sequences, *Fd-GOGAT* and *NADH-GOGAT* genes were screened via an EcoTILLING approach. The different combinations of specific genomic primers (A and B) reported in Table S2 were amplified in the genotypes of the collection and analyzed for the presence of polymorphisms.

In this work, each combination was amplified as reported in the Materials and Methods section, and SNPs and indels polymorphisms were determined by duplex ether formation and digestion with the *CelI* enzyme. The duplex was obtained from each single genotype in the collection and the cv Svevo was used as control.

Digestion probes of different primer combinations for *Fd-GOGAT* genes allowed the identification of six SNPs polymorphism. Out of the six SNPs found, three were located in intronic regions (Introns 5, 10 and 31), and three were located in the exon region (Exons 6, 31 and 32). None of them determined a change in aminoacidic predicted sequences.

With the aim of determining the association between SNP markers found in *Fd-GOGAT* genes and the protein content of the caryopsis, regression analysis was conducted between the SNPs markers and the percentages of grain protein of the trials previously described. As reported in Table 3, the six SNPs identified three allelic variants for both *Fd-GOGAT-A2* and *Fd-GOGAT-B2* genes, named as a, b, and c allelic variants, respectively. Out of the three identified variants in the *Fd-GOGAT-A2* gene, the *Fd-GOGAT-A2a* allelic variant showed a highly significant correlation with GPC. This polymorphism was a C/T transition, significant at $p \geq 0.001$ in all three environments. Phenotypic variation ranged from 6.9 to 14.6% in Valenzano 2010, suggesting its potential effect on GPC. On the other hand, none of the three allelic variants identified for *Fd-GOGAT-B2* showed any correlation with GPC.

Table 3. Regression analysis between Fd-GOGAT allelic variants and GPC (% DW^{-1}) in a tetraploid wheat collection evaluated in three different environments.

Gene	Allele	Frequency	Environments								
			Foggia 2009			Valenzano 2009			Valenzano 2010		
			$-\log_{10}(p)$	Effect	R^2	$-\log_{10}(p)$	Effect	R^2	$-\log_{10}(p)$	Effect	R^2
Fd-GOGAT-A2a	C/T	204–17	4.1 ***	−1.47	6.9	6.7 ***	−2.48	11.7	8.4 ***	−3.05	14.6
Fd-GOGAT-A2b	C/T	115–99	0.2	−0.09	0.1	0.0	−0.02	0.0	0.2	−0.14	0.1
Fd-GOGAT-A2c	A/G	100–117	0.3	0.12	0.2	0.0	0.02	0.0	0.2	0.13	0.1
Fd-GOGAT-B2a	A/G	181–38	0.9	−0.41	1.1	2.3	−0.98	3.5	2.7 *	−1.19	4.3
Fd-GOGAT-B2b	A/G	37–181	1.1	0.47	1.4	2.6 *	1.06	4.0	3.0 *	1.28	4.9
Fd-GOGAT-B2c	A/G	117–101	0.3	−0.15	0.3	0.0	0.00	0.0	0.2	−0.14	0.1

* and *** = significant at $p \geq 0.05$ and $p \geq 0.001$, respectively, using the Bonferroni threshold (P/18) to control for multiple testing. R^2 = Phenotypic variation (%).

The same approach was followed in order to find out polymorphisms in *NADH-GOGAT* genes. A very similar scenario was detected for this gene: six different SNPs were detected in the screened genotypes, and specifically three SNPs for each genome resulting in three allelic variants for each homoeologous gene, named as a, b and c allelic variants, respectively (Table 4). The results of regression analysis showed that, out of the three allelic variants for *NADH-GOGAT-A3*, only the T/G transversion, identified as the *NADH-GOGAT-A3b* allelic variant, showed a very high and significant association at $p \geq 0.001$ with GPC in the three considered environments. Phenotypic variation ranged from 7.0 to 11.1% in Valenzano 2010. *NADH-GOGAT-A3a*, a C/T transition, was found to be significantly associated in one environment only, Foggia 2009, at $p \geq 0.05$ with a phenotypic variation of 4.4%.

As found in the *NADH-GOGAT-A3* gene, the 3B homoeologous gene showed three allelic variants, identified as a, b, and c and corresponding to a G/T transversion, and C/T and A/G transitions, respectively. None of them showed a strong correlation nor a phenotypic effect as the *NADH-GOGAT-A3b* allelic variant, but two of them, *NADH-GOGAT-B3a* and *NADH-GOGAT-B3c*, had a significant association at $p \geq 0.05$ with GPC in the Foggia 2009 environment only, with an R^2 of 4.6% and 4.2%, respectively.

Gene expression and post-transcriptional modification have a great effect on enzyme activity. However, mutations located both in exonic and intronic regions can affect gene expression levels. Insertion, deletions, and point mutations (single nucleotide polymorphisms) in introns can, for instance, introduce novel splice sites, activate novel promoters, or introduce/eliminate enhancer activity.

A recent work of Zeng et al. [38] showed how a single nucleotide polymorphism, leading to an aminoacid substitution in rice *Fd-GOGAT* genes, resulted in an increased GPC, confirming its important role as a potential candidate in NUE improvement.

Table 4. Regression analysis between NADH-GOGAT allelic variants and GPC (% DW^{-1}) in a tetraploid wheat collection evaluated in three different environments.

Gene	Allele	Frequency	Environments								
			Foggia 2009			Valenzano 2009			Valenzano 2010		
			$-\log_{10}(p)$	Effect	R^2	$-\log_{10}(p)$	Effect	R^2	$-\log_{10}(p)$	Effect	R^2
NADH-GOGAT-A3a	C/T	145–70	2.7 *	0.68	4.4	1.3	0.58	1.9	1.6	0.73	2.4
NADH-GOGAT-A3b	G/T	42–176	6.1 ***	−1.24	10.6	4.1 ***	−1.36	7.0	6.4 ***	−1.87	11.1
NADH-GOGAT-A3c	A/G	111–103	0.1	−0.04	0.0	0.7	−0.36	0.8	1.3	−0.60	1.9
NADH-GOGAT-B3a	G/T	71–147	2.8 *	−0.68	4.6	1.5	−0.61	2.1	1.8	−0.76	2.6
NADH-GOGAT-B3b	C/T	115–103	0.1	−0.06	0.0	0.9	−0.43	1.1	1.6	−0.68	2.3
NADH-GOGAT-B3c	A/G	146–72	2.6 *	0.65	4.2	1.2	0.54	1.6	1.5	0.68	2.1

* and *** = significant at $p \geq 0.05$ and $p \geq 0.001$, respectively, using the Bonferroni threshold (P/18) to control for multiple testing. R^2 = Phenotypic variation (%).

The *NADH-GOGAT* gene was also identified as a major candidate gene for cereal NUE by a cross-genome ortho-meta QTL study of NUE [39]. QTL for GPC has also been reported on chromosome 3AL in in the homoeologous position of the *NADH-GOGAT-3B* gene [40] A proof of the central role of *GOGAT* genes was obtained in rice; the suppression of both GOGAT genes reduced yield per plant and thousand kernel weight, phenotypic indications of nitrogen starvation [41].

4. Conclusions

GPC is one of the most important wheat agronomic trait, in relation to both nutritional and technological properties. GPC is a typical quantitative trait, controlled by several genes and influenced by environmental factors. Different physiological processes also influence GPC, such as nitrogen uptake, assimilation, and remobilization to the grain. Several studies, carried out on different genetic materials, have reported the influence of the homoeologous chromosome groups 2 and 3 on GPC control. In 1990, important QTLs for GPC on group 2 chromosomes were firstly reported on durum wheat [42]. QTLs for GPC were also found on the short arms of homoeologous group 2 chromosomes in both bread and durum wheat [40,43]. Stable QTLs for GPC were also identified on 2A and 2B chromosomes in Canadian durum wheat populations [44]. Several authors have focused on deciphering GPC and NUE quantitative traits, and genetic diversity at candidate genes have lately been considered for this purpose. Both *GS2* and *Fd-GOGAT* genes have been mapped in homoeologous group 2 chromosomes and have been found to co-localize with important QTLs for GPC in durum wheat [10,27]. Homoeologous group 3 was also found to be important for GPC control. An otho-Meta QTLs analysis for NUE identified *NADH-GOGAT* as one of the major effectors of NUE in wheat, rice, sorghum, and maize [39] A QTL for GPC was also reported in the homoeologous position of chromosome 3AL [40]. Another important aspect to be considered in exploiting GPC and NUE is the importance of genotypic variation in gene and QTL expressions. Nigro et al. [12] showed how the genotype plays an important role in *GS* expression, especially at different N regimes. Previously, other authors have reported the importance of genotypic variation in NUE and final grain nitrogen content, determining that the genotype was one of the causes of variation in analyzed traits, after the N-rate and the growth stage [15].

The above data suggest that the genomic region surrounding *GS2* and *GOGAT* genes are involved in grain protein accumulation, and the identification of new useful alleles for marker-assisted selection is valuable for breeding wheat varieties with improved agronomic performance and N-use efficiency.

Supplementary Materials: The following are available online at www.mdpi.com/1424-2818/9/4/52/s1, Table S1: Genome specific primer combination for GS2 genes, Table S2: Genome specific primer combination for GOGAT genes, Table S3: List of tetraploid accession used in the study.

Acknowledgments: This work was supported by Puglia Region, Italy, projects PSR "SAVEGRAIN" and "Intervento cofinanziato dal Fondo di Sviluppo e Coesione 2007–2013"—APQ Ricerca Regione Puglia "Programma regionale a sostegno della specializzazione intelligente e della sostenibilità sociale ed ambientale—FutureInResearch". The authors declare no conflicts of interest.

Author Contributions: Domenica Nigro, Agata Gadaleta and Antonio Blanco conceived and designed the experiments; Domenica Nigro, Stefania Fortunato and Stefania Lucia Giove performed the experiments; Giacomo Mangini, Ines Yacoubi and Antonio Blanco analyzed the data; Domenica Nigro, Rosanna Simeone and Agata Gadaleta wrote the paper.

Conflicts of Interest: The authors declare no conflict of interest.

References

1. Lawlor, D.W. Carbon and nitrogen assimilation in relation to yield: Mechanisms are the key to understanding production systems. *J. Exp. Bot.* **2002**, *53*, 789–799. [CrossRef]
2. Triboi, E.; Triboi-Blondel, A.M. Productivity and grain or seed composition: A new approach to an old problem. *Eur. J. Agron.* **2002**, *16*, 163–186. [CrossRef]
3. Lea, P.J.; Azevedo, R.A. Nitrogen use efficiency. 2. Amino acid metabolism. *Ann. Appl. Biol.* **2007**, *151*, 269–275. [CrossRef]
4. Hirel, B.; le Gouis, J.; Ney, B.; Gallais, A. The challenge of improving nitrogen use efficiency in crop plants: Towards a more central role for genetic variability and quantitative genetics within integrated approaches. *J. Exp. Bot.* **2007**, *58*, 2369–2387. [CrossRef] [PubMed]
5. Obara, M.; Sato, T.; Sasaki, S.; Kashiba, K.; Nagano, A.; Nakamura, I.; Ebitani, T.; Yano, M.; Yamaya, T. Identification and characterization of a QTL on chromosome 2 for cytosolic glutamine ynthetase content and panicle number in rice. *Theor. Appl. Genet.* **2004**, *110*, 1–11. [CrossRef] [PubMed]
6. Swarbreck, S.M.; Defoin-Platel, M.; Hindle, M.; Saqi, M.; Habash, D.Z. New perspectives on glutamine synthetase in grasses. *J. Exp. Bot.* **2011**, *62*, 1511–1522. [CrossRef] [PubMed]
7. Bernard, S.M.; Møller, A.L.B.; Dionisio, G.; Kichey, T.; Jahn, T.P.; Dubois, F.; Baudo, M.; Lopes, M.S.; Tercé-Laforgue, T.; Foyer, C.H.; et al. Gene expression, cellular localization and function of glutamine synthetase isozymes in wheat (*Triticum aestivum* L.). *Plant. Mol. Biol.* **2008**, *67*, 89–105. [CrossRef] [PubMed]
8. Thomsen, H.C.; Eriksson, D.; Moller, I.S.; Schjoerring, J.K. Cytosolic glutamine synthetase: A target for improvement of crop nitrogen use efficiency? *Trends Plant. Sci.* **2014**, *19*, 656–663. [CrossRef] [PubMed]
9. Habash, D.Z.; Bernard, S.; Schondelmaier, J.; Weyen, J.; Quarrie, S.A. The genetics of nitrogenuse in hexaploid wheat:N utilisation, development and yield. *Theor. Appl. Genet.* **2007**, *114*, 403–419. [CrossRef] [PubMed]
10. Gadaleta, A.; Nigro, D.; Giancaspro, A.; Blanco, A. The glutamine synthetase (GS2) genes in relation to grain protein content of durum wheat. *Funct. Integr. Genom.* **2011**, *11*, 665–670. [CrossRef] [PubMed]
11. Gadaleta, A.; Nigro, D.; Marcotuli, I.; Giancaspro, A.; Giove, S.L.; Blanco, A. Isolation and characterization of cytosolic glutamine synthetase (GSe) genes and association with grain protein content in durum wheat. *Crop Pasture Sci.* **2014**, *65*, 38–45. [CrossRef]
12. Nigro, D.; Fortunato, S.; Giove, S.L.; Paradiso, A.; Gu, Y.Q.; Blanco, A.; de Pinto, M.C.; Gadaleta, A. Glutamine synthetase in Durum Wheat: Genotypic Variation and Relationship with Grain Protein Content. *Front. Plant Sci.* **2016**, *7*, 971. [CrossRef] [PubMed]
13. Forde, B.G.; Lea, P.J. Glutamate in plants: Metabolism, regulation, and signalling. *J. Exp. Bot.* **2007**, *58*, 2339–2358. [CrossRef] [PubMed]
14. Cren, M.; Hirel, B. Glutamine Synthetase in Higher Plants Regulation of Gene and Protein Expression from the Organ to the Cell. *Plant Cell Physiol.* **1999**, *40*, 1187–1193. [CrossRef]
15. Tabuchi, M.; Abiko, T.; Yamaya, T. Assimilation of ammonium ions and reutilization of nitrogen in rice. *J. Exp. Bot.* **2007**, *58*, 2319–2327. [CrossRef] [PubMed]
16. Cánovas, F.M.; Avila, C.; Cantón, F.R.; Cañas, R.A.; de la Torre, F. Ammonium assimilation and amino acid metabolism in conifers. *J. Exp. Bot.* **2007**, *58*, 2307–2318. [CrossRef]
17. Leegood, R.C.; Lea, P.J.; Adcock, M.D.; Hausler, R.E. The regulation and control of photorespiration. *J. Exp. Bot.* **1995**, *46*, 1397–1414. [CrossRef]

18. Ferrario-Méry, S.; Hodges, M.; Hirel, B.; Foyer, C.H. Photorespiration dependent increases in phosphoenolpyruvate carboxylase, isocitrate dehydrogenase and glutamate dehydrogenase in transformed tobacco plants deficient in ferredoxin-dependent glutamine-alpha-ketoglutarate aminotransferase. *Planta* **2002**, *214*, 877–888. [CrossRef] [PubMed]

19. Ferrario-Méry, S.; Valadier, M.H.; Godefroy, N.; Miallier, D.; Hirel, B.; Foyer, C.H.; Suzuki, A. Diurnal changes in ammonia assimilation in transformed tobacco plants expressing ferredoxin-dependent glutamate synthase mRNA in the antisense orientation. *Plant Sci.* **2002**, *163*, 59–67. [CrossRef]

20. Lancien, M.; Martin, M.; Hsieh, M.H.; Leustek, T.; Goodman, H.; Coruzzi, G.M. Arabidopsis *glt1*-T mutant defines a role for NADH-GOGAT in the non-photorespiratory ammonium assimilatory pathway. *Plant J.* **2002**, *29*, 347–358. [CrossRef] [PubMed]

21. Sakakibara, H.; Kawabata, S.; Takahashi, H.; Hase, T.; Sugiyama, T. Molecular cloning of the family of glutamine synthetase genes from maize: Expression of genes for glutamine synthetase and ferredoxin-dependent glutamate synthase in photosynthetic and non-photosynthetic tissues. *Plant Cell Physiol.* **1992**, *33*, 49–58.

22. Zehnacker, C.; Becker, T.W.; Suzuki, A.; Carrayol, E.; Caboche, M.; Hirel, B. Purification and properties of tobacco ferredoxin-dependent glutamate synthase, and isolation of corresponding cDNA clones. Light-inducibility and organspecificity of gene transcription and protein expression. *Planta* **1992**, *187*, 266–274. [CrossRef] [PubMed]

23. Coschigano, K.T.; Melo-Oliveira, R.; Lim, J.; Coruzzi, G.M. Arabidopsis *gls* mutants and distinct *Fd.-GOGAT* genes: Implications for photorespiration and primary nitrogen assimilation. *Plant Cell* **1998**, *10*, 741–752. [CrossRef] [PubMed]

24. Avila, C.; Márquez, A.J.; Pajuelo, P.; Cannell, M.E.; Wallsgrove, R.M.; Forde, B.G. Cloning and sequence analysis of a cDNA for barley ferredoxin-dependent glutamate synthase and molecular analysis of photorespiratory mutants deficient in the enzyme. *Planta* **1993**, *189*, 475–483. [CrossRef] [PubMed]

25. Boisson, M.; Mondon, K.; Torney, V.; Nicot, N.; Laine, A.L.; Bahrman, N.; Gouy, A.; Daniel-Vedele, F.; Hirel, B.; Sourdille, P.; et al. Partial sequences of nitrogen metabolism genes in hexaploid wheat. *Theor. Appl. Genet.* **2005**, *110*, 932–940. [CrossRef] [PubMed]

26. Nigro, D.; Gu, Y.Q.; Huo, N.; Marcotuli, I.; Blanco, A.; Gadaleta, A.; Anderson, O.D. Structural analysis of the wheat genes encoding NADH-dependent glutamine-2-oxoglutarate amidotransferases genes and correlation with grain protein content. *PLoS ONE* **2013**, *8*, e73751. [CrossRef] [PubMed]

27. Nigro, D.; Blanco, A.; Anderson, O.D.; Gadaleta, A. Characterization of Ferredoxin-Dependent Glutamine-Oxoglutarate Amidotransferase (Fd-GOGAT) Genes and Their Relationship with Grain Protein Content QTL in Wheat. *PLoS ONE* **2014**, *9*, e103869. [CrossRef] [PubMed]

28. Sharp, P.J.; Kreis, M.; Shewry, P.R.; Gale, M.D. Location of b-amylase sequences in wheat and its relatives. *Theoret. Appl. Genet.* **1988**, *75*, 286–290. [CrossRef]

29. Freed, R.; Eisensmith, S.P.; Goetz, D.; Reicosky, D.; Smail, V.W.; Wolberg, P. *MSTAT-C: A Microcomputer Program for the Design, Management, and Analysis of Agronomic Research Experiments*; Michigan State University: East Lansing, MI, USA, 1991.

30. Laido, G.; Mangini, G.; Taranto, F.; Gadaleta, A.; Blanco, A.; Cattivelli, L.; de Vita, P. Genetic diversity and population structure of tetraploid wheats (*Triticum turgidum* L.) estimated by SSR, DArT and pedigree data. *PLoS ONE* **2013**, *8*, e67280. [CrossRef] [PubMed]

31. Marcotuli, I.; Houston, K.; Schwerdt, J.G.; Waugh, R.; Fincher, G.B.; Burton, R.A.; Gadaleta, A. Genetic diversity and genome wide association study of β-glucan content in tetraploid wheat grains. *PLoS ONE* **2016**, *11*, e0152590. [CrossRef] [PubMed]

32. Colasuonno, P.; Lozito, M.L.; Marcotuli, I.; Nigro, D.; Giancaspro, A.; Mangini, G.; Simeone, R. The carotenoid biosynthetic and catabolic genes in wheat and their association with yellow pigments. *BMC Genom.* **2017**, *18*, 122. [CrossRef] [PubMed]

33. Nigro, D.; Laddomada, B.; Mita, G.; Blanco, E.; Colasuonno, P.; Simeone, R.; Blanco, A. Genome-wide association mapping of phenolic acids in tetraploid wheats. *J. Cereal Sci.* **2017**, *75*, 25–34. [CrossRef]

34. Li, X.P.; Zhao, X.Q.; He, X.; Zhao, G.Y.; Li, B.; Liu, D.C.; Li, Z.S. Haplotype analysis of the genes encoding glutamine synthetase plastic isoforms and their association with nitrogen-use-and yield-related traits in bread wheat. *New Phytol.* **2011**, *189*, 449–458. [CrossRef] [PubMed]

35. Guttieri, M.J.; Frels, K.; Regassa, T.; Waters, B.M.; Baenziger, P.S. Variation for nitrogen use efficiency traits in current and historical great plains hard winter wheat. *Euphytica* **2017**, *213*, 87. [CrossRef]

36. Fontaine, J.X.; Ravel, C.; Pageau, K.; Heumez, E.; Dubois, F.; Hirel, B.; Le Gouis, J. A quantitative genetic study for elucidating the contribution of glutamine synthetase, glutamate dehydrogenase and other nitrogen-related physiological traits to the agronomic performance of common wheat. *Theor. Appl. Genet.* **2009**, *119*, 645–662. [CrossRef] [PubMed]

37. Bogard, M.; Allard, V.; Martre, P.; Heumez, E.; Snape, J.W.; Orford, S.; Griffiths, S.; Gaju, O.; Foulkes, J.; Le Gouis, J. Identifying wheat genomic regions for improving grain protein concentration independently of grain yield using multiple inter-related populations. *Mol. Breed.* **2013**, *31*, 587. [CrossRef]

38. Zeng, D.D.; Qin, R.; Li, M.; Alamin, M.; Jin, X.L.; Liu, Y.; Shi, C.H. The ferredoxin-dependent glutamate synthase (*OsFd-GOGAT*) participates in leaf senescence and the nitrogen remobilization in rice. *Mol. Genet. Genom.* **2017**, *292*, 385. [CrossRef] [PubMed]

39. Quraishi, U.M.; Abrouk, M.; Murat, F.; Pont, C.; Foucrier, S.; Desmaizieres, G.; Confolent, C.; Riviere, N.; Charmet, G.; Paux, E.; et al. Cross-genome map based dissection of a nitrogen use efficiency ortho-meta QTL in bread wheat unravels concerted cereal genome evolution. *Plant J.* **2011**, *65*, 745–756. [CrossRef] [PubMed]

40. Blanco, A.; Mangini, G.; Giancaspro, A.; Giove, S.; Colasuonno, P.; Simeone, R.; Signorile, A.; de Vita, P.; Mastrangelo, A.M.; Cattivelli, L.; et al. Relationships between grain protein content and grain yield components through QTL analyses in a RIL population derived from two elite durum wheat cultivars. *Mol. Breed.* **2012**, *30*, 79–92. [CrossRef]

41. Lu, Y.; Luo, F.; Yang, M.; Li, X.; Lian, X. Suppression of glutamate synthase genes significantly affects carbon and nitrogen metabolism in rice (*Oryza sativa* L.). *China Life Sci.* **2011**, *54*, 651–663. [CrossRef] [PubMed]

42. Joppa, L.R.; Cantrell, R.G. Chromosomal location of genes for grain protein content of wild tetraploid wheat. *Crop Sci.* **1990**, *30*, 1059–1064. [CrossRef]

43. Prasad, M.; Varshney, R.K.; Kumar, A.; Balyan, H.S.; Sharma, P.C.; Edwards, K.J.; Gupta, P.K. A microsatellite marker associated with a QTL for grain protein content on chromosome arm 2DL of bread wheat. *Theor. Appl. Genet.* **1999**, *99*, 341–345. [CrossRef]

44. Suprayogi, Y.; Pozniak, C.J.; Clarke, F.R.; Clarke, J.M.; Knox, R.E.; Singh, A.K. Identification and validation of quantitative trait loci for grain protein concentration in adapted Canadian durum wheat populations. *Theor. Appl. Genet.* **2009**, *119*, 437–448. [CrossRef] [PubMed]

45. Barraclough, P.B.; Lopez-Bellido, R.; Hawkesford, M.J. Genotypic variation in the uptake, partitioning and remobilization of nitrogen during grain-filling in wheat. *Field Crops Res.* **2014**, *156*, 242–248. [CrossRef] [PubMed]

diversity

MDPI

Article

The Phylogeny and Biogeography of *Phyla nodiflora* (Verbenaceae) Reveals Native and Invasive Lineages throughout the World

Caroline L. Gross [1,*], Mohammad Fatemi [1,2], Mic Julien [3,4], Hannah McPherson [5] and Rieks Van Klinken [3]

[1] Ecosystem Management, University of New England, Armidale 2351, NSW, Australia
[2] Current Address: Biology, Golestan University, Beheshti St., Gorgan, Iran; fatemi980@yahoo.com
[3] Ecosystem Sciences, CSIRO, EcoSciences Precinct, Dutton Park 4102, QLD, Australia; rieks.vanklinken@csiro.au
[4] Current Address; 64 Tyalgum Creek Rd, Tyalgum 2484, NSW, Australia; mic.julien@hotmail.com
[5] National Herbarium of New South Wales, Royal Botanic Gardens and Domain Trust, Mrs Macquaries Road, Sydney 2000, NSW, Australia; hannah.mcpherson@rbgsyd.nsw.gov.au
* Correspondence: cgross@une.edu.au; Tel.: +61-2-6773-3759

Academic Editors: Rosa Rao and Giandomenico Corrado
Received: 3 March 2017; Accepted: 4 May 2017; Published: 10 May 2017

Abstract: *Phyla nodiflora* is an herbaceous perennial and an enigmatic species. It is indigenous to the Americas but is considered a natural component of the flora in many areas and a weed in others. Our aim was to circumscribe the native range of *P. nodiflora*, to explore dispersal mechanisms and routes and to test the hypothesis that *P. nodiflora* is native outside of the Americas. Determining whether distributions are natural or human-induced has implications for decisions regarding weed control or conservation. We undertook phylogenetic analyses using sequence variation in nuclear DNA marker ITS (Internal Transcribed Spacer) for a global sample of 160 populations of *P. nodiflora* sourced from Asia, Australia, central America, the Mediterranean, southern North America, South America and Africa. Analyses included maximum likelihood, maximum parsimony, a Bayesian estimation of phylogeny and a parsimony network analysis which provided a genealogical reconstruction of ribotypes. We evaluated phylogenies against extensive historical and biogeographical data. Based on the sequences, 64 ribotypes were identified worldwide within *P. nodiflora* and considerable geographic structure was evident with five clades: one unsupported and the remaining weakly supported (bootstrap support ranging from 52% to 71%). Populations from central and southern North America formed the core area in the indigenous range and we have detected at least three native lineages outside of this range. Within Australia *P. nodiflora* is represented by at least one native lineage and several post-European introductions. *Phyla nodiflora* is one of the few species in the family Verbenaceae to have a pan-tropical native distribution, probably resulting from natural dispersal from America to Africa then to Australasia. However, it has also undergone human-mediated dispersal, which has obscured the native-origin of some ribotypes. These introductions present a risk of diluting the pan-tropical structure evident in this species and therefore they have important conservation implications.

Keywords: biogeography; nuclear DNA ITS variation; native versus alien status; *Phyla nodiflora*; *Phyla canescens*; phylogeny; ribotype analyses

1. Introduction

Cosmopolitan species provide opportunities to test hypotheses on dispersal processes (vicariance versus long distance dispersal) and with modern phylogenetic tools, to clarify the status of species as native or alien to an area. Native (indigenous) plants have been defined as "Taxa that have originated in a given area without human involvement or that have arrived there without intentional or unintentional intervention of humans from an area in which they are native" [1]. For some species, strong evidence suggests that their wide distribution is due to non-human dispersal mechanisms (e.g., by ocean currents *Cakile edentula*, [2]; ancient tectonic plate movements *Pteridium*, [3]. In other species it is clear that humans have aided the movement of plants (e.g., sweet potato, [4]) and there is a third class in which the distribution of the species is cryptogenic [5]—the basis to the biogeography is puzzling and yet to be resolved.

Phyla nodiflora (L.) Greene (Verbenaceae), a perennial mat-forming herb in neotropical *Phyla* (Verbenaceae), is a cryptogenic species. In the Americas *P. nodiflora* occurs from lower North America (including Florida) to northern South America [6] with a disjunct occurrence in Brazil [6,7]. In Australia, *P. nodiflora* occurs in tropical and sub-tropical regions with several disjunct temperate occurrences [8]. With its distribution on all vegetated continents and many archipelagos [9] occurring over a 9000 km latitudinal range (39.5° N to 44.8° S), it has long intrigued biologists. Depending on their definitions of native status, different authors have considered *P. nodiflora* to be either native to the Americas and dispersed by humans elsewhere [6,8] or with a natural distribution that extends to the old world tropics [10]. The hypothesis that *P. nodiflora* is native outside of the Americas therefore requires testing.

Phyla species radiated within the Americas [11] and *P. nodiflora* is assumed to have originated from there [6]. Recent phylogenetic analyses [11] suggest that the family Verbenaceae has had up to six migration events from the Americas leading to subsequent radiations in Africa (*Lantana*, *Lippia*, *Verbena*, *Chascanum*, *Coelocarpum* and possibly *Stachytarpheta*). A similar natural migration by *Phyla nodiflora* from the Americas to Africa may therefore not be unexpected. Natural migration of species further east to Australia may, however, be exceptional. In fact, Munir [8]) considers that the species was "probably introduced into Australia during the last century and has now become naturalised in most mainland states". However, there is some evidence that it could be native in eastern Australia as the earliest collection was made in 1802 by Robert Brown at Shoalwater Bay (Qld) while on expedition with Captain Matthew Flinders. The Shoalwater Bay area was not occupied by Europeans until 1853 [12] and the only previous European contact in the area was a brief visit by Captain Cook in May 1770.

Our objective was to elucidate the global biogeography of *Phyla nodiflora* and to determine its native/alien status outside of the Americas and particularly for Australasia. This is important because determining whether distributions are natural or human-induced has implications for decisions regarding weed control e.g., providing insights for management [13] and conservation. We constructed a molecular phylogeny of *P. nodiflora* and combined this with phytogeographical and ecological evaluations of the species. We applied these data to a set of criteria we collated from Webb [14] and Bean [15] (Table 1) to determine invasive or native status. Our aim was to circumscribe the native range of *P. nodiflora*, to explore dispersal mechanisms and routes and to test the hypothesis that *P. nodiflora* is native outside of the Americas.

2. Materials and Methods

2.1. Study Species

Phyla nodiflora is an herbaceous perennial from the Americas [16]. It was considered to be a separate species from *P. canescens* (Kunthe) Greene in taxonomic revisions by Kennedy [6] and Munir [8], but most recently O'Leary and Múlgura [7] made a new combination and listed these taxa as varieties of *P. nodiflora* (*P. nodiflora* var. *nodiflora* and *P. nodiflora* var. *minor* (Hook) N. O'Leary & Múlgura). Their distinction was based on leaf shape, leaf size, leaf apex characters and leaf indumentum. In this paper we follow the taxonomy of Kennedy [6] and Munir [8] because (i) the species are morphologically distinctive [6,8,17] (Figure S1a,b); (ii) they appear to have a mostly non-overlapping distribution in their native range [6] and Australia [6,17]; (iii) have distinctive climatic preferences [8,17] and (iv) the species vary in their breeding biology (Fatemi, unpub data, see below; [18]).

Table 1. Criteria and characteristics of alien and native (indigenous) species, particularly for Australia, adapted and modified from Bean [15].

Criterion	Characteristics Likely to be Associated with a Native Species	Characteristics Likely to be Associated with an Invasive Species	Equivocal Information
1	(a) consistently occurs in intact unmodified habitat	(b) species known only from croplands, roadsides and other frequently disturbed sites	
2	(a) is not persistently invasive in its area of occurrence	(b) persistently invades or encroaches upon natural communities	
3	(a) is attended by a range of pests or diseases	(b) species that is pest- and disease-free	(c) Damage to herbarium material may have occurred post collecting
4	(a) displays a range of phenotypic or genetic diversities	(b) phenotypically or genetically uniform populations, probably derived from a single introduction	
5	(a) does not display any post-settlement expansion of geographical range within the region	(b) species that has a known or inferred expansion in its range over the past 100–150 years	
6	(a) any discontinuities of distribution of the species within the region are related to climatic and edaphic factors	(b) a species with a patchy distribution correlated with human settlement patterns is probably alien	
7	(a) a species is probably indigenous if closely related species occur as natives in Australia or nearby (e.g., New Guinea, Timor, Java, New Zealand)	(b) where the closest relatives occur on another continent, the species is likely to be an alien	
8		(b) a species known to be alien in areas outside the region must be under suspicion of being alien within the region	
9		(b) a species is probably not indigenous to the region if the nearest occurrence outside the region represents a major disjunction	
10		(b) the initial herbarium record dates well after the first European settlement of the region	(c) alternatively, if the initial herbarium record precedes or is soon after a European settlement, no useful conclusions can be drawn
11	(a) plant has an established ethnobotanical use by indigenous peoples	(b) a written record exists of the introduction or importation of a species in a journal, nursery catalogue, or botanic gardens listing	(c) alternatively, if no written record exists, no useful conclusions can be drawn

This is further supported by results from this study (see results section). O'Leary and Múlgura [7] do not cite the monograph by Kennedy [6]. However, it would appear that Kennedy [6] and O'Leary and Múlgura [7] also differ in the circumscription of species limits as the same specimens from southern South America are allocated to different species by them (e.g., *MacBride 5880*, US, is recognised as *P. canescens* [6] or *P. nodiflora* var *nodiflora* [7]).

2.2. Phylogenetics

2.2.1. Plant Material Collection

Leaf material of *P. nodiflora* was sourced as comprehensively as possible from across its global distribution using herbarium samples, supplemented by additional field surveys conducted in Australia, Iran, eastern Mexico and Florida, USA. In addition, representative *P. canescens* and *Phyla* spp. (specimens that were morphologically similar to either *P. canescens* or *P. nodiflora* but could not be definitively identified) populations were included from a parallel global survey (Fatemi, unpublished data).

For the nuclear DNA assay, we gathered leaf material for *P. nodiflora*, *P. canescens* and *Phyla* spp. from herbarium specimens or living plants from 445 samples (living accessions or herbarium specimens from 20 herbaria, see Acknowledgements) worldwide and from these we successfully extracted DNA from 181 samples, representing 179 populations. These samples were distributed as follows; *P. nodiflora* (160 populations, 161 samples) from the Americas (1 sample from each of $n = 35$ populations), Australasia ($n = 65$), the Mediterranean ($n = 3$), Africa (including the Middle East and Indian oceanic islands, $n = 31$, with two samples from one population in Iran), Asia ($n = 24$) and Hawaii ($n = 2$); *P. canescens* (19 populations, 20 samples) from South America ($n = 6$ populations, with two samples from one population in Bolivia), Hawaii ($n = 1$) and Australia ($n = 6$) and *Phyla* sp. from Australia and South America ($n = 6$). Most of the 181 samples were unequivocally from single plants in discrete populations, i.e., we did not repeat sample at that location, or if we did (Iran and Bolivia) we spaced 20 m between collections. We consider it unlikely that these samples were from the same genet. Effort was directed at sampling areas of taxonomic uncertainty including populations from the Balranald-Mildura region of southern NSW, northern Victoria and south-eastern South Australia ($n = 4$) and northern Argentina ($n = 5$).

2.2.2. DNA Extraction, Amplification and Sequencing

DNA was obtained from small amounts of leaf material (c. 20 mg) taken from herbarium sheets (see above) or from fresh leaves collected during field studies ($n = 22$ populations, Table S1). Our sampling from herbarium sheets included contemporary and very old specimens (*Wilkes s.n.*, 1838, Philippines, NY1239756) and overall we had mixed success with extracting DNA from them. The oldest specimen from which we successfully extracted DNA was from 1899 (*Tracy 6472*, Biloxi, Harrison County, Mississippi, 23 June 1899, US). Many contemporary collections failed to yield useable DNA (e.g., *Okebiro 701*, Kenya, Kiserian-L. Baringo; 19 March 1989, K). Fresh leaf samples were desiccated by placing them in plastic bags containing silica gel and herbarium vouchers for them were deposited with CANB, NE or SI. DNA was readily extracted from all fresh samples. From the 181 samples that yielded DNA of sufficient quality, we examined sequence variation in nuclear ribosomal internal transcribed spacers (nrDNA-ITS, ITS1, 5.8S rDNA and ITS2). Our pilot work with *rbcL* and *trnL-F* and more recent extensive investigation with *trnL-F* and *PetB* using leaf material from 262 populations (Data S1) did not yield variable regions and we abandoned investigations with these markers. Population locations, methods and results for chloroplast investigations are available in Supporting Information (Appendix A, Data S1).

Total genomic DNA was isolated using DNeasy Plant Mini Kit (Qiagen, Chadstone, Vic, Australia) or Wizard SV Genomic DNA Purification System (Catalogue number A2361, Promega, Sydney, NSW, Australia). The entire internal transcribed spacer region (ITS1/5.8S/ITS2) was amplified using primers

ITS4 and ITS5 [19]. Each 25 µL of amplification reaction contained 16 mM $(NH_4)_2SO_4$, 67 mM Tris-HCl (pH 8.8), 0.01% Tween-20, 10% DMSO, 4 mM $MgCl_2$, 0.4 mM of each dNTP, 0.5 unit *Taq* Polymerase (GoTaq®Flexi DNA Polymerase, Promega), 0.5 µM each ITS4 and ITS5 primer and 20 ng genomic DNA. Dimethyl sulfoxide (DMSO, 10%) was included in the reaction mix to exclude the amplification of low stability templates and to avoid preferential amplification of paralogous gene copies or nonfunctional pseudogenes [20–22]. Each PCR product was electrophoresed in a 1% agarose gel, stained with SYBR safe (Catalogue number S33102, Invitrogen, Thermo Fisher Scientific Australia, Scoresby, Vic, Australia), then excised and eluted using Wizard SV Gel and PCR Clean-Up System (Catalogue number A2361, Promega). PCR fragments were sequenced using both forward and reverse primers by SUPAMAC (The University of Sydney, Australia).

2.2.3. Sequence Data Preparation

DNA sequences were proofed and edited in FinchTV version 1.4 (available from www.geospiza. com/finchtv). nrDNA-ITS sequences generated in this study have been deposited in the Genbank database and accession numbers are listed in Table S1. Forward and reverse reads were analyzed for all sequences. The guidelines for obtaining reliable nrDNA-ITS sequences in plants [23] were followed to minimise the pitfalls of using ITS sequences to infer phylogenetic patterns. This included the addition of DMSO, repeated PCR reactions, sequencing complementary strands to verify the variable sites and verifying the secondary structure stability of sequences. DNA sequences were aligned using ClustalX [24] with default parameters for gap penalty and extension. One sequence for *Lantana camara* L. (Genbank AF437853) from Taiwan and two sequences for *Lippia alba* (Mill.) N.E.Br. (Genbank EU761076 and EU761078) from Colombia were used as out-groups. The beginning and end of the ITS regions were determined with reference to the *Lantana camara* sequence from Genbank (AF437853). Prior to the phylogenetic analyses, accessions with identical sequences were detected and merged using Jalview 2.4.0.b2 [25]. In total 80 ribotypes were included in the analyses as an ingroup (Table S1).

2.2.4. Consensus Tree Analyses

Consensus tree analyses included maximum likelihood (ML) and maximum parsimony (MP) using PAUP*4.0 beta 10 [26] and MEGA 5.0 [27] and the Bayesian estimation of phylogeny using the MrBayes version 3.1.2 [28]. Maximum parsimony was performed with gap states treated as a fifth character [29] and all character transformations weighted equally. Bootstrap analyses [30] were performed by resampling the data matrix 1000 times. To check for consistency with the results of the parsimony analysis (see below) we first determined the model of nucleotide substitution with ModelTest v. 3.7 [31] and ran two million generations of four simultaneous Markov Chains to approximate the posterior probabilities of trees. A single tree from every 200 generations was retained for the construction of the Bayesian consensus tree. To infer evolutionary divergence between ribotypes within *P. nodiflora*, evolutionary distances were measured in the units of the number of base pair substitutions per site [32] and calculated by using the Maximum Composite Likelihood model [33] incorporated into MEGA 5 [27]. Gaps were treated as complete deletions and standard errors for evolutionary divergence measures were obtained by a bootstrap procedure with 1000 replicates. Values in the distance matrix indicate evolutionary divergence between two ribotypes. A zero indicates that the difference between the two ribotypes is only in a gap or deletion.

2.2.5. Parsimony Network Analyses

We used parsimony networking methods to visualise character incongruence caused by reticulation. Network methods can incorporate population processes in building and refining relationships and also allow a more detailed display of population-level information than bifurcating trees [34]. Parsimony networks were obtained with the software TCS [35], which uses statistical parsimony and genealogical reconstruction algorithms from Templeton et al. [36]. To minimise the effect of gap coding on the accuracy of the parsimony network and to overcome the problem of order dependent collapsing of sequences into haplotypes, as described by Joly et al. [37], we reshuffled the data matrix and re-analysed the data until all individual sequences were assigned to their respective ribotypes.

2.3. Biogeography—Delimitation of Native-Range Distributions of P. nodiflora Clades

We gathered ecological information on the populations of *P. nodiflora* from journal articles, books, reports, archived newspapers (National Library of Australia, www.//trove.nla.gov.au) and from 336 herbarium vouchers in addition to those sampled for DNA (AD, BRI, CANB, CBG, DNA, HO, MEL, NSW and PERTH). Care was taken with historical searches to exclude *Aloysia citriodora* Ortega ex Pers. (Lemon Verbena) as it has been variously known as 'Lippia' owing to an earlier combination as *Lippia citriodora* H.B. & K. We evaluated our data against Bean's [15] topology, which we broke down into eleven criteria (Table 1), for *Phyla nodiflora* and for each major clade within the species (see results). We gathered information for each criterion where available. For criterion five (post-settlement expansion) and using data from the Australian Virtual Herbarium (sourced 4 April 2012), we compared the number of herbarium vouchers collected over time for *P. nodiflora* ($n = 311$) with (a) endemic *Scutellaria humilis* R.Br (Lamiaceae) ($n = 292$), a small trailing herb found in moist places across five states in Australia and (b) the unequivocally alien shrub in Australia, *Lantana camara* L. (Verbenaceae) ($n = 595$). Our expectation here was that all species would show an increase in the number of voucher records with time since settlement as a consequence of heightened survey effort, but an alien species would show a major range expansion over time compared with the native species, *S. humilis* (and see [15]). Duplicate samples were excluded in the analyses. While herbarium collections can have inherent biases reflecting non-random sampling by botanists (e.g., [38]), we assumed that these biases would be similar across the three species.

3. Results

3.1. Phylogenetics

3.1.1. Characteristics of the ITS Sequences and Ribotype Discovery

The aligned ITS data matrix was 630 bp long. There were 101 variable characters, of which six were indels and 95 were base substitutions. A high number of sites in ITS1 were variable (58 sites), whereas only 10 and 33 variable sites were observed in 5.8S and ITS2 regions respectively (Table S2). Based on nrDNA-ITS sequences of 181 samples (179 populations), a total of 64 ribotypes were identified worldwide within *P. nodiflora*, 11 for *P. canescens* and five for undetermined *Phyla* species (Table S1). Our global sampling of *P. nodiflora* ($n = 160$ populations) included 35 populations from the Americas where we found 10 ribotypes, a discovery level of 29%; for Australia we sampled 65 populations which yielded 26 ribotypes, a discovery level of c. 40% and from the rest of the world we sampled 61 populations where we found 38 ribotypes at a discovery level of 62%.

3.1.2. Consensus Trees

Five equally maximum parsimonious trees 806 steps long, with a consistency index of 0.75 and retention index of 0.78 were resolved. The strict and semi-strict consensus trees were identical. The Bayesian consensus tree of the 10,000 trees sampled in the analyses was highly congruent with the MP and ML analyses (Figure 1) with slight differences in branching topology among poorly supported nodes. The Bootstrap (BS) and Bayesian analyses provided 100% support for the *Phyla* clade. Within the *Phyla* clade three major *Phyla* lineages were recognized each with at least 97% BS support: a *Phyla* spp. clade, *Phyla canescens* and *Phyla nodiflora* (Figure 1). The *Phyla* spp. clade was recorded in South America, the Gulf of Mexico and southern Australia (Figure 2) and may represent several taxa (analysis not shown). *Phyla canescens* was found in central South America, Hawaii and Australia (Figure 2).

In the Americas *P. nodiflora* was restricted to central and North America and northern South America, but it was the most wide-spread and commonly sampled species elsewhere in the world (Figure 2).

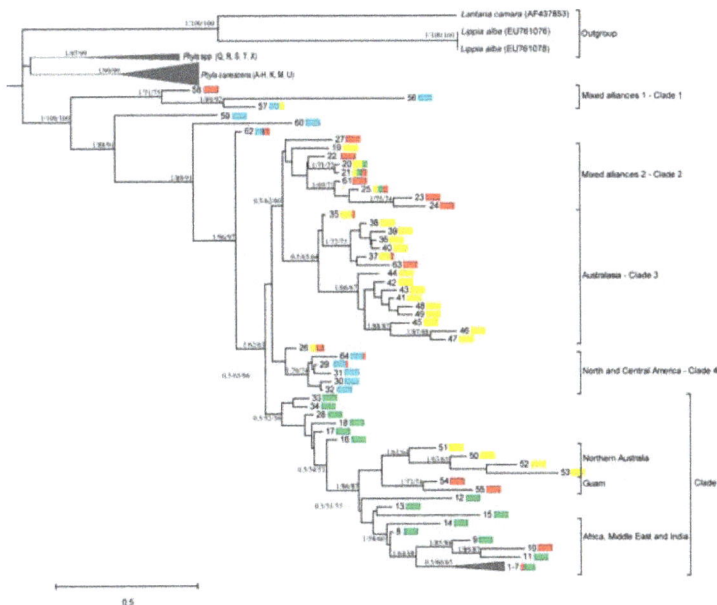

Figure 1. The maximally parsimonious tree based on nrDNA-ITS sequence data. The tree is drawn to scale, with branch lengths in the same units as those of the evolutionary distances used to infer the phylogenetic tree. Numbers on branches indicate posterior probabilities or bootstrap values for Bayesian analyses/parsimony analysis/maximum likelihood. Ribotypes are terminal taxa and details for them are given in Table S1 and Figures 2 and 3. Colour coding matches the global colour scheme in Figure 3 and the bandwidths are not informative. AF437853 = *Lantana camara*. Five clades, 1–5 are recognised.

There was considerable structure evident within the *P. nodiflora* clade. We recognize five clades, one that is unsupported (Clade 2) and the remaining four clades weakly supported with bootstrap support ranging from 52% to 71% (Figure 1). In addition, ribotypes 59 (Utah), 60 (New Mexico) and 62 (Vietnam, Mexico) were not grouped with any major clade within *P. nodiflora* (Figure 1).

Clade 1 contained three distinct ribotypes which together spanned three continents (Asia, North America and Australia) (Figures 1 and 2). It included two ribotypes from Central America, one of which was also recorded in eastern Australia and one only recorded in Asia (China).

Clade 2 was distributed globally, including in Africa, Australia, Asia, Europe and two Oceanic Islands, although it was notably absent from the Americas (Figure 1). This clade contains widely distributed ribotypes 21 and 25, each present across three continents (Figures 1 and 2). Ribotype 21 has the largest latitudinal spread from southwest Western Australia through to China (Figure 2.)

Clade 3 was largely restricted to Australia (and the island of New Guinea and East Timor) (Figure 1). Fifteen of the 16 ribotypes were recorded from Australia. Ribotype 37 was widely distributed across northern Australia, Timor and the island of New Guinea. Only one ribotype, 63 from Palau, was not found within Australasia (Figure 2).

Clade 4 was largely restricted to the Americas, although one ribotype was also present in Taiwan (64) and one in Hawaii (29) (Figure 1). In the Americas clade 4 ribotypes were recorded in south-eastern USA, Mexico, South America (coastal Ecuador and Venezuela) and the Caribbean (Figure 2).

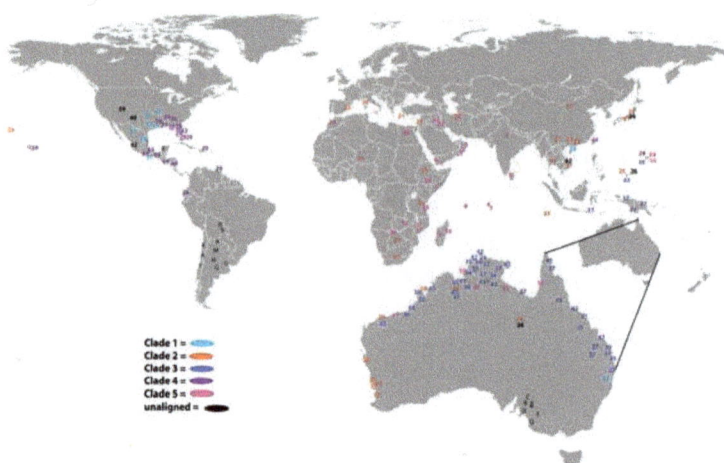

Figure 2. The global distribution of the 64 ribotypes discovered for *Phyla nodiflora* (numbers), the eleven ribotypes discovered for *P. canescens* (A-H, K, M, U) and five ribotypes (unaligned) from an undetermined taxa or taxon within *Phyla* (Q-T, X). Underlined numbers represent multiple collections of that ribotype from that location area. *Phyla nodiflora* ribotypes with clade support are coloured according to clade number on Figure 1 or if unaligned they are in black notation. Further details of ribotypes are given in Table S1.

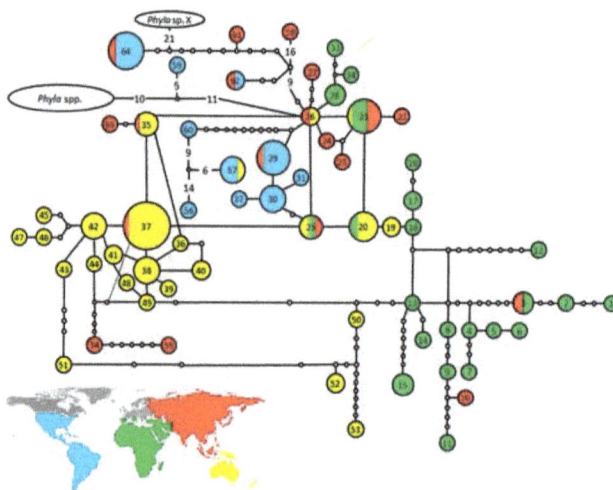

Figure 3. A parsimony network displaying the relationships among the 64 *Phyla nodiflora* nrDNA-ITS ribotypes. Each link between the ribotypes represents one mutational difference. Unlabelled nodes indicate inferred steps not found in sampled populations. Loops in the network are the result of homoplasies. The angle of bifurcation and the length of links between ribotypes have no significance. Each circle is proportional to the ribotype frequency. The colour of each ribotype relates to the biogeographical region in the insert map. Circles with multiple colours are ribotypes present in two or more biogeographic regions. Ribotype numbers and letters are explained in Table S1.

Ribotype 26 was an unsupported sister to clade 4 and was not detected in the Americas but was found on Palau and in Australia.

Considerable genetic and geographic structure was evident in Clade 5 (Figure 1). Most ribotypes were restricted to Africa, the Middle East, the subcontinent and islands in the Indian Ocean (20 ribotypes). However, there were also sub-clades within Clade 5 restricted to Australia (4 ribotypes) and Guam (2 ribotypes).

3.1.3. Parsimony Network

A parsimony network of the ribotypes derived from the ITS region in *P. nodiflora* revealed a complex structure within this species (Figure 3) and is in broad congruence with the phylogenetic analyses. Ribotypes of *P. nodiflora* were connected with 15 or more evolutionary steps to the ribotypes of other species of *Phyla*. Several loops (homoplasies or recombination) were detected within the parsimony network, indicating low resolution in the ITS sequence data or recombination.

Strong biogeographic structure was revealed by the parsimony network analysis (Figures 2 and 3). No ribotype was found on every continent. Only ten of the 64 *P. nodiflora* ribotypes occurred in two or more biogeographical regions (ribotypes 3, 20, 21, 25, 26, 29, 35, 62, 64 and 57). Three of these were present in the Americas (62, 64 and 57). Ribotype 57, a ribotype found extensively in the indigenous American range, was also discovered in eastern Australia in a floodplain disturbed by cattle grazing. Ribotype 25 was the most widely distributed type detected in our study being found in the Middle East, France, Spain, South-East Asia and Australia, but not the Americas. Ribotype 26 was found in Asia (Japan, Palau) and Australia. Ribotype 21 was detected in the Mediterranean, Asia, Africa and Australia. Ribotype 20 was found in Iran and Western Australia. Ribotype 37, the most common ribotype discovered by our sampling in Australia, is shared between Australia (Northern Territory and Queensland), Timor, PNG and West Papua. Ribotype 37 differs by one nucleotide from common ribotypes 25 and 35 and these are the closest links outside of Australia.

Twenty-five of the 64 ribotypes (c. 39%) discovered world-wide were found in Australia and 18 of these (c. 72%) are unique to Australia (ribotypes; 19, 36, 38–53, Figure 3). Ribotype 37, followed by ribotype 42, are the most common endemic types in Australia and they span the tropics of northern Australia (Figure 2). Ribotype 35 is found on the east coast of Australia and in Guam and differs by one nucleotide from the common eastern hemisphere ribotype 26.

Three ribotypes were present on remote islands in the Indian Ocean. Two ribotypes, 4 and 7, found in the Indian Ocean archipelago of Chagos, differed in one deletion and two substitutions respectively at positions 56, 152 and 222 of ITS1. These ribotypes are closely related to ribotypes 5 and 6, found in nearby Yemen, which differ from each other in two substitutions at positions 10 and 182 of ITS1. Ribotype 8, from the Seychelles in the Indian Ocean is closely related to ribotype 9, found in nearby Somalia. They differ in three substitutions at positions 6, 8 and 15 of ITS1 (Table S2).

Ribotypes 54 and 55 in Clade 5 were only found on the island of Guam but differed in five substitutions in their ITS profile (two in ITS1, two in 5.8S and one in ITS2 regions). They were found in disturbed areas near a military base, which would offer opportunities for multiple introductions (e.g., ribotype 35 from Clade 3, ribotype 29 Clade 4 were also found on Guam).

The largest genetic distance observed between two ribotypes was between ribotypes 1 (Oman) and 56 (Mexico) with a divergence value of 0.0826 (Supporting Information, Data S2)—this difference was due to 50 substitutions and three deletions. The differences between ribotype 56 and 11 (Somalia) was also relatively large at 0.0788. Whereas, divergences within regions were generally not pronounced. For example, the distance between ribotypes 29 and 32 was only 0.0016 and ribotype 47 compared with 37, both from Australia, was 0.0065.

3.2. Biogeography—Delimitation of Native-Range Distributions of P. nodiflora Clades

Each of the five clades was associated mostly with one or two biogeographic regions. We attempted to distinguish the likely native-range limits of the clades from human-assisted dispersal, based on our hypothesised phylogeny and additional evidence reviewed against a typology presented in Table 1. In the native range, we would expect a range of genetic diversities to be present with genotypes structured geographically. Furthermore, a natural distribution would be reflected in the species occurring in intact unmodified habitats, accompanied by a lack of invasive encroachments, a lack of post-European expansion in geographical range and any discontinuities of distribution would be related to climatic, geological and edaphic factors. In an alien species, patterns of genotypic diversity would not be expected to be structured geographically. Human-mediated dispersal would also be indicated by the species being found mainly in disturbed areas, with invasive encroachments on native communities, an expansion of range in the last 200 years and a patchy distribution correlated with human settlement.

We will refer to the criteria in Table 1 as C1 to C11 throughout this section. We found that a constraint to utilising Bean's [15] criterion 1 fully (C1a intact habitat versus C1b frequently disturbed sites) is that *Phyla nodiflora* is a pioneer species in its native range of lower North America to northern South America, where, Kennedy [6] notes that it is weedy. Kennedy [6] also notes that *P. nodiflora* is tolerant to a wide variety of environmental situations and is often associated with water sources. In the context of *P. nodiflora*, we restricted C1a to cover pristine or near pristine habitats and C1b to highly disturbed areas with other signs of gross disturbance including the presence of invasive species.

3.2.1. Clade 1—Mixed Alliances 1: Americas (North, Central, Northern South America)

Ribotypes from this clade occur in the native range of *P. nodiflora* (sensu Kennedy [6]). General ecological information on the species from this region (which could also apply to Clade 4 populations, see below) includes that *Phyla nodiflora* is known to be a pioneer and a weedy species [6]. Plants from ribotypes 56 and 57 grow in river and in irrigation channels, floodplains and alluvial fans and therefore match C1b in Table 1. However, plants have been found in native plant communities (C1a, Table 1) as well as townships (C1b, Table 1) but not as an invasive species (C2a, Table 1). The butterfly *Phyciodes phaon* Edwards (the Phaon crescent), is restricted to the Americas and uses *P. nodiflora* as a food plant for its larva in Florida [39]. This is strong evidence of *P. nodiflora*'s nativity to lower North America (C3a). Furthermore, Genc et al. [39] report that several species of native Lepidoptera use the leaves of *P. nodiflora* as a food plant in this region. Bean [15] argues that a range of phenotypes (C4a) would be indicative of a native species and the corollary that uniform diversity would be suggestive of an alien lineage (C4b). In the Americas however, where *P. nodiflora* is indigenous, Kennedy [6] notes the extensive variation in vegetative characters but that the variation does not have a geographical basis. In Bentham's [40] *Flora Australiensis* and in regard to the species' global diversity, he notes that *P. nodiflora* "is very variable in the breadth of the leaves, the size of the spikes and the flowers, the points and teeth of the bracts, &c". Ribotype 58 was restricted to China and the specimen was found growing on coastal sands. Ribotype 57 was also detected in a single population in eastern Australia (*Nelson, s.n. NE*) in a flood plain disturbed by cattle. This isolated occurrence in Australia (Figure 2), location and clade position (Figure 1) and network position (Figure 3) accords with Bean's [15]) criteria for characteristics associated with an alien species (C1b, C2b, C5b, C6b, C9b, C10b) and we suggest that this ribotype has recently arrived in Australia from the Americas.

3.2.2. Clade 2—Mixed Alliances 2: Asia, Africa, Australia

Ribotypes from this clade are highly mobile with three of them occurring in two or more continents (Figures 1 and 3). We tentatively propose that this clade originates from Asia where it is common. Supporting this and against Criterion 7, the Verbenaceae have been found in fossil pollen records for China (pollen resembling *P. nodiflora*, Miocene Heilongjing Formation, [41]). In Asia, plants have been found along the seashore (Japan) and very disturbed roadsides (China). Further east, on Midway Atoll, the plants were recorded in a drainage ditch and as a new record in 1999 on Sand Island (*BISH 659789*). *Phyla nodiflora* is listed as a food plant for several species of butterfly in Asia including *Junonia almana* L., *Junonia atlites* L., *Junonia lemonias* L. and *Junonia orithya* L. [42] suggesting an evolutionary relationship between these herbivores and host-plant. In Australia and Africa, Clade 2 ribotypes match many of the alien criteria of Bean's (2007 [15]) topology (Table 1; C1b, C2b, C5b, C6b, C9b, C10b) as they grow in very disturbed areas associated with farming and irrigation and co-occur with invasive species such as *Salvinia molesta* D.Mitch. (Kununurra, W.A., *Thompson s.n. 26 July 2007, CANB*). Location data from specimens of *P. nodiflora* were mapped against region and collection time revealing that Clade 3 is disjunct from the main distribution of *P. nodiflora* in Australia (Figure 4, C9b) with the majority of collections occurring after 1980 (C10b). However, in 1899 E.J. Bickford President of the Mueller Botanical Society of W.A. lists 'Lippia' as a wildflower of the districts and sand plains north of Perth (Western Mail, 9 December 1899). We reject the possibility that this early record was *Phyla canescens* as that species has never been collected from this region in W.A., nor is it found in intact ecosystems. Furthermore the earliest gatherings of *P. nodiflora* in Western Australia were made by Augustus Oldfield (*Oldfield s.n. 1850–1859, MEL583739A, MEL583740A*) between 1850–1859 around the Murchison River area near Kalbarri which was being explored for mining and agriculture from 1839 [43], although not well explored until the 1850s (p. 35, [44]). We cannot determine the ribotype of these Oldfield specimens although ribotype 20 from the historic Grey's Well at Kalbarri is a close geographic match. A ribotype from Clade 2 may have existed naturally in the Kalbarri region.

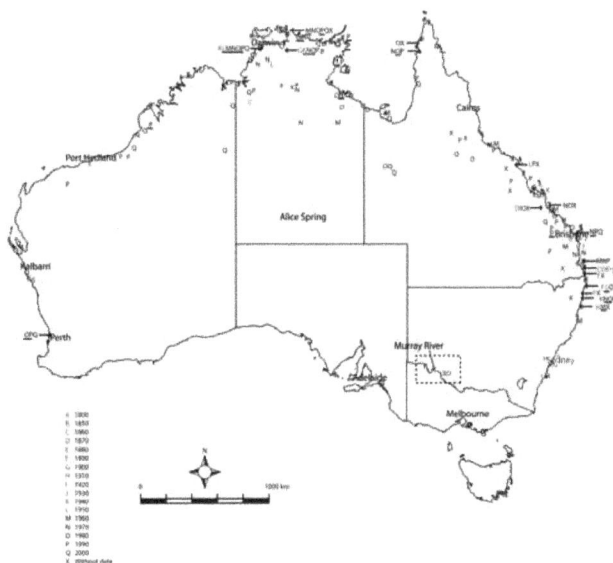

Figure 4. The location and decade of Australian herbarium-collections for *P. nodiflora* in Australia. An underlined letter indicates that multiple collections exist for that location in that decade. Red letters indicate specimens that were collected prior to 1900. The rectangle insert indicates specimens listed in herbaria as *P. nodiflora* but have been determined by us to be *P. canescens*.

3.2.3. Clade 3: Australasia

We propose that this clade is native to Australia. The ribotypes are found in areas that are relatively undisturbed by human activities (C2a, see also C10, Table 1). For example, *Phyla nodiflora* is scantily distributed on the floodplains of Oenpelli, a remote location in the Northern Territory (A. Mitchell, pers. comm. 21 February 2011, *Specht 1181, 13 October 1948, AD; Cowie 2119, 13 November 1991, MEL*). On the mainland of Australia we failed to find data (either from herbarium records or from our own field work) where *P. nodiflora* has intense infestations, although plants can be found in disturbed areas co-occurring with invasive species (*Vachellia farnesiana* (L.) Wight & Arn, Bullo River Station, NT, *Kerrigan 1040, 10 March 2006, DNA*). Ribotype 44 was found on mainland Australia and on Croker Island (Arafura Sea, Northern Territory) where *P. nodiflora* has invaded the seasonally inundated Paludal plains and where the invasion has been facilitated by feral animals overgrazing the area [45]. Clade 3 *Phyla nodiflora* found in both modified and unmodified habitats in Australia (C1a and C1b Table 1) including rubbish tips, drainage channels, roadsides and built recreational areas with lawns (WA, Karratha, R.D. van Klinken s.n. unpub. data). In concert with the behaviour of *P. nodiflora* in its native range, we found habitat preferences suggestive of both a native and alien species. Against Criterion 2, Clade 3 *P. nodiflora* are mostly an occasional or sometimes a common component of natural communities (Ord River, W.A., R.D. van Klinken s.n. 24 July 2007 unpub. data)—but apparently not to the extent that it dominates the community. For Criterion 3 herbarium label data were not informative for the presence or absence of pests and diseases in *P. nodiflora*. Plant collectors may avoid diseased or imperfect specimens when collecting and thus screening herbarium specimens for pests and diseases may not be informative. We have seen an image of the first specimen collected in Australia (*BM, Bennett No. 2335*, image at NE, see Criterion 10) and about half of the leaves are imperfect with holes and other insect damage. This could have happened after the specimen was collected. We found listings for the butterfly *Junonia villida* Fabricius using *P. nodiflora* as a food plant but it also forages on the alien and invasive *P. canescens* (e.g., [46]). We thus do not have unequivocal information of an evolutionary relationship between *P. nodiflora* and native herbivores in Australia.

Bean (2007 [15]) argues that a range of phenotypes would be indicative of a native species and the corollary that uniform diversity would be suggestive of an alien lineage (C4). For Australia, Munir [8]) and Macdonald [17]) note the extensive variability within *P. nodiflora* as well as within the alien and invasive *P. canescens*. Due to the variability expressed in the Americas (see above), we conclude that the variability observed in *P. nodiflora* in Australia is not a useful indicator for its status in Australia. For Criterion 5 the collection dates on herbarium specimens were collated and plotted as a percentage of the collecting effort for each species and against decades for *P. nodiflora*, the native herb *Scutellaria humilis* and the alien *Lantana camara* (Figure 5). As expected *L. camara* has had a rapid expansion of collection records, especially from the 1950s. Overall *S. humilis* was collected more often than *P. nodiflora* up until the 1970s from when *P. nodiflora* was collected more frequently (Figure 4). The data for *P. nodiflora* is somewhat intermediate between the patterns observed for native *S. humilis* and alien *L. camara* and we tentatively suggest a post-settlement expansion in Australia has occurred for *P. nodiflora* especially since 1970.

Against Criterion 6 and Clade 3 in Australia, most of the Australian populations are associated with tropical habitats and freshwater or coastal dune systems. For Criterion 7, *Phyla nodiflora* is most closely related to the South American *P. canescens* [6]. Kennedy [6] also considers the North American species *P. cuneifolia* (Torr.) Greene to be closely related to *P. nodiflora* while O'Leary and Múlgura [7] consider that *P. nodiflora* is the taxon most similar morphologically to *P. lanceolata* (Michx.) Greene. *Phyla canescens* is a highly invasive herb in Australia [18,47–49] whereas *P. cuneifolia*, *P. fructicosa* and *P. lanceolata* are not listed as occurring in Australia (Australia's Virtual Herbarium, accessed 26 February 2017). The taxa outside of Australia with the closest native range to *P. nodiflora* include the Latin American *P. canescens*, *P. betulifolia* (Kunth) Greene and *P. fructicosa* (= *P. nodiflora* var. *reptans* (Kunth) Moldenke, [7]) but none of these taxa is native to Australia's near neighbours of New Guinea, Timor, Java or New Zealand. For Criterion 8 (Table 1) several islands in Asia (e.g., Guam) hold alien populations of *P. nodiflora* and are potential source areas for distribution into Australia. Gupta et al. [10] consider that *P. nodiflora* is native to the whole of Africa, temperate and tropical Asia, Australasia, Europe and tropical America and that it is wide-spread and locally common in "open and wet places near streams, ponds, paddy fields, ditches, backwaters, brackish water". Location data taken from specimens of *P. nodiflora* were mapped against region and collection time revealing that *P. nodiflora* has a disjunct distribution in Australia (Figure 4). Populations are found in the north of Australia, with sparse occurrences along the east coast (C9b, Table 1). For Criterion 10 (Table 1) the earliest collection of *P. nodiflora* from Australia was made in 1802 by Robert Brown at Shoalwater Bay Qld (*BM, Bennett No. 2335*, image at NE) and is within the Clade 3 geographic distribution. At least a further 16 collections from the mid to late 1800s have been lodged in Australian herbaria, including collections from south-eastern Queensland. None of these areas in southern Queensland was unequivocally pristine at the time these specimens were collected. Captain James Cook for example, had made land in the Shoalwater Bay area in 1770 (late May-early June) where an expedition was made to Pier Headland, in the same vicinity that Robert Brown collected from some 32 years later in 1802. There is no record of a Banks and Solander collection of *P. nodiflora* from 1770 (specimen data from the Herbarium of New South Wales, Royal Botanic Gardens and Domain Trust, 21 February 2011) although botanical gatherings were made in the Shoalwater Bay area by them [50]. Three collections were made in the 1890s along the north coast of New South Wales. One herbarium specimen however, provides strong evidence for a pre-European occurrence of *P. nodiflora* in northern Australia (*Mueller s.n. MEL 583741A*). Mueller was the Botanist on the Gregory Expedition into Northern Australia (1855–1856) and in 1856 he wrote that he collected a 'Lippia' on that expedition [51], Appendix II in [52]. This is a significant finding as the Victoria River area that these explorers traversed at this stage of the Gregory Expedition was not settled and was unchartered territory at the time [53]. We did not find any records of its use as a medicinal plant by the Australian Aboriginal peoples.

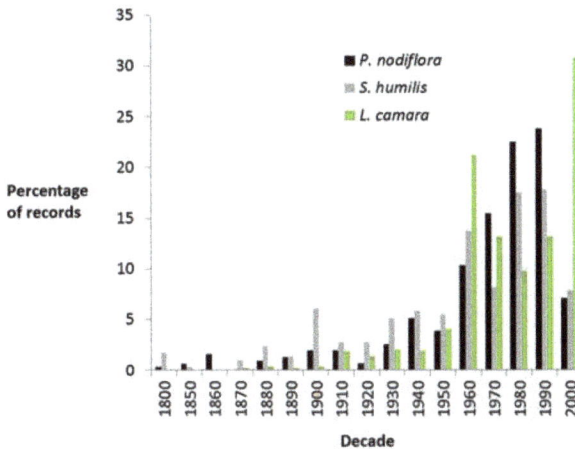

Figure 5. The distribution of Australian collection-records from eight herbaria (AD, BRI, CANB, CBG, HO, MEL, NSW, PERTH) plotted as a percentage of the collecting effort for each species and against time (decadal periods) for *Phyla nodiflora* (*n* = 311) *Scutellaria humilis* (*n* = 292) and *Lantana camara* (*n* = 595).

3.2.4. Clade 4—North and Central America with Asia

We suggest that these ribotypes are native to the Americas with the Asian representatives being exotic to Taiwan and Guam. In the distributional range of Clade 4 in the Caribbean, *Phyla nodiflora* is a host food plant for the butterflies *Junonia genoveva* Cramer and *Junonia evarete* Cramer [42]. In Latin America plants from ribotypes 29 were found along rivers, mangroves, open savannahs, beaches and coastal dunes. In lower North America ribotype 29 plants were found in disturbed areas related to the built environment including airforce bases, boat launches roadsides and empty lots. Ribotype 29 includes a specimen used as a landscaping plant on Guam (*Fosberg 46296 13 July 1965, NY*). We found direct evidence of *P. nodiflora* being accidentally transferred between biogeographic regions. In Hawaii ribotype 29 was grown from soil scraped from a soldier's boot—the soldier was newly arrived from Guam (*BISH 728774*).

3.2.5. Clade 5—Northern Australia, Guam, Africa, Middle East and India

We suggest that this clade is native to Africa but that there has been recent movement from there to Australia and Asia. *Phyla* type pollen fossils have been found in Africa from the Holocene [54] and *P. nodiflora* has a long history as a medicinal plant in Nepal [55], India [56] Taiwan [57] and the Middle East [58]. *Phyla nodiflora* has been recorded from different habitats ranging from pristine dune systems in eastern Africa to black soil flood plains and forests in the Middle East and the Indian subcontinent and from desert oases in Egypt to lake sides in Australia [59–61]. In Africa and the Middle East, ribotypes were found along stream edges, desert oases and natural lakes (C1a) but also in disturbed areas on farms, military camps and air strips (C1b). The ecological data accompanying herbarium specimens for ribotypes 50–53 accords with them being recent arrivals to Australia. All of them were found near coastal settlements and in ruderal habitats such as disturbed areas around developments. In Guam, the samples are also associated with highly disturbed areas (landfill, roadsides on an air base).

4. Discussion

4.1. Molecular Markers

Chloroplast DNA exhibits uni-parental inheritance and a slow rate of mutation [62,63]. We found insufficient sequence variability in three chloroplast regions for resolving relationships within *Phyla nodiflora* or *Phyla canescens*. This is not an unusual finding (but see [64]). For example, in a whole chloroplast genome study of 71 rainforest taxa, Rossetto et al. [65] found 75% of the trees sequenced (up to 18 individuals per species) had less than 25 single nucleotide polymorphisms across an average of 122,373 bp sequenced. Diazgranados & Barber [66], working with 110 species in the subtribe Espeletiinae Cuatrec., Asteraceae, found that most chloroplast regions screened were uninformative or the variability was too low to resolve phylogenetic relationships. Low variation in many chloroplast regions highlights that they may be less than ideal for global studies of closely related lineages. ITS regions were more variable and informative with the large clades of *Phyla nodiflora* and *Phyla canescens* strongly supported. Clades within *P. nodiflora* were moderately or weakly supported or unsupported, however the geographic structure of the five clades and the parsimony network of ribotypes discussed below provide a strong argument for the global expansion of *P. nodiflora*.

4.2. The Global Expansion of Phyla nodiflora Involves Native and Invasive Lineages

We set out to determine the native range of *P. nodiflora*, a species with a pantropical distribution. In the Americas, *Phyla nodiflora* is found in unmodified and disturbed habitats and this complicates the determination of its native/alien status in other parts of the world where it also occupies disturbed and undisturbed areas. Our study supports a Neotropical origin for the genus [11] and for *Phyla nodiflora* [6]. Our results also concur with Kennedy's [6] that *P. nodiflora* is indigenous to southern USA, the Caribbean, Mexico and Central America. However, we also found substantial molecular and phytogeographical evidence that *P. nodiflora* has been dispersed widely around the world through both natural and human-assisted spread of propagules. We suggest a North American origin, with subsequent natural dispersal to Africa, Asia and finally Australasia. As such we demonstrate an additional colonisation of Africa to the 5–6 events already determined for Verbenaceae by Marx et al. [11]. In addition, available evidence suggests that human-assisted dispersal has resulted in trans-continental movement of all five clades that we recognized.

Most lineages that we recognized within *P. nodiflora* were present in at least two continents. However, the native range of most could be identified using biogeographic evidence. Two distinct clades were probably native to the Americas, with all five ribotypes from one clade and two of the three ribotypes from another, being recorded from there. The exception was a relatively distinct ribotype that was only recorded from China. Further sampling is required to test its origin, although fossilized pollen evidence from the Miocene Heilongjing Formation (5.3–23 Ma, [41]) does suggest the species, if not necessarily this clade, was already present there at that time.

The status of *P. nodiflora* in Australia is generally considered as naturalised after an introduction to the continent [8], despite it being first recorded there in 1802, soon after European settlement and during first European contact in north-western Australia [51], Appendix II in [52]. However, we identify a lineage that is clearly native to Australia (and possibly Papua New Guinea, West Papua and East Timor). Genetic variation was considerable, although no geographic structure was evident. Only two records of this lineage were found outside Australia, both in Micronesia where all indications are that *P. nodiflora* is alien (see above).

The native-range of the two remaining clades is more ambiguous. The most genetically divergent clade appears to be native throughout Africa, including a subclade that has radiated in and around the Arab Peninsula. A long history of *P. nodiflora* in Africa is also supported by Holocene fossil evidence from Africa [54]. However, a further two sub-clades were only recorded from Australia and Guam respectively, despite available evidence suggesting that they may be alien there and therefore of unknown origin (see below). The final clade, clade 5, was the most ubiquitous, with ribotypes present in Africa (from South Africa to the Mediterranean), Europe, Asia and Australasia and was also the least well supported phylogenetically. Available evidence suggests that it is native to Eurasia, but that it may also have been extensively moved around by humans. For example, it is likely to be alien in Australia where provenances are used as lawns (see below).

Phyla nodiflora grows in surprisingly diverse habitats, in coastal dune systems, along sandy lake shores, shore muds, grasslands, roadsides and salt marshes [59–61,67]. Our sampling included populations growing in seasonally flooded black-soil plains, dune systems and urban environments (such as lawns, roadsides and rubbish tips). However, we found no evidence for ecotypes with differing habitat preferences despite considerable genetic variation within the species.

4.3. Natural Dispersal

All evidence suggests that the genus *Phyla* radiated in the Americas, so *P. nodiflora* most likely dispersed and subsequently radiated from there. Our phylogeny is congruent with movement from the Americas, to Africa and Eurasia and into Australia, although other pathways can't be ruled out. In any case *P. nodiflora* is one of very few non-halophytic and terrestrial plant species known to have achieved such a pantropical distribution (see [3]). However, many taxa have moved naturally across the Atlantic between the Americas and Africa. Renner [68] compiled a list of 110 angiosperm genera in 53 families that contain species from both sides of the tropical Atlantic, with wind and sea currents being identified as the most likely vectors. Dispersal of plant taxa from Africa to Australia has also been documented ([69] and references therein) which Mummenhoff et al. [69] suggest for *Lepidium* L. (Brassicaceae) is congruent with bird migration patterns between Africa and Australia.

We can only speculate on when and how *P. nodiflora* dispersed globally through natural means. Waterbirds are one possibility. *Phyla* seeds have been found in the stomachs of migratory birds in North America [70,71] and seeds are considered to be dispersed by endo and ectozoic means in India [72]. In northern Australia, *Phyla nodiflora* has been found in the stomachs of the Comb Crested Jacana (*Iredipurru gallinacea* Temminck) [73], a water bird common to Asia and Australia that is dispersive among water bodies. Sea currents are thought to be responsible for the pantropical distribution of *Hibiscus tiliaceus* L. [74]; and it may also be a pathway for *P. nodiflora* as seeds are tolerant to inundation and saline conditions [75] and *P. nodiflora* grows well in coastal sand dunes in Africa [59] and in Australia (*Carter s.n. 1 July 2007, CANB*).

4.4. Human Assisted Dispersal

There is evidence that all five clades that we recognize have been moved by humans between biogeographic regions, although some clades appear to have been moved more than others. *Phyla nodiflora*, has long been used and therefore potentially dispersed, by people as a medicinal plant in the tropics of the eastern hemisphere (e.g., [56]), although we failed to find any data on *P. nodiflora* use by indigenous Australians. More recently there are documented cases of *P. nodiflora* being imported into Australia. For example, 'Lippia' was imported into Tasmania in c. 1902 by the then curator of the Botanical Gardens, Mr Francis Abbot (Western Mail, 16 March 1917). The plant was considered 'an absolute failure' and passed on to Victoria where it was 'condemned as worthless'. Its dismal performance as a lawn substitute in Tasmania and Victoria suggests that the taxon evaluated in 1902 was not *Phyla canescens* and it may have been *P. nodiflora*. It was also imported (as "*Phyla nudiflora*") to Australia in 1948 (Commonwealth Plant Introductions, G. Cook, pers. comm.) from an unknown location and for an unknown purpose. Plant collectors continue to trade "*Phyla nodiflora*" within and among continents (e.g., *Phyla nodiflora* is listed for sale in a Victorian nursery catalogue in 1886, [76]; plants are globally traded (e.g. [77] including as a lawn or ground cover [78]; Cocos (Keeling) Island [79]). However, *P. canescens* plants are commonly mis-labelled and sold as *P. nodiflora*, at least in Australia [17] and so such records should be treated with caution. Nonetheless, we found some *P. nodiflora* ribotypes widely used as ornamentals or as lawns. For example, ribotype 20 was commonly associated with lawns in arid Australian towns (van Klinken, pers. obs) and is one of the ribotypes in that clade that has become naturalized near urban centres in south-western Australia ([80]; Keighery pers. comm. October 2009). On Guam *P. nodiflora* is used as a landscaping plant (*Fosberg 46296 13 July 1965, NY*). There is also evidence of *P. nodiflora* ribotypes being moved unintentionally between continents. For example, the evidence of unintentional movement by humans with the detection of seeds transported from Guam to Hawaii from the soil scraped from a soldier's boot (*BISH 728774, 18 July 2006*).

4.5. Native Lineages Outside of the Americas Have Had Secondary Emigration Events

A grouping of ribotypes (50–53) in Clade 5 from northern Australia is closely aligned with African material and we suggest that these are secondary dispersal events from native African populations to Australia. Each of these ribotypes found in Australia is represented by one specimen. Ribotype 51 was collected on a rubbish tip, ribotype 50 from a swamp and 52 and 53 are from agricultural properties. We tentatively assign these ribotypes as alien to Australia, but further collections from Africa could help resolve their status. This clade also has ribotypes 54 and 55, each with one sample, from disturbed areas on a Military base on Guam. Ribotype 44 from Croker Island, Arafura Sea, Australia, was the accession closest to ribotypes 54 and 55 from Guam and ribotype 37, a common type from northern Australia was also found in Guam and the intermediate step of Papua New Guinea. Croker Island receives Australian and US military vessels from time to time (e.g., [81]). Clade 5 also has 14 ribotypes that are centred on land near the Red Sea, the Persian Gulf and the Arabian Sea. Most of the collections were from disturbed habitats.

4.6. Taxonomic Considerations

Our genetic analyses strongly support the treatment of *P. canescens* and *P. nodiflora* as separate species, a position that is in contrast to the recent morphological work by O'Leary & Múlgura [7] where these taxa are listed as varieties under *P. nodiflora*. We did not sample from or view the type specimen of *Phyla nodiflora*. We accept Munir's [8] position on the nomenclature for the taxon as he viewed microfiche images of syntypes available to Linnaeus and he also accepted Verdcourt's [67] lectotypification of *P. nodiflora* [67] cited in [8]. Verdcourt [67] does not utilise an earlier lectotypification of *P. nodiflora* by Townsend [82] probably due to that specimen being from cultivated material and instead he employed specimens available to Linnaeus. This contrasts with the position taken by O'Leary and Múlgura [7] who note that the lectotype used by Verdcourt may well be the original material for the name but that the earlier lectotypification by Townsend [82] was effectively published.

O'Leary & Múlgura [7] considered varietal status "for taxa when we observed a group of organisms with a gradual variation of characters, which would suggest an incomplete segregation of incipient species sharing the same geographical area". O'Leary & Múlgura [7] did not use a molecular approach but they describe the considerable variation that separates the two taxa in terms of leaf morphology, indumentum and flower colour and to this we would add the succinct differences in inflorescence structure that the two taxa maintain (Figure S1a,b).

Our work in progress has shown that the two species have different breeding systems; *P. nodiflora* is capable of autogamous seed production (83% of bagged flowers set seed automatically, *n* = c. 250 flowers, Fatemi, unpub. data, 2006 whereas *P. canescens* is not capable of autogamous seed production, [18]). Our research has found that there is little geographic overlap in the species in either the native or Australian range (Figure 2) which concurs with Kennedy ([6] Figures 8.2 and 8.9) and MacDonald [17] respectively. Our work also revealed additional species that are closely related to *P. canescens* and *P. nodiflora* and these will be discussed elsewhere (M. Fatemi, unpublished data). In most cases our species placements agreed with those of recent taxonomic treatments [6,8], confirming the circumscription of this taxon by these authorities. However, there were some important exceptions. Two specimens from the Caribbean that were determined as *P. nodiflora* by Kennedy [6] were assigned by our analyses to *P. canescens*. The circumscription of *P. nodiflora* in Australia by Munir [8] includes material from the Murray River region (*Lucas 71, Balranald, 1878, MEL583748A*) which would be an outlying population for this species (see Figure 5). Plants from this region show phenotypic variation that warrants further investigation and indeed duplicate specimens of Conn 4043 (*Conn 4043, Kings Billabong, E of Mildura, 16 January 1994, AD*) are variously curated as *P. nodiflora* (AD) or *P. canescens* (BRI, CANB, NSW, MEL). Our ITS studies showed that *Conn 4043* is *P. canescens*. We did not sample from Lucas 71 but we did successfully sample from six herbarium specimens from different populations in the Murray River area and all but one of these was resolved as *P. canescens*. One specimen from Lakala South Australia (Ribotype Q) may be a new taxon for Australia, based on phylogenetic distance, although we are unable to determine which species of *Phyla* it is at present.

4.7. Implications for Conservation

The detection of both native and exotic lineages of *P. nodiflora*, in Africa, Eurasia and Australasia is of conservation significance. First, it confirms that it is native in many parts of the world where its status had previously been considered exotic (e.g., Australia, [8]; South Africa, [83]). Second, exotic lineages could become invasive in their own right, as *P. canescens* has already done in Australia [48,49], although there is currently limited evidence of it doing so. Finally, native and exotic lineages could potentially interbreed, threatening the integrity of native lineages. For example, Saltonstall [84] showed that the American common reed, *Phragmites australis*, had been largely replaced by the cryptic non-native lineages from Europe or Asia over the past 150 years. Cryptic invasion may also contribute to the biological invasion through hybridization with native lineages and introgression [85–87]. Sympatry between *P. nodiflora* and *P. canescens* has not been recorded in Australia although the species are found in the same general regions of the south coast of NSW (*P. canescens Michael s.n. 23-October-1969, CANB, Storrie 08/04, 13 February 2004, NSW; P. nodiflora Boorman s.n., 28 February 1900, NSW*) and Perth (*P. canescens Reynolds s.n. PERTH; Hocking s.n., 2 May 1953 PERTH; P. nodiflora Swarbrick 10679, 11 April 1993 BRI*). Further work is required to understand whether or not introgression between native and alien lineages of *P. nodiflora* is occurring and the evolutionary and ecological significance of any such events. *Phyla canescens* is increasingly progressing into regions where *P. nodiflora* occurs in Australia [17] and further work is required to determine the potential for hybridization between the two especially as Kennedy [6] notes that many apparently intermediate forms of *P. nodiflora* and *P. canescens* exist in the Americas.

5. Conclusions

Phyla nodiflora is a cosmopolitan pioneer species of seasonally damp places and sand dunes where it behaves as a weed of minor importance including within its indigenous range of lower North America to northern South America. This is the key to the enigmatic behaviour of *P. nodiflora*—as a pioneer species, it has a set of life history characteristics that promotes dispersal and establishment of propagules to new areas. We have detected at least three native lineages outside of the indigenous range for *P. nodiflora*, in Australasia, Africa and Eurasia, although this is further confounded by subsequent intercontinental human-movement of several clades. Our results highlight the need for genetic studies, combined with ecological and historical observations, to help elucidate the limits of species' distributions and to identify cases of cryptic invasions facilitated by humans.

Supplementary Materials: The following are available online at www.mdpi.com/1424-2818/9/2/20/s1. Table S1: Specimens sequenced for nr-DNA and the phylogeography of *Phyla nodiflora*, with voucher information and GenBank accession numbers. Herbarium acronyms stand for: AD = State Herbarium of South Australia; BISH = Herbarium Pacificum; BRI = Queensland Herbarium; CANB = Australian National Herbarium; DNA = Northern Territory Herbarium; FLAS = Florida Museum of Natural History; Gorgan = University of Gorgan; K = Royal Botanic Gardens Kew; MEL = National Herbarium of Victoria; MO = Missouri Botanical Garden Herbarium; MPU = Université Montpellier II; NE = N. C. W. Beadle Herbarium; NSW = Royal Botanic Gardens, Sydney; PERTH = Western Australian Herbarium; PRE = South African National Biodiversity Institute; SI = Museo Botánico; UC = University of California; US = Smithsonian Institution; XAL = Instituto de Ecología, A.C. s.n. = without number. Table S2: The 101 variable sites of the aligned sequences of the ITS1, 5.8S and ITS2 regions yielding 64 ribotypes in *Phyla nodiflora*. A dot (.) indicates that the character states are the same as for ribotype 1 and a dash (-) indicates an insertion in the aligned sequences. Data S1. The mostly univariate sites of the aligned sequences of Petb and trnL-F in *Phyla nodiflora*. A dot (.) indicates that the character states are the same as for ribotype 1, N stands for any base and a dash (-) indicates an insertion in the aligned sequences. Data S2. Estimates of evolutionary divergence between ribotypes (ribotype number is the first column and row). The number of base substitutions per site between sequences is shown. Standard error estimates are shown above the diagonal. Analyses were conducted using the Maximum Composite Likelihood model (Tamura et al., 2004 [33]) implemented in MEGA5 (Tamura et al., 2011 [27]). The analysis involved 64 DNA sequences and a total of 630 positions in the final dataset. 0 = ribotypes are identical and 1 = ribotypes differ in all positions. Figure S1a: Floral and leaf characters for *Phyla nodiflora* (ribotype 35, GenBank HM194056) from Corindi Beach NSW, Australia (30.0155° S, 153.1855° E) and *Phyla canescens* from Bendigo, Victoria, Australia (36.7570° S, 144.2794° E). Images taken by M. Fatemi 2006. Figure S1b: Habit and spent flower heads of *Phyla nodiflora* (ribotype 26, GenBank HM194113) from Mt Isa, Qld (20.7256° S, 139.4927° E) and *Phyla canescens* from Kialami, New South Wales, Australia (30.4503°S, 151.5325°E). Images taken by A. White (CSIRO) 2007 (lower left) and C.L. Gross 2008 (lower right).

Acknowledgments: This study was supported in part by the Australian Weeds Research Centre (AWRC) and by UNE. We thank A. Haddadchi, M. Macdonald, P. Nelson, C. Xu, T. Woods, S. Smith, S.J. Hadavi, N.A.H. Abbasi, S. Sosa, R. Seguraand M. Martinez for providing leaf samples. Herbarium samples were generously provided by AD, BISH, BRI, CANB, DNA, FLAS, Gorgan, K, MEL, MO, MPU, NE, NSW, NY, PERTH, PRE, SI, UC, US and XAL, and we thank the Directors of these institutions. Louisa Murray is thanked for her assistance with specimens at NSW, B. Conn for informative discussions on *Phyla*, I. Simpson for photographs at NE and J. Bruhl and R. Southwood for facilitating loans to NE from overseas institutions. We thank D. Crayn, K. Schulte, L. Vary, M. Rossetto, J. Bruhl, Wal Whalley, C. Xu, P. Prentice, D. Mackay, A. Bean and M. Harrington for their comments on an earlier version of the manuscript or our approach to analyses. The original idea to investigate the biogeography of *Phyla nodiflora* started with Green (1899) and we were encouraged to embark on the work by John A. Duggin.

Author Contributions: This project was designed by C.L.G., R.V.K., M.F., M.J. The molecular laboratory work was conducted by M.F., the ITS analyses were conducted by M.F., the plastid analysis were conducted by H.M. and C.L.G. with interpretations from all authors. The geography data were collated by C.L.G. with contributions from R.V.K. The paper was written by C.L.G., R.V.K., H.M. with contributions from M.F. & M.J.

Conflicts of Interest: The authors declare no conflict of interest.

Appendix A. Methods for the Chloroplast Regions

Total genomic DNA was isolated using DNeasy Plant Mini Kit (Qiagen, Chadstone, Vic, Australia) or Wizard SV Genomic DNA Purification System (Catalogue number A2361, Promega). Polymerase chain reaction amplification was performed on three intergenic spacer regions from the single-copy portion of the chloroplast genome ([88–90]); these three primers were selected for amplification after testing a set of seven primers (3′rps16/5′trnK UUU, trnL-trnF, trnH-psbA, atpB-rbcL, petB intron, petL-psbE, psbB-psbF). The template specific primers were tailed with universal M13 primers to attach fusion primers (see Table A1).

Our first-round PCR to amplify intergenic regions followed a standard program, adjusted only for annealing temperatures: (1) Initial denaturation was conducted for 2 min at 95 °C. (2) Thirty cycles of denaturation were conducted for 30 s at 94 °C, annealing for 30 s at 56–57 °C temperature depending on the region and elongation for 1 min at 72 °C (3) A final elongation step of 20 min at 72 °C was performed.

Second-round PCR was conducted in 15 µL volume consisting of 7.5 µL 2 × mastermix (GoTaq®G2 Hot Start Colorless Mastermix, Promega), 1 µL PCR product from first-round PCR as template DNA, 5 µM of each fusion primers A and B (Table A1) and PCR grade H2O to make the final volume. The following cycling program was used: (1) Initial denaturation was conducted for 2 min at 95 °C. (2) Twenty cycles of denaturation were conducted for 30 s at 94 °C, annealing for 30 s at 55 °C and elongation for 1 min at 72 °C. (3) A final elongation step of 20 min at 72 °C was performed.

Before sequencing, concentration of the PCR products was determined using QIAxcel multi-capillary electrophoresis system (Qiagen) and then equimolar volumes of each PCR products were pooled into two pools: 144 samples in each (Data S1). The PCR products were then purified using the Wizard®SV Gel and PCR Clean-Up System kit (Promega Corporation, Madison, WI, USA) following manufacturer's instructions. Sequencing of the PCR products was performed on a Roche 454 Genome Sequencer FLX (Roche Diagnostics) with a Titanium XL+ sequencing kit at Macrogen (Seoul, Korea).

Sequences obtained from the 454 Genome Sequencer were assigned to their respective accessions according to MID tags. Fastq files for each of 262 accessions were imported into CLC Genomics Workbench 7.0.4 (CLC). Since reference sequences were not available, a de novo assembly of all samples pooled together was performed with minimum contig length 100, length fraction 0.5 and similarity fraction 0.8. Twenty-five contigs were produced. These were further assembled in Geneious 6.1.7 (Geneious) resulting in 5 contigs. A Blastn search (performed on Genbank) confirmed contig 4 as trnL-trnF, contig 5 as petB intron and 3′rps16/5′trnK UUU was not retrieved. Contigs 4 and 5 were used as reference sequences. Reads for each of 288 accessions were trimmed to a quality value of 0.05 (an error probability calculated in CLC equivalent to a minimum average Phred count of 13 for each sequence) and a minimum length of 50 bp and adaptors were removed. Trimmed sequences were mapped to the reference sequences in CLC and consensus sequences for each accession saved (with low coverage areas, <5×, removed). Consensus sequences were aligned in Geneious in order to compare variation across samples.

Table A1. Primer pairs to study phylogeography based on chloroplast intergenic sequences.

Region	Forward		Reverse		Tm	Length	Source
3′rps16/5′trnK	(UUU) F	AAAGTGGGTT TTTATGATCC	(UUU) R	TTAAAAGCCG AGTACTCTACC	56	631	[89]
trnL-trnF	UniE	GGTTCAAGTC CCTCTATCCC	UniF	ATTTGAACTGG TGACACGAG	57	273–392	[90]
petB intron	petB intron F	AGAGATGGTT CTACTTCGTC	petB intron R	ACTTTCATCT CGTACAGCTC	57	552	[88]

[88] Nishizawa T, Watano Y (2000) Primer pairs suitable for PCR-SSCP analysis of chloroplast DNA in angiosperms. *Journal of Phytogeography and Taxonomy* 48:63–64; [89] Shaw J, Lickey E, Schilling E, Small R (2007) Comparison of whole chloroplast genome sequences to choose noncoding regions for phylogenetic studies in angiosperms: The tortoise and the hare III. Am J Bot 94:275–288; [90] Taberlet P, L. Gielly, G. Pautou, J. Bouvet (1991) Universal primers for amplification of three non-coding regions of chloroplast DNA. Plant Mol Biol 17:1105–1109.

References

1. Pyšek, P.; Richardson, D.M.; Rejmánek, M.; Webster, G.L.; Williamson, M.; Kirschner, J. Alien plants in checklists and floras: Towards better communication between taxonomists and ecologists. *Taxon* **2004**, *53*, 131–143. [CrossRef]

2. Gormally, C.; Donovan, J.H.A. Genetic structure of a widely dispersed beach annual, *Cakile edentula* (Brassicaceae). *Am. J. Bot.* **2011**, *98*, 1657–1662. [CrossRef] [PubMed]

3. Der, J.P.; Thomson, J.A.; Stratford, J.K.; Wolf, P.G. Global chloroplast phylogeny and biogeography of bracken (*Pteridium*; Dennstaedtiaceae). *Am. J. Bot.* **2009**, *96*, 1041–1049. [CrossRef] [PubMed]

4. Horrocks, M.; Rechtman, R.B. Sweet potato (*Ipomoea batatas*) and banana (*Musa* sp.) microfossils in deposits from the Kona Field System, Island of Hawaii. *JAS* **2009**, *36*, 1115–1126. [CrossRef]

5. Carlton, J.T. Biological invasions and cryptogenic species. *Ecology* **1996**, *77*, 1653–1655. [CrossRef]

6. Kennedy, K.L. A Systematic Study of the Genus *Phyla* Lour (Verbenaceae: Verbenoideae, Lantanae). Ph.D. Thesis, The University of Texas at Austin, Austin, TX, USA, 1992.

7. O'Leary, N.; Múlgura, M.E. A Taxonomic Revision of the Genus *Phyla* (Verbenaceae). *Ann. Mo. Bot. Gard.* **2012**, *98*, 578–596. [CrossRef]

8. Munir, A.A. A taxonomic revision of the genus *Phyla* Lour.(Verbenaceae) in Australia. *J. Adel. Bot. Gard.* **1993**, *15*, 109–128.

9. Greene, E. Neglected generic types I. *Pittonia* **1899**, *4*, 45–51.

10. Gupta, A.; Sadasivaiah, B.; Bhat, G. *Phyla nodiflora*. Version 2016-3. Available online: www.iucnredlist.org (accessed on 24 February 2017).

11. Marx, H.E.; O'Leary, N.; Yuan, Y.-W.; Lu-Irving, P.; Tank, D.C.; Múlgura, M.E.; Olmstead, R.G. A molecular phylogeny and classification of Verbenaceae. *Am. J. Bot.* **2010**, *97*, 1647–1663. [CrossRef] [PubMed]

12. Cosgrove, B. *Shoalwater Bay: Settlers in a Queensland Wilderness*; Central Queensland University Press: Rockhampton, Australia, 1996.

13. Hawkins, J.; Boutaoui, N.; Cheung, K.; van Klinken, R.; Hughes, C. Intercontinental dispersal prior to human translocation revealed in a cryptogenic invasive tree. *New Phytol.* **2007**, *175*, 575–587. [CrossRef] [PubMed]

14. Webb, D. What are the criteria for presuming native status? *Watsonia* **1985**, *15*, 231–236.

15. Bean, A.R. A new system for determining which plant species are indigenous in Australia. *Aust. Syst. Bot.* **2007**, *20*, 1–43. [CrossRef]

16. Steven, W.; Ulloa, C.; Pool, A.; Montiel, O. Flora de Nicaragua. In *Monographs in Systematic Botany from the Missouri Botanical Garden*; Missouri Botanical Garden: St. Louis, MO, USA, 1978.

17. Macdonald, M.J. Ecology of *Phyla canescens* (Verbenaceae) in Australia. Ph.D. Thesis, University of New England, Armidale, Australia, 2008.

18. Gross, C.; Gorrell, L.; Macdonald, M.; Fatemi, M. Honeybees facilitate the invasion of *Phyla canescens* (Verbenaceae) in Australia–no bees, no seed! *Weed Res.* **2010**, *50*, 364–372. [CrossRef]

19. Roalson, E.; Friar, E. Phylogenetic relationships and biogeographic patterns in North American members of *Carex* section *Acrocystis* (Cyperaceae) using nrDNA ITS and ETS sequence data. *Plant Syst. Evol.* **2004**, *243*, 175–187. [CrossRef]

20. Buckler, E.S.; Ippolito, A.; Holtsford, T.P. The evolution of ribosomal DNA divergent paralogues and phylogenetic implications. *Genetics* **1997**, *145*, 821–832. [PubMed]

21. Consaul, L.L.; Gillespie, L.J.; Waterway, M.J. Evolution and polyploid origins in North American Arctic *Puccinellia* (Poaceae) based on nuclear ribosomal spacer and chloroplast DNA sequences. *Am. J. Bot.* **2010**, *97*, 324–336. [CrossRef] [PubMed]

22. Mäder, G.; Zamberlan, P.M.; Fagundes, N.J.; Magnus, T.; Salzano, F.M.; Bonatto, S.L.; Freitas, L.B. The use and limits of ITS data in the analysis of intraspecific variation in *Passiflora* L. (Passifloraceae). *Genet. Mol. Biol.* **2010**, *33*, 99–108. [CrossRef] [PubMed]

23. Feliner, G.N.; Rosselló, J.A. Better the devil you know? Guidelines for insightful utilization of nrDNA ITS in species-level evolutionary studies in plants. *Mol. Phylogen. Evol.* **2007**, *44*, 911–919. [CrossRef] [PubMed]

24. Chenna, R.; Sugawara, H.; Koike, T.; Lopez, R.; Gibson, T.J.; Higgins, D.G.; Thompson, J.D. Multiple sequence alignment with the Clustal series of programs. *Nucleic Acids Res.* **2003**, *31*, 3497–3500. [CrossRef] [PubMed]

25. Waterhouse, A.M.; Procter, J.B.; Martin, D.M.; Clamp, M.; Barton, G.J. Jalview Version 2—A multiple sequence alignment editor and analysis workbench. *Bioinformatics* **2009**, *25*, 1189–1191. [CrossRef] [PubMed]

26. Swofford, D.L. *PAUP*. Phylogenetic Analysis Using Parsimony (* and Other Methods)*, Version 4; Sinauer Associates: Sunderland, MA, USA, 2003.

27. Tamura, K.; Peterson, D.; Peterson, N.; Stecher, G.; Nei, M.; Kumar, S. MEGA5: Molecular evolutionary genetics analysis using maximum likelihood, evolutionary distance, and maximum parsimony methods. *Mol. Biol. Evol.* **2011**, *28*, 2731–2739. [CrossRef] [PubMed]

28. Huelsenbeck, J.P.; Ronquist, F. MRBAYES: Bayesian inference of phylogenetic trees. *Bioinformatics* **2001**, *17*, 754–755. [CrossRef] [PubMed]

29. Ogden, T.H.; Rosenberg, M.S. How should gaps be treated in parsimony? A comparison of approaches using simulation. *Mol. Phylogen. Evol.* **2007**, *42*, 817–826. [CrossRef] [PubMed]

30. Felsenstein, J. Confidence limits on phylogenies: An approach using the bootstrap. *Evolution* **1985**, *39*, 783–791. [CrossRef]

31. Posada, D.; Crandall, K.A. Modeltest: Testing the model of DNA substitution. *Bioinformatics* **1998**, *14*, 817–818. [CrossRef] [PubMed]
32. Kimura, M. *Population Genetics, Molecular Evolution, and the Neutral Theory: Selected Papers*; University of Chicago Press: Chicago, IL, USA, 1994; p. 704.
33. Tamura, K.; Nei, M.; Kumar, S. Prospects for inferring very large phylogenies by using the neighbor-joining method. *Proc. Natl. Acad. Sci. USA* **2004**, *101*, 11030–11035. [CrossRef] [PubMed]
34. Posada, D.; Crandall, K.A. Intraspecific gene genealogies: Trees grafting into networks. *Trends Ecol. Evol.* **2001**, *16*, 37–45. [CrossRef]
35. Clement, M.; Posada, D.; Crandall, K.A. TCS: A computer program to estimate gene genealogies. *Mol. Ecol.* **2000**, *9*, 1657–1659. [CrossRef] [PubMed]
36. Templeton, A.R.; Crandall, K.A.; Sing, C.F. A cladistic analysis of phenotypic associations with haplotypes inferred from restriction endonuclease mapping and DNA sequence data. III. Cladogram estimation. *Genetics* **1992**, *132*, 619–633. [PubMed]
37. Joly, S.; Stevens, M.I.; van Vuuren, B.J. Haplotype networks can be misleading in the presence of missing data. *Syst. Biol.* **2007**, *56*, 857–862. [CrossRef] [PubMed]
38. Tobler, M.; Honorio, E.; Janovec, J.; Reynel, C. Implications of collection patterns of botanical specimens on their usefulness for conservation planning: An example of two neotropical plant families (Moraceae and Myristicaceae) in Peru. *Biodivers. Conserv.* **2007**, *16*, 659–677. [CrossRef]
39. Genc, H.; Nation, J.L.; Emmel, T.C. Life history and biology of Phyciodes phaon (Lepidoptera: Nymphalidae). *Fla. Entomol.* **2003**, *86*, 445–449. [CrossRef]
40. Bentham, G. *Flora Australiensis: A Description of the Plants of the Australian Territory*; Reeve & Co.: London, UK, 1870; Volume 5, pp. 31–70.
41. Song, Z.C.; Wang, W.M.; Huang, F. Fossil pollen records of extant angiosperms in China. *Bot. Rev.* **2004**, *70*, 425–458.
42. Robinson, G.S.; Ackery, P.R.; Kitching, I.J.; Beccaloni, G.W.; Hernández, L.M. HOSTS—A Database of the World's Lepidopteran Hostplants. Available online: http://www.nhm.ac.uk/hosts (accessed on 23 February 2017).
43. Grey, G. *Journals of Two Expeditions of Discovery in North-West and Western Australia, during the Years 1837, 38, and 39*; T and W Boone: London, UK, 1841; Volume II.
44. Knibbs, G.H. *Commonwealth Bureau of Census and Statistics Melbourne*; Australian Government: Melbourne, Australia, 1909.
45. Day, K.J.; Forster, B.A. *Report on the land units of Croker Island, N.T.*; Department of Northern Australia, Land Conservation Section: Canberra, Australia, 1975.
46. Edwards, E.D.; Newland, J.; Regan, L. *Lepidoptera: Hesperioidea, Papilionoidea*; CSIRO Publishing: Melbourne, Australia, 2001; Volume 31.6, pp. 545, 615.
47. Leigh, C.; Walton, C.S. *Lippia (Phyla canescens) in Queensland*; Land Protection, Department of Natural Resources, Mines, and Energy: Brisbane, Australia, 2004.
48. Price, J.; Macdonald, M.; Gross, C.; Whalley, R.D.; Simpson, I. Vegetative reproduction facilitates early expansion of Phyla canescens in a semi-arid floodplain. *Biol. Invasions* **2011**, *13*, 285–289. [CrossRef]
49. Xu, C.Y.; Julien, M.H.; Fatemi, M.; Girod, C.; Van Klinken, R.D.; Gross, C.L.; Novak, S.J. Phenotypic divergence during the invasion of Phyla canescens in Australia and France: Evidence for selection-driven evolution. *Ecol. Lett.* **2010**, *13*, 32–44. [CrossRef] [PubMed]
50. Cilento, R.; Lack, C. *Triumph in the Tropics: An Historical Sketch of Queensland/compiled and edited by Sir Raphael Cilento; with the assistance of Clem Lack; for the Historical Committee of the Centenary Celebrations Council of Queensland*; Smith & Paterson: Brisbane, Australia, 1959; p. 446.
51. Mueller, F. North Australian botany, observations on, by Dr Frederick (sic) Mueller, botanist to the NW Australian Government Expedition, under the command of Mr Surveyor Gregory; in a letter to Sir WJ Hooker (published with the sanction of the Colonial Office). *Hooker's J. Bot. Kew Gard. Misc.* **1856**, *8*, 321–331.
52. Mueller, F. Botanical report on the North Australian exploring expedition, under the command of A.C. Gregory Esq. *J. Proc. Linn. Soc. Lond. Bot.* **1858**, *2*, 137–144. [CrossRef]
53. Knibbs, G.H. *Official Year Book of the Commonwealth of Australia Containing Authoritative Statistics for the Period 1901–1908 and Corrected Statistics for the Period 1788 to 1900. No. 2*; Melbourne, Australia, 1909.

54. Dupont, L.; Behling, H.; Kim, J.-H. Thirty thousand years of vegetation development and climate change in Angola (Ocean Drilling Program Site 1078). *CliPa* **2008**, *4*, 107–124.

55. Dangol, D.; Gurung, S. Ethnobotany of the Tharu tribe of Chitwan district, Nepal. *Int. J. Pharm.* **1991**, *29*, 203–209. [CrossRef]

56. Shukla, S.; Patel, R.; Kukkar, R. Study of phytochemical and diuretic potential of methanol and aqueous extracts of aerial parts of *Phyla nodiflora* Linn. *Int. J. Pharm. Pharm. Sci.* **2009**, *1*, 85–91.

57. Wang, Y.-C.; Huang, T.-L. Screening of anti-*Helicobacter pylori* herbs deriving from Taiwanese folk medicinal plants. *FEMS Immunol. Med. Microbiol.* **2005**, *43*, 295–300. [CrossRef] [PubMed]

58. Ahmed, A.B.A.; Gouthaman, T.; Rao, A.S.; Rao, M.V. Micropropagation of *Phyla nodiflora* (L.) Greene: An important medicinal plant. *Iran. J. Biotechnol.* **2005**, *3*, 186–190.

59. Abuodha, J.; Musila, W.; van der Hagen, H.; van der Meulen, F. Floristic composition and vegetation ecology of the Malindi Bay coastal dune field, Kenya. *J. Coast. Conserv.* **2003**, *9*, 97–112. [CrossRef]

60. Estes, J.R.; Brown, L.S. Entomophilous, intrafloral pollination in *Phyla incisa*. *Am. J. Bot.* **1973**, *60*, 228–230. [CrossRef]

61. Pascual, M.; Slowing, K.; Carretero, E.; Mata, D.S.; Villar, A. Lippia: Traditional uses, chemistry and pharmacology: A review. *J. Ethnopharmacol.* **2001**, *76*, 201–214. [CrossRef]

62. Smith, D.R. Mutation rates in plastid genomes: They are lower than you might think. *Gen. Biol. Evol.* **2015**, *7*, 1227–1234. [CrossRef] [PubMed]

63. Wolfe, K.H.; Li, W.-H.; Sharp, P.M. Rates of nucleotide substitution vary greatly among plant mitochondrial, chloroplast, and nuclear DNAs. *Proc. Natl. Acad. Sci. USA* **1987**, *84*, 9054–9058. [CrossRef] [PubMed]

64. Call, A.; Sun, Y.X.; Yu, Y.; Pearman, P.B.; Thomas, D.T.; Trigiano, R.N.; Carbone, I.; Xiang, Q.Y.J. Genetic structure and post-glacial expansion of *Cornus florida* L. (Cornaceae): integrative evidence from phylogeography, population demographic history, and species distribution modeling. *J. Syst. Evol.* **2015**, *54*, 136–151. [CrossRef]

65. Rossetto, M.; McPherson, H.; Siow, J.; Kooyman, R.; Merwe, M.; Wilson, P.D. Where did all the trees come from? A novel multispecies approach reveals the impacts of biogeographical history and functional diversity on rain forest assembly. *J. Biogeogr.* **2015**, *42*, 2172–2186. [CrossRef]

66. Diazgranados, M.; Barber, J.C. Geography shapes the phylogeny of frailejones (Espeletiinae Cuatrec., Asteraceae): A remarkable example of recent rapid radiation in sky islands. *PeerJ* **2017**, *5*, e2968. [CrossRef] [PubMed]

67. Verdcourt, B. Flora of Tropical East Africa. In *Verbenaceae*; CRC Press: London, UK, 1992.

68. Renner, S. Plant dispersal across the tropical Atlantic by wind and sea currents. *Int. J. Plant Sci.* **2004**, *165*, S23–S33. [CrossRef]

69. Mummenhoff, K.; Linder, P.; Friesen, N.; Bowman, J.L.; Lee, J.-Y.; Franzke, A. Molecular evidence for bicontinental hybridogenous genomic constitution in *Lepidium sensu stricto* (Brassicaceae) species from Australia and New Zealand. *Am. J. Bot.* **2004**, *91*, 254–261. [CrossRef] [PubMed]

70. Pettingill, O.S. Additional information on the food of the American woodcock. *Wilson Bull.* **1939**, *51*, 78–82.

71. Rundle, W.D.; Sayre, M.W. Feeding ecology of migrant Soras in southeastern Missouri. *J. Wildl. Manag.* **1983**, *47*, 1153–1159. [CrossRef]

72. Razi, B.A. A contribution towards the study of the dispersal mechanisms in flowering plants of Mysore (south India). *Ecology* **1950**, *31*, 282–286. [CrossRef]

73. Dostine, P.; Morton, S.R. Seasonal abundance and diet of the comb-crested Jacana *Irediparra gallinacea* in the tropical Northern Territory. *Emu-Austral Ornithol.* **2000**, *100*, 299–311. [CrossRef]

74. Takayama, K.; Kajita, T.; Murata, J.; Tateishi, Y. Phylogeography and genetic structure of *Hibiscus tiliaceus*—Speciation of a pantropical plant with sea-drifted seeds. *Mol. Ecol.* **2006**, *15*, 2871–2881. [CrossRef] [PubMed]

75. Baldwin, A.H.; McKee, K.L.; Mendelssohn, I.A. The influence of vegetation, salinity, and inundation on seed banks of oligohaline coastal marshes. *Am. J. Bot.* **1996**, *83*, 470–479. [CrossRef]

76. Brookes, M.; Barley, R. *Plants Listed in Nursery Catalogues in Victoria 1855–1889*; Published for the Ornamental Plant Collections Association Inc., Royal Botanic Gardens: Melbourne, Australia, 1992.

77. Lorenz's OK Seeds. Available online: http://www.lorenzsokseedsllc.com/?s=lippia (accessed on 26 February 2017).

78. Moldenke, H. *Verbenaceae*; Amerind Publishing Co. Pvt. Ltd.: New Delhi, India, 1983.

79. Barker, R.; Telford, I. Oceanic Islands 2. In *Flora of Australia*; Australian Biological Resources Study and the Australian Government Publishing Service: Canberra, Australia, 1993; Volume 50, p. 415.

80. Hussey, B.; Keighery, G.; Dodd, J.; Lloyd, S.; Cousens, R. *Western Weeds: A Guide to the Weeds of Western Australia*; The Plant Protection Society of Western Australia: Victoria Park, Australia, 2007.

81. HMAS Inverell. Available online: http://www.navy.gov.au/hmas-inverell (accessed on 26 February 2017).

82. Townsend, C. Verbenaceae. In *Flora of Turkey and the East Aegean Islands*; Davis, P., Ed.; Edinburgh University Press: Edinburgh, UK, 1982; pp. 31–35.

83. Germishuizen, G.; Meyer, N. *Plants of Southern Africa: An Annotated Checklist*; National Botanical Institute Pretoria: Pretoria, South Africa, 2003.

84. Saltonstall, K. Cryptic invasion by a non-native genotype of the common reed, *Phragmites australis*, into North America. *Proc. Natl. Acad. Sci. USA* **2002**, *99*, 2445–2449. [CrossRef] [PubMed]

85. Ellstrand, N.C.; Schierenbeck, K.A. Hybridization as a stimulus for the evolution of invasiveness in plants? *Proc. Natl. Acad. Sci. USA* **2000**, *97*, 7043–7050. [CrossRef] [PubMed]

86. Mason-Gamer, R.J. Reticulate evolution, introgression, and intertribal gene capture in an allohexaploid grass. *Syst. Biol.* **2004**, *53*, 25–37. [CrossRef] [PubMed]

87. Miura, O. Molecular genetic approaches to elucidate the ecological and evolutionary issues associated with biological invasions. *Ecol. Res.* **2007**, *22*, 876–883. [CrossRef]

88. Nishizawa, T. Primer pairs suitable for PCR-SSCP analysis of chloroplast DNA in angiosperms. *J. Phytogeogr. Taxon* **2000**, *48*, 63–66.

89. Shaw, J.; Lickey, E.B.; Schilling, E.E.; Small, R.L. Comparison of whole chloroplast genome sequences to choose noncoding regions for phylogenetic studies in angiosperms: The tortoise and the hare III. *Am. J. Bot.* **2007**, *94*, 275–288. [CrossRef] [PubMed]

90. Taberlet, P.; Gielly, L.; Pautou, G.; Bouvet, J. Universal primers for amplification of three non-coding regions of chloroplast DNA. *Plant Mol. Biol.* **1991**, *17*, 1105–1109. [CrossRef] [PubMed]

diversity

MDPI

Editorial

The Contribution of Professor Gian Tommasso Scarascia Mugnozza to the Conservation and Sustainable Use of Biodiversity

Mario Augusto Pagnotta [1,*] (ID) and Arshiya Noorani [2,*]

[1] Department of Agricultural and Forestry Sciences (DAFNE), Università degli Studi della Tuscia, Via S.C. de Lellis, 01100 Viterbo, Italy
[2] Plant Production and Protection Division, FAO, Viale delle Terme di Caracalla, 00153 Rome, Italy
*   Correspondence: pagnotta@unitus.it (M.A.P.); arshiya.noorani@yahoo.com (A.N.); Tel.: +39-0761-357-423 (M.A.P.)

Received: 23 November 2017; Accepted: 12 December 2017; Published: 16 January 2018

Abstract: During his lifetime, Professor Scarascia Mugnozza contributed significantly to the field of population genetics, his research ranging from wheat breeding in arid and semi-arid regions, to the conservation of forest ecosystems. He promoted regional networks across the Mediterranean, linking science and policy at national and international levels, focusing on the conservation and sustainable use of genetic diversity. In addition, he worked intensely on improvement of knowledge bases, raising awareness on how research could inform international agreements, and thus lead to evidence-based policies. The loss of biodiversity and the resulting implications for environmental, socio-economic, political, and ethical management of plant genetic resources were of major concern, and he highlighted the absolute necessity for conservation of genetic diversity, stressing the importance of building positive feedback linkages among ex situ, in situ, on-farm conservation strategies, and participatory approaches at the community level. His work emphasized the importance of access to diverse plant genetic resources by researchers and farmers, and promoted equitable access to genetic resources through international frameworks. Farmers' rights, especially those in centres of origin and diversity of cultivated plants, were a key concern for Professor Scarascia Mugnozza, as their access to germplasm needed to be secured as custodians of diversity and the knowledge of how to use these vital resources. Consequently, he promoted the development of North-South cooperation mechanisms and platforms, including technology transfer and the sharing of information of how to maintain and use genetic resources sustainably.

Keywords: biodiversity; germplasm; genetic resources; International Treaty; conservation

1. Introduction

The current president of the Italian Science Academy also called "The Academy of the Forty" declared, during the Scarascia Mugnozza commemoration day, that *"It is difficult to fully remember the work of Gian Tommaso Scarascia Mugnozza, a man of charismatic personality, brilliant intelligence, great culture, with an extraordinary capacity of translate his ideas and intuitions into concrete projects"* [1]. The contribution of Professor Scarascia Mugnozza in promoting plant genetic resources diversity for mitigating and adapting to environmental changes, and its importance in addressing malnutrition and food security, is substantial and should be not be lost in time (see Figure 1 for a timeline highlighting the major developments at international and national levels).

At the 19th McDougall Memorial Lecture (October 1995) at the Food and Agriculture Organization of the United Nations (FAO) in Rome, Professor Scarascia Mugnozza began his speech by asserting "loss of biodiversity is often presented as an ecological problem, but the fundamental underlying causes are socio-economic and political" [2], and it is this that guided his lifetime's work and achievements.

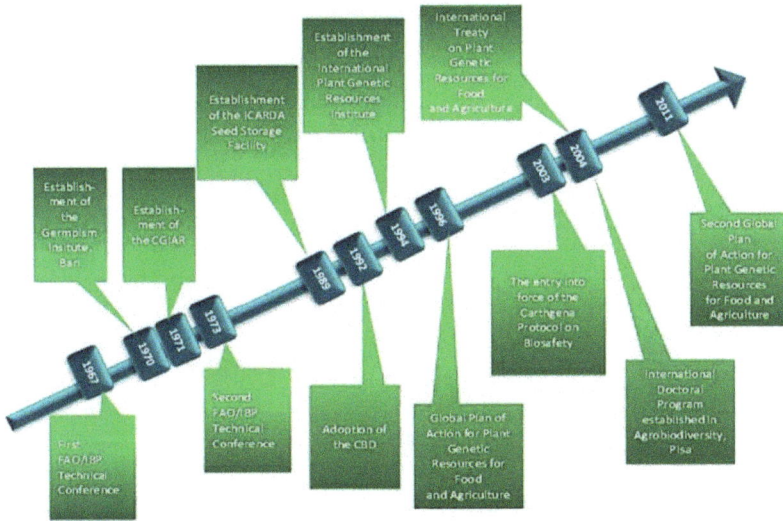

Figure 1. Timeline highlighting the major contributions of Prof. G.T. Scarascia Mugnozza for the conservation and sustainable use of plant diversity.

In this paper, which was published in the Special Issue launched to honour the memory of Professor Gian Tommaso Scarascia Mugnozza, we provide an overview of his work within the national and global technical and policy landscapes of the conservation and use of genetic resources for food and agriculture.

2. Conservation Activities

2.1. Ex Situ Conservation and International Frameworks

In 1945 in Quebec, Canada, the Food and Agriculture Organization of the United Nations (FAO) was founded with initial support from 42 countries, the mandate of which was increased food security through improved agricultural production [3]. In the 1960s, the FAO began to address issues that related not only to production, but also to the conservation of genetic resources. This was addressed at the FAO Technical Conference held in Rome in 1967, three seed banks were proposed for Northern Europe (Sweden), Central Europe (Germany), and for the Mediterranean region. Professor Scarascia Mugnozza, then Professor of Agricultural Genetics at Bari University, proposed that the Committee for Agricultural Sciences of the National Research Council Institute (CNR) establish a national genebank in Bari through the establishment of the Germplasm Institute. He became the first director of the Germplasm Institute and took part in numerous collection missions that were conducted in the 1970s and 1980s by the Institute and FAO [4,5].

2.2. Consultative Group on International Agricultural Research

In 1971, he attended the international meeting in Beltsville, United States of America (USA), which would give rise to the Institutes and Centres of the Consultative Group on International Agricultural Research (CGIAR). The CGIAR is a network of 15 international agricultural research centres, which manages approximately 600,000 agricultural seed samples [6]. It was established in 1971, and launched at the World Bank, with the sponsorship of UNDP, FAO, the World Bank, and an initial group of just over twenty donors. In 1972, Professor Scarascia Mugnozza participated in the Union Nations Conference on Human Environment organized in Stockholm, working out guidelines for safeguarding the genetic resources of plants. Following this development, Scarascia Mugnozza joined the Technical Advisory Committee of CGIAR. Finally, in 1994, most of the crop germplasm that was held in CGIAR genebanks was placed under the auspices of the FAO to be held in trust by the world community (see Figure 2 for the geneflows among the diverse stakeholders in the CGIAR System) [7].

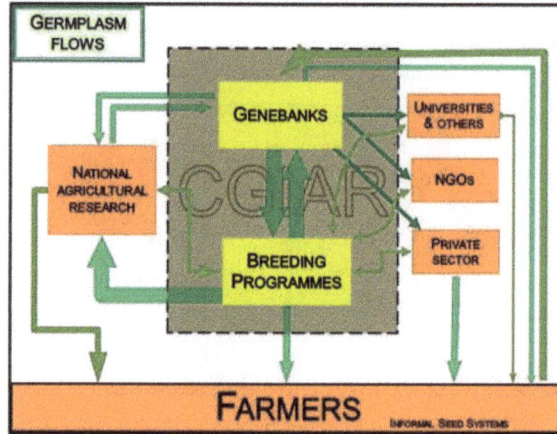

Figure 2. Germplasm flows in and out of the Consultative Group on International Agricultural Research (CGIAR) [7].

Professor Scarascia Mugnozza believed firmly that conservation of germplasm had to be as inclusive as possible. He liaised with prof. M.S. Swaminathan in India and helped to promote community-based conservation approaches, involving programmes for conserving farmers' varieties and landraces, and providing compensation and access to the stored PGR [8]. The "Scarascia Mugnozza Community Genetic Resources Centre", established in Chennai, India, with funding from the Government of Italy (Figure 3), allows for the exchange of genetic resources among farmers and promoting participatory plant breeding [9,10].

Figure 3. Professor Scarascia Mugnozza inaugurating the Community ex situ Genebank of the SMCGRC at MSSRF in Chennai. The picture is taken from the historical archives of the Accademia delle Scienze Detta dei XL (Rome, Italy).

2.2.1. International Plant Genetic Resources Institute

The need for international coordination of conservation of plant genetic resources activities was an area to which Professor Scarascia Mugnozza contributed substantially. In 1974, the International Board for Plant Genetics (IBPGR) was established at FAO in Rome to coordinate an international plant genetic resources program, including emergency collecting missions, and building and expanding national, regional, and international genebanks.

Professor Scarascia Mugnozza was member of the IBPGR Board of Trustees in 1991, when it became the International Plant Genetic Resources Institute (IPGRI) [11]. He assisted in identifying a location in Rome, where IPGRI was established, and, together with other international institutes dealing with agriculture and food, started in 1988. In January 1994, IPGRI began to operate independently as a CGIAR centre, and, at the request of CGIAR, took over the governance and administration of the International Network for the Improvement of Banana and Plantain (INIBAP).

2.2.2. International Centre for Agriculture Research in the Dry Areas

In 1977, CGIAR established the International Centre for Agriculture Research in the Dry Areas (ICARDA) with headquarters in Aleppo in Syria. Professor Scarascia Mugnozza was part of the Board of Trustees for two terms over three years, engaging in delineating research lines and operational aspects. ICARDA works on addressing the needs of resource-poor farmers through research on improving productivity, incomes, and livelihoods, especially in those areas where water is a limiting factor for agricultural production. He oversaw the renovation of an experimental station, which was built by Denmark in the Bekah valley, and its conversion into a seeds storage facility, inaugurated in 1989. The ICARDA Genebank now holds over 135,000 accessions from over 110 countries: traditional varieties, improved germplasm, and wild crop relatives. These include wheat, barley, oats, and other cereals; food legumes, such as faba bean, chickpea, lentil, and field pea; forage crops, rangeland plants, and wild relatives of each of these species.

3. Rolling Global Plan for Plant Genetic Resources for Food and Agriculture

Professor Scarascia Mugnozza participated actively in negotiations that led to the development of the Global Plan of Action for the Conservation and Sustainable Use of Plant Genetic Resources for Food and Agriculture (GPA), adopted in 1996 by 150 countries at the International Technical Conference on Plant Genetic Resources in Leipzig, Germany [12]. The resulting Leipzig Declaration asserted, "Our primary objective must be to enhance world food security through conserving and sustainably using plant genetic resources" [13].

Today, the Second Global Plan of Action for Plant Genetic Resources for Food and Agriculture (Second GPA), which was adopted by the FAO Council on 29 November 2011, updates the GPA [14]. The Second GPA was prepared through a series of regional consultations, with the participation of 131 countries and representatives of the international research community, the private sector and civil society. It plays an important role in the international policy framework for world food security, as a supporting component of the International Treaty for Plant Genetic Resources for Food and Agriculture.

4. International Treaty on Plant Genetic Resources for Food and Agriculture

In his speech, delivered at the 19th McDougall Memorial Lecture in 1995 [2], Professor Scarascia Mugnozza focused on the need to develop an international mechanism to allow for the fair and equitable exchange of Plant Genetic Resources for Food and Agriculture (PGRFA). He followed up on this by contributing to the development of the Madras Declaration, which was undersigned by diverse stakeholders from 76 countries (Figure 4) and was presented at the World Food Summit, November 1996, in Rome [8]. The Declaration highlighted the need for increased investment in agricultural research and rural development in order to guarantee food security and social equity by the establishment of an international fund.

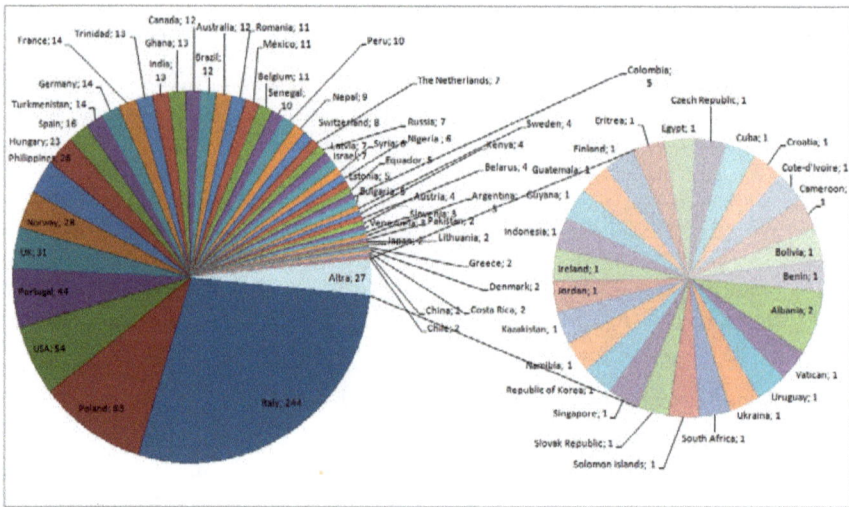

Figure 4. Number of scientists and stakeholders signed the Scarascia-Swaminata petition.

Prior to this, the International Undertaking on Plant Genetic Resources (IUPGR), which was adopted in 1983, was the first and main international instrument addressing the conservation and sustainable use of PGRFA. This was then revised in accordance with the Convention on Biological Diversity (CBD), which entered into force in 1992 [15]. However, it was not comprehensive and legally binding enough to address the needs highlighted in the Madras Declaration.

In response to the global demand for a more detailed instrument, the International Treaty for Plant Genetic Resources (the Treaty) was developed and adopted at the Thirty-first Session of the FAO Conference 2001 in Rome. Professor Scarascia Mugnozza collaborated extensively with the Treaty Secretariat to promote the linkages between research and the contribution of agricultural biodiversity to food and nutrition security and to sustainable agriculture.

The Treaty entered into force 90 days after 40 signatory States ratified it on 29 June 2004 [16]. In accordance with the Convention on Biological Diversity (CBD), the Treaty established a Multilateral System (MLS) of access and benefit-sharing for the most important crops for agriculture (64 crops, representing 52 genera and 29 forage genera), listed Annex 1 of the Treaty, to address the interdependency of countries for access to germplasm (www.planttreaty.org). Thus, contracting parties are obliged to provide access to PGRFA listed under the MLS when requested to do so by another Party (or a legal or natural person under the jurisdiction of a Party, or by an international institute that has signed an agreement with the governing body). Contracting Parties are also obliged to provide access when such PGRFA has been acquired under these same terms [17]. The scope of the Treaty, including its articles on conservation and sustainable use, is all PGRFA, while the MLS is specific to the crops listed under Annex 1 of the Treaty.

5. In Situ Conservation of Natural Ecosystems and Biodiversity

Professor Scarascia Mugnozza played a key role in ecosystem management at the national level, developing a set of tools to monitor biodiversity and ecosystem management. The 1970s and 1980s were decades where public awareness drove adaptation and mitigation strategies in research to address air and water pollution that threatened the survival of forest ecosystems. Hundreds of thousands of acres of woodlands in Scandinavia, Poland, Germany, France, and Switzerland, showed severe symptoms of defoliation due to the phenomenon of acid rain. This was also seen in areas of Italy, which, in addition to being severely affected by pollutions, was compounded by coastal erosion due to deforestation.

Professor Scarascia Mugnaozza coordinated case studies that were undertaken carried out in two locations: the 6000 ha forests of San Rossore, Tuscany (Italy), and those of Castelporziano, close to Rome. These lands had been managed at low levels of intensity and intervention, having been used traditionally for activities, such as hunting, firewood, and the breeding of local equine and bovine breeds [18]. He headed a commission to study the degradation of San Rossore's vegetation, and, in 1993, was appointed to chair the Technical Scientific Commission for monitoring the Castelporziano estate.

In collaboration with researchers from diverse disciplines, he set up a Territorial Information System consisting of seven working groups for collecting data on the atmosphere, soil, hydrogeology, vegetation, fauna, and anthropogenic impacts. This resulted in the review of those criteria that had inspired the guidelines of previous activities, and established, among other things, changes in the management of woodlands with a view to promote natural forest regeneration and conserve the biodiversity that was present within those ecosystems. In accordance with the European Community Action Plan (Mediterranean Action Plan), the San Rossore and Castelporziano estates are now used for environmental research, and to assess the impacts of different woodland management regimes, with the goal to upscale the sustainable management of forests to other areas.

6. Biotechnology Role in the Conservation of Biodiversity and Sustainable Production Systems

Professor Scarascia Mugnozza was at the forefront of promoting new technologies that were addressing development in the sustainable use of genetic resources. As a member of the Italian delegation, in 1955 he participated in the Conference of Geneva on the 'Peaceful Uses of Atomic Energy'. Scarascia Mugnozza established contacts with the delegates of American Atomic Commission during the conference and obtained, under "Atoms for Peace", a cobalt reactor. This was used for the irradiation of plants and seeds in so-called 'gamma fields' in a continuous manner (Figure 5). This inspired him to set up the Farm Laboratory at the Centro Casaccia of the National Centre for Nuclear Energy (CNEN), established in 1960 near Rome [10]. One of his most important contributions was as the scientific secretary of the advisory commission in 1961, and the 'Plant Genetics' Laboratory became the 'Laboratory for Nuclear Applications in Agriculture' under his guidance. Scarascia Mugnozza identified four research lines of interest for Italian research in agriculture, and this led to the development of the following fields of research in the Domestic Radiation Applications Laboratory:

1. the induction of mutations in improving agricultural crops;
2. a new means of biological control: the technique of sterilizing insects with gamma irradiation;
3. the application of the radioisotopes method to the study of soil-fertilizer-plant relations;
4. the irradiation of foods in order to ensure their preservation [19].

Figure 5. "Gamma fields" at the National Centre for Nuclear Energy (CNEN) in 1962.

The laboratory results were in the fields of:

- Applied mutagenesis to the improvement of agrarian plants.

The objective was to use mutagenesis to induce genetic variability useful for the improvement of autogamous plants (cereals, horticultural), and to increase the frequency of somatic mutations in vegetative propagation plants (fruit trees and flowering plants). Several crop varieties were produced with significant reduced plant height, including the durum wheat varieties Castelfusano and Castelporziano that were selected from Cappelli mutants (Figure 6). The variety, Creso, registered in 1974, was the major variety used in Italian agriculture, and, in 1984, was the seed certified variety that was used in more than 60% of the field and more than 25% of the overall durum wheat area (certified and uncertified seeds) with an utilization on more than 400,000 ha in 1982. Moreover, its cultivation was present over more than 30 years. Several other important varieties have been created utilizing Creso in their pedigree.

- The sterile insect larvae technique for the fruit fly (*Ceratitis capitata*).

Male insects were sterilized by radiation and released to the environment. Being sterile, the insect population was drastically reduced without any environmental impact. The diffusion of sterile insects were first trialled on island environments before being used on the Italian mainland.

- Technique of radioisotopes (N, P, K) used as tracers for the study of physiology and nutrition of plants.
- Preserving foodstuffs for the extension of market life, particularly fruit and vegetables, and the preservation of grain and other agricultural products, such as: potatoes, onions, strawberries, oranges, grapes, and in radio-disinfestation of dried figs and packs of flowers (carnations).

Figure 6. Professor Scarascia Mugnozza (right) demonstrating Durum wheat mutant lines with agronomic value. The picture is taken from the historical archives of the Accademia delle Scienze Detta dei XL (Rome, Italy).

Professor Scarascia Mugnozza also promoted the use of biotechnology for human benefit [20,21]. In 2001, as the president of a commission of scientists, he sought to clarify the scientific basis of the potential and impact of plant biotechnologies. This commission described how, instead of relying on random recombination between a large number of genes by conventional crossing of plants (including different species), the molecular method allows for the inclusion of sequences of DNA carrying specific characters in the genetic information of an organism (genome), creating a transgenic organism, also known as Genetically Modified Organism (GMO) or Living Modified Organism (LMO). This can reduce the timing of selection, preserve beneficial characteristics of the original genotype, and add individual genes where genotypes are deficient, thus making it possible to precisely and minimally alter the genome. The method also allows for the exchange of genes between sexually incompatible organisms, thus drastically increasing the potential of using natural biological diversity.

His approach was that rather than cultivating new land, thereby destroying forests, biodiversity, and other elements of possible climate change mitigation, it was possible to decrease impacts on natural ecosystems, while increasing the productivity of existing agro-ecosystems. Additionally, the careful use of biotechnology, and, in particular, genetic engineering, could contribute to innovative processes allowing the development of new, more resilient and productive crop varieties, while decreasing the levels of inputs that are needed (pesticides, fertilizers, herbicides). Nevertheless, the use of new varieties may also pose risks. Thus, both transgenic and conventional varieties should be subjected to risk analysis, and transgenics accepted when, in relation to expected benefits, they are less dangerous than those that are obtained by conventional techniques. The Cartagena Protocol on Biosafety (entered into force in 2003) defines a LMO as "any living organism that possesses a novel combination of genetic material, obtained through the use of modern biotechnology". Policies have been put into force to safeguard natural systems from potential environmental risks from such LMOs. The Protocol established that an advance informed agreement (AIA), would ensure that countries be provided with the documentation necessary to make informed decisions before agreeing to imports and agricultural production. The Protocol recommends a precautionary approach, and is based on Principle 15 of the Rio Declaration on Environment and Development.

He was recognized as an authority on biotechnology, and was requested to contribute an article to the prestigious Italian Enciclopedia Treccani [22], and in 2004, he received an honorary degree in biotechnology from the University of Naples for his position and activities on the role of biotechnology highlighted in the paper of Scarascia Mugnozza and De Pace [23].

7. Building Upon Global Frameworks of PGRFA

In order to enlarge the research of conservation and evaluation of different genetic resources, Professor Scarascia Mugnozza founded the International Doctoral Program in Agrobiodiversity, established in 2004 at the Scuola Superiore Sant'Anna of Pisa. The first of its kind, it attests to Scarascia's commitment to the training of young people, in particular from developing countries. The school received initial contributions from the Italian government. The course aims to promote scientific and policy research in the field of genetic diversity of agricultural and forestry plants, fostering sustainable agricultural practices to address issues of food insecurity and malnutrition.

In response to the high rates of genetic erosion created by the rapid loss of crop diversity, and by its consequences on agricultural growth and food security, all the activities and institutions he participated in emphasized the very close relationship between technological advancement and basic research, and the economic and social problems that are posed by disadvantaged areas. This distinguished gentleman passed away on 28 February 2011 at the age of 85, and will be remembered as a world authority on plant genetics, and a strong supporter of the scientific community's role in the conservation of biodiversity.

Acknowledgments: Authors acknowledge the Accademia delle Scienze Detta dei XL (Rome, Italy) for the assistance in the bibliography.

Author Contributions: The authors equally contributed to wrote the paper.

Conflicts of Interest: The authors declare no conflict of interest.

References

1. Chiancone, E. Intervento. In *Rendiconti dell'Accademia Nazionale delle Scienze detta dei XL, Memorie di Scienze Fisiche e Naturali*; Aracne Editrice: Rome, Italy, 2012; pp. 75–85.
2. Scarascia Mugnozza, G.T. The Protection of Biodiversity and the Conservation and Use of Genetic Resources for Food and Agriculture: Potential and Perspective. In *19th Mc Dougall Memorial Lecture*; FAO: Rome, Italy, 1995.
3. Phillips, R.W. *FAO: Its Origins, Formation and Evolution 1945–1981*; FAO: Rome, Italy, 1981. Available online: http://www.fao.org/3/a-p4228e.pdf (accessed on 22 November 2017).
4. Scarascia Mugnozza, G.T.; Perrino, P. State, Use, Problems of Ex Situ Plant Germplasm Collections. In *Rendiconti dell'Accademia Nazionale delle Scienze detta dei XL, Memorie di Scienze Fisiche e Naturali*; Accademia Nazionale delle Scienze detta dei XL: Rome, Italy, 2000; Volume 24, pp. 57–100. Available online: http://media.accademiaxl.it/memorie/S5-VXXIV-P1-2-2000/ScarasciaMugnozza-Perrino57-100.pdf (accessed on 22 November 2017).
5. Scarascia-Mugnozza, G.T.; Perrino, P. The History of Ex Situ Conservation and Use of Plant Genetic Resources. In *Managing Plant Genetic Diversity*; Engels, J.M.M., Rao, V.R., Eds.; CABI: Wallingford, UK, 2002; pp. 1–22.
6. Renkow, M.; Byerlee, D. The impacts of CGIAR research: A review of recent evidence. *Food Policy* **2010**, *35*, 391–402. [CrossRef]
7. López Noriega, I.; Halewood, M.; Galluzzi, G.; Vernooy, R.; Bertacchini, E.; Gauchan, D.; Welch, E. How policies affect the use of plant genetic resources: the experience of the CGIAR. *Resources* **2013**, *2*, 231–269. [CrossRef]
8. Scarascia Mugnozza, G.T.; Swaminathan, M.S. Appello agli Scienziati di tutto il mondo per la conservazione e l'utilizzazione della biodiversità e delle risorse genetiche essenziali per l'agricoltura e la produzione agroalimentare. [Appeal to the scientists of the world for the maintenance and use of biodiversity and genetic resources important for food and agriculture]. In *Rendiconti dell'Accademia delle Scienze detta dei XL, Memoria di Scienze Fisiche e Naturali*; Accademia delle Scienze detta dei XL: Rome, Italy, 1996; pp. 27–31. Available online: http://media.accademiaxl.it/memorie/S5-VXX-P1-2-1996/ScarasciaMugnozza-Swaminathan27-31.pdf (accessed on 22 November 2017).
9. Bala Ravi, S.; Rani, M.G.; Swaminathan, S. Conservation of Plant Genetic Resources at the Scarascia Mugnozza Community Genetic Resources Centre. In *Rendiconti dell'Accademia delle Scienze detta dei XL, Memoria di Scienze Fisiche e Naturali*; Aracne Editrice: Rome, Italy, 2010; Volume 24, pp. 47–58.
10. Porceddu, E. Intervento. In *Rendiconti dell'Accademia delle Scienze detta dei XL, Memorie di Scienze Fisiche e Naturali*; Aracne Editrice: Rome, Italy, 2012; pp. 127–132.
11. IPGRI. *The Mulino at Maccarese*; International Plant Genetic Resources Institute: Rome, Italy, 2001.
12. FAO. *Global Plan of Action for the Conservation and Sustainable Utilization of Plant Genetic Resources for Food and Agriculture*; FAO: Rome, Italy, 1996. Available online: http://www.fao.org/tempref/docrep/fao/meeting/015/aj631e.pdf (accessed on 22 November 2017).
13. Leipzig Declaration. In Proceedings of the International Technical Conference on Plant Genetic Resources, Leipzig, Germany, 17–23 June 1996. Available online: http://www.fao.org/FOCUS/f/96/06/more/declar-f.htm (accessed on 22 November 2017).
14. FAO. *The Second Global Plan of Action for Plant Genetic Resources for Food and Agriculture*; FAO: Rome, Italy, 2012.
15. Khoury, C.; Laliberté, B.; Guarino, L. Trends in ex situ conservation of plant genetic resources: A review of global crop and regional conservation strategies. *Genet. Resour. Crop Evol.* **2010**, *57*, 625–639. [CrossRef]
16. Moore, G.K.; Tymowski, W. *Explanatory Guide to the International Treaty on Plant Genetic Resources for Food and Agriculture (No. 57)*; IUCN: Gland, Switzerland, 2005.
17. Anishetty, M. Conservation and utilization of plant genetic resources for food and agriculture: Strengthening local capacity for food security. In *Protecting and Promoting Traditional Knowledge: Systems, National Experiences and International Dimensions*; United Nations: New York, NY, USA; Geneva, Switzerland, 2004; pp. 33–39.
18. Giordano, E. Il contributo del prof. Scarascia Mugnozza per la conservazione degli ecosistemi della Regione mediterranea. In *Rendiconti dell'Accademia delle Scienze detta dei XL, Memorie di Scienze Fisiche e Naturali*; Aracne Editrice: Rome, Italy, 2012; pp. 101–104.
19. Rossi, L.; Donini, B.; Bozzini, A. CNEN/ENEA Laboratorio Applicazioni in Agricoltura. In *Rendiconti dell'Accademia delle Scienze detta dei XL, Memorie di Scienze Fisiche e Naturali 133*; Parte II, Tomo, I; Accademia Nazionale delle Scienze detta dei XL: Rome, Italy, 2015; Volume XXXIX, pp. 173–193. Available online: http://www.accademiaxl.it/wp-content/uploads/2011/09/Rossi-Donini-Bozzini-173-193.pdf (accessed on 22 November 2017).

20. Sonnino, A. Internazionalizzazione della ricerca e cooperazione scientifica internazionale—L'attualità dell'insegnamento di Gian Tommaso Scarascia Mugnozza. In *Rendiconti dell'Accademia delle Scienze detta dei XL, Memorie di Scienze Fisiche e Naturali 133*; Parte II, Tomo I; Accademia Nazionale delle Scienze detta dei XL: Rome, Italy, 2015; Volume XXXIX, pp. 203–216. Available online: http://media.accademiaxl.it/memorie/S5-VXXXIX-P2-2015/Sonnino203-216.pdf (accessed on 22 November 2017).

21. Sonnino, A. International Instruments for Conservation and Sustainable Use of Plant Genetic Resources for Food and Agriculture: An Historical Appraisal. *Diversity* **2017**, *9*, 50. [CrossRef]

22. Scarascia Mugnozza, G.T. *Biotecnologie—Il Vocabolario Treccani*; Istituto della Enciclopedia Italiana: Rome, Italy, 2003.

23. Scarascia Mugnozza, G.T.; De Pace, C. Biotecnologie: Ricerche e applicazioni nel comparto agricolo-alimentare e ambientale. In *Storia Dell'agricoltura Italiana, L'età Contemporanea*; Accademia dei Georgofili: Firenze, Italy, 2003; Volume 3, pp. 259–322.

MDPI
St. Alban-Anlage 66
4052 Basel
Switzerland
Tel. +41 61 683 77 34
Fax +41 61 302 89 18
www.mdpi.com

Diversity Editorial Office
E-mail: diversity@mdpi.com
www.mdpi.com/journal/diversity

www.ingramcontent.com/pod-product-compliance
Lightning Source LLC
Chambersburg PA
CBHW051847210326
41597CB00033B/5808